WALTER DE GRUYTER, INC.
3 Westchester Plaza
ELMSFORD, N.Y. 10523

Ralf Steudel

Chemistry of the Non-Metals

With an Introduction to Atomic Structure
and Chemical Bonding

English Edition by

F. C. Nachod, J. J. Zuckerman

Walter de Gruyter Berlin · New York 1977

Title of the Original Edition

Chemie der Nichtmetalle
mit einer Einführung in die Theorie der Atomstruktur
und der Chemischen Bindung
© Copyright 1973 by Walter de Gruyter & Co., Berlin

Author

Ralf Steudel, Dr. rer. nat.
Professor of Inorganic Chemistry
Technische Universität Berlin

English Edition by

F. C. Nachod, Ph. D., D. Sc.
Sterling-Winthrop Research Institute
Rensselaer, N.Y. 12144
and Adjunct Professor of Chemistry
Rensselaer Polytechnic Institute
Troy, N.Y. 12181

J. J. Zuckerman, Ph. D., Ph. D., D. Sc.
Professor, Chairman of the Department of Chemistry
The University of Oklahoma
Norman, OK. 13019

With 123 figures and 52 tables

Library of Congress Cataloging in Publication Data

Steudel, Ralf
  Chemistry of the non-metals.

  Translation of Chemie der Nichtmetalle.
  Includes index.
  1. Nonmetals. 2. Atomic theory. 3. Chemical bonds.
  I. Title.
  QD161.S7313   546'.7   76-17273
  ISBN 3-11-004882-5

CIP-Kurztitelaufnahme der Deutschen Bibliothek

Steudel , Ralf
Chemistry of the non-metals: with an introd. to
atomic structure and chem. bonding. – 1. Aufl. –
Berlin, New York: de Gruyter, 1976.
  (De Gruyter textbook)
  ISBN 3-11-004882-5

Preface

This text was written at the suggestion of the publishers from the notes of lectures which I had given to advanced students for several years. It presupposes fundamental knowledge of general inorganic chemistry and attempts to present the chemical behavior of the nonmetallic elements in condensed form. The primary focus is on simple, easily-visualized compounds and reactions. Molecular structure is discussed on the basis of current models of the chemical bond.

The important theoretical ideas of atomic structure and chemical bonding are treated in the first part of the book. Toward the end of this part empirical properties of chemical bonds are introduced which lead on to the second, topically-oriented part. Here the indispensable information about compounds and reactions becomes the main theme, and theoretical concepts are enlarged and illustrated by numerous examples which are interspersed in the representation of the factual material. The simple theory of chemical bonding based on numerous approximations is shown to be quite adequate for the interpretation of molecular structure, however, of little help so far in the explanation of chemical reactions. Hence, a balanced presentation of theory and factual material is required.

Berlin, September 1976   *R. Steudel*

Translators' Preface

The marvelous variety and infinite subtlety of the non-metallic elements, their compounds, structures and reactions, is not sufficiently acknowledged in the current teaching of chemistry. It was our desire to bring to the English-speaking student the new information available on this important class of elements, by preparing an English translation of Professor Ralf Steudel's very successful text. The work opens with a brief exposition of modern ideas of chemical bonding followed by a discussion of hydrogen and the 22 elements which occupy the upper-right quadrant of the long form of the periodic table. These elements, found above the zig-zag line conventionally drawn between aluminium, germanium, antimony and polonium and their higher congeners, are outnumbered 4:1 by the metals, but form many more compounds. The elements germanium and polonium are included among the non-metals. The emphasis throughout is on structure and bonding, rather than the compilation of an encyclopaedic listing of reactions.

In commending the book to its readers, the translators wish to acknowledge the collaboration of the author Professor Ralf Steudel, and Dr. Rudolf Weber of Walter de Gruyter in Berlin.

Albany, October 1976

J. J. Zuckerman
F. C. Nachod

Table of Contents

Part I: Atomic Structure and Chemical Bonding

1. Introduction

Atoms are structured particles. This was first concluded from experiments on the conductivity of electrolytes and discharge through dilute gases which showed that electrons were a common constituent of all atoms. A further indication of the complex nature of heavier atoms came from the discovery of radioactivity by A. H. Bequerel (1896).

P. Lenard discovered in 1903 that fast electrons can penetrate several layers of atoms in a metal foil without substantial change in direction. He concluded that atoms are not homogeneous spherical masses as had been assumed in the kinetic theory of gases, but rather that they are mostly empty space.

Experiments by E. Rutherford, H. Geiger and E. Marsden (1906–1909) on the scattering of α-particles during the penetration of gold foils supported Lenard's conclusion. The mass of the atom must be concentrated in a very small region, since, while most α-particles went through the foil without scattering, some were scattered through large angles. Rutherford showed from the angular dependence of α-particle scattering that the atom had a small, massive nucleus of radius 10^{-12} cm, and he proposed a detailed model in 1911.

The atom consists of a very small, positively-charged nucleus, representing most of the mass of the atom, which is surrounded by moving electrons. The movement of the electrons is determined by the equivalence of centrifugal and Coulombic attractive forces. The numbers of nuclear charges and electrons are equal, insuring electric neutrality. From the α-particle scattering, Rutherford calculated the nuclear charge number which agreed with the atomic number of the respective element in the periodic system proposed in 1869 by D. I. Mendeleev and independently by L. Meyer. The radii of atomic nuclei were calculated to be of the order of $\leq 10^{-12}$ cm. In comparison with the atomic radius of 10^{-8} cm obtained from kinetic theory, the result indicated that the largest portion of the atomic volume was made up of almost massless electrons.

N. Bohr, using Planck's quantum theory calculated the planetary model of the atom exactly for hydrogen, and gave a detailed explanation of its atomic spectrum.

2. Elementary Particles

Elementary particles are units which are not further sub-divisible, but which may be transmuted into other elementary particles. At present some hundred elementary particles are known, but only a few are important to the chemist (cf. Table 1).
A. Einstein in 1905 formulated the equivalence of mass and energy in his special theory of relativity

$$E = mc^2 \tag{1}$$

$$c = \text{light velocity} = 2.998 \cdot 10^8 \text{ m/s}$$

Thus, a particle may be characterized by its mass or its energy. While physicists prefer energy (electron volts), the chemist is more familiar with classification according to mass. According to Einstein, however, the mass of a particle is a function of its velocity:

$$m = \frac{m_0}{\sqrt{1 - \left(\dfrac{v}{c}\right)^2}} \tag{2}$$

$$c = \text{light velocity}$$

where m_0 ist the mass at $v = 0$, or rest mass. As long as v is less than 10% of c, m differs from m_0 by less that 0.5%. Yet for $v = 50\%$ c, $m = 1.15\, m_0$.
Forms of energy which like photons or γ-quanta have a rest mass of zero are listed among elementary particles. The energy of photons, according to M. Planck (1900), is proportional to their frequency, v:

$$E = h v \tag{3}$$

$$h = \text{Planck's constant}$$
$$6.626 \cdot 10^{-34} \text{ J} \cdot \text{s}$$

The properties of some elementary particles are summarized in Table 1.

The scattering of high energy electrons by protons and neutrons shows that the latter two are not point masses, but are structured as well. The radii of electrons, protons, and neutrons are ca. 10^{-13} cm. Protons and neutrons are bound together by nuclear forces which have an effective range of only 10^{-13} cm. The electrons are attracted to the nucleus by Coulombic forces which are also responsible for the chemical bond.
Neutrons are unstable outside the nucleus and decompose with a half-life of ca. 12 minutes

into a proton, an electron, and an anti-neutrino ($\bar{\nu}$)

$$n \rightarrow p + e^- + \bar{\nu} \tag{4}$$

The anti-neutrino has the same mass, spin and lifetime as the neutrino, but differs in reactions with other elementary particles.

Table 1   Properties of Some Elementary Particles

Particle	Symbol	Rest mass[a]	Charge[b]	Spin[c]	Comments
Photon	γ	0	0	1	stable
Electron	e^-	0.0005486	$-e$	1/2	stable
Proton	p	1.007276	$+e$	1/2	stable
Neutron	n	1.008665	0	1/2	$t_{1/2} = 12$ min.
Neutrino	ν	0	0	1/2	stable

a in atomic units p. 20
b in atomic charge units, p. 20
c discussion of spin, cf. pp. 44 and 51–2.

3. Units and Conversion Factors

Recent international agreements have given rise to new units for important parameters in the scientific literature. The units and symbols which are employed in this text, called International Units or S.I. (from *Système International*) units, are surveyed here and pertinent conversion factors are given.

The base unit of *length* is the meter [m]; derived units are: $1 m = 10^2 cm = 10^3 mm = 10^6 \mu m$ (micrometer) $= 10^9 nm$ (nanometer) $= 10^{12} pm$ (picometer). One $a_0 = 0.5292 \cdot 10^{-10} m$ the radius of the Bohr hydrogen atom is used as the atomic unit. In addition, the Angstrom $1 \text{ Å} = 10^{-10} m = 100 pm$ is still used.

The base unit of *mass* is the kilogram [1 kg]. $1 kg = 1000$ grams [g]. The atomic mass unit [u] is $1 u = 1.6605 \cdot 10^{-27}$ kg which corresponds to $\frac{1}{12}$ of the mass of the ^{12}C nuclide.

The base unit for *time* is the second [s].

The base unit for *energy* is the Joule [J]. $1 J = 2.39 \cdot 10^{-4}$ kcal, or, conversely, $1 kcal = 4.187 \cdot 10^3 J$ or $4.187 kJ$. The electron volt (eV) is used as the atomic unit of energy, the energy acquired by 1 electron in traversing a potential field of 1 volt; $1 eV = 3.83 \cdot 10^{-23} kcal = 1.602 \cdot 10^{-12} erg = 8066 cm^{-1} = 1.602 \cdot 10^{-19} J$. The energy units cal, kcal, erg, and cm^{-1} are, according to international agreement, not to be used further.

The unit of *temperature* is the Kelvin [K]. On this scale, the triple point for water $(0°C)$ is 272.16 K, the boiling point of water 372.16 K. As a special name for the Kelvin, the use of degree Centigrade (or Celsius) is still permitted. However, degree and °K are no longer to be used.

The base unit of *pressure* is the Pascal [Pa] or Bar [bar]. $1 bar = 10^5 Pa = 750 Torr = 0.987$ atm. The pressure unit Torr remains in use; $1 Torr = 133.3 Pa = 1.333 mbar$ (millibar).

The unit of *current* is the Ampere [A] and of *charge* (amount of electricity) the Coulomb [C]. As the unit of atomic charge, $1 e = 1.602 \cdot 10^{-19} C$.

The unit of *substance* is the Mol [mol]. 1 mol is the amount of substance of a system consisting of atoms, molecules, electrons, etc., which consists of as many particles as the number of atoms in 12 grams of the nuclide ^{12}C, or $6.0225 \cdot 10^{23}$ particles.

4. Atomic Models

4.1. Structure and Properties of the Electron Cloud

4.1.1. The Hydrogen Atom

The Bohr Atomic Model

The hydrogen atom consists of a proton and an electron. According to Bohr the proton is the nucleus of the atom and the electron rotates about it on a circular orbit in a spherically symmetrical field. The atom thus has the shape of a circular cut of a sphere.

The energy of the atom is composed of two parts, firstly the potential energy, U, the result of Coulombic attraction between opposite charges:

$$U = -\int_{\infty}^{r} K \, dr = +\int_{\infty}^{r} \frac{e^2}{r^2} \, dr = \frac{-e^2}{r} = E_{pot} \qquad (5)$$

$$K = \frac{(+e) \cdot (-e)}{r^2}$$

r = distance between nucleus and electron

The energy of the proton and electron separated to infinity is by definition zero (E = 0). However, if an electron approaches a proton to the distance r, the system changes to a more favored energetic state (E < 0). The energy thus gained, $+\dfrac{e^2}{r}$, is partly liberated and partly changed into kinetic energy.

The atom achieves stability when the electron and proton rotate about the common center of gravity. Owing to the much smaller mass of the electron, the center of gravity is near the center of the nucleus, which is considered at rest. This leads to:

$$E_{kin} = \tfrac{1}{2} \cdot \theta \cdot \omega^2$$

$$E_{kin} = \tfrac{1}{2} \cdot m \omega^2 r^2 \qquad (6)$$

$$\theta = m \cdot r^2 \quad \text{moment of inertia}$$

$$\omega = \frac{\varphi}{t} \quad \text{angular velocity;} \qquad \varphi \text{ is the angle for the time interval, t.}$$

The kinetic energy, unlike the potential, is always positive. The total energy:

$$E = E_{pot} + E_{kin} \qquad (7)$$

in a stable state, is negative, i.e., energy is liberated in the formation of an atom from a nucleus and electron. Hence: $|E_{pot}| > E_{kin}$

The calculation according to Bohr for the hydrogen atom, gives:

$$E_{kin} = \frac{e^2}{2r} = \tfrac{1}{2}|E_{pot}| \tag{8}$$

For stable rotation of an electron about the nucleus, the Coulomb attraction (centripetal force) and centrifugal force, $m\omega^2 r$, must be equivalent:

$$\frac{e^2}{r^2} = m\omega^2 r \tag{9}$$

Since e and m are constant, $\omega^2 \sim 1/r^3$, i.e., the rotation velocity increases with decreasing radius. Since a continuous variation of r between the values 0 and ∞ is inconsistent with the evidence, Bohr postulated certain stationary states by the condition that angular momentum, p_1, of the electron must be an integer multiple, n, of the momentum, $h/2\pi$:

$$p_1 = \theta \cdot \omega = m\omega r^2 = n \cdot \frac{h}{2\pi} \tag{10}$$

p_1 = angular momentum
h = Planck's constant
n = quantum number

According to Equation 10, the angular momentum is quantized (Bohr's quantum condition), and n is the principal quantum number. The state n = 1 is the ground state, and the excited states, n > 1. The electron moves in circular orbitals which fulfill condition (10) without loss of energy in the form of electromagnetic radiation, although a time-dependent dipole according to classical theory should continuously radiate energy. Bohr's theory gave no explanation for this contradiction.

The radius of the hydrogen atom is obtained from equations (9) and (10):

$$r = n^2 \cdot \frac{h^2}{4\pi^2 \cdot e^2 \cdot m} = n^2 \cdot a_0 \tag{11}$$

For n = 1, r = 0.5292 Å, which is abbreviated as a_0 ($= 1$ Bohr) and is the atomic unit of length.

The total energy, E, can be calculated from Equations (5) and (6) by the use of Equations (7) and (9):

$$E = E_{pot} + E_{kin} = -\frac{e^2}{r} + \frac{e^2}{2r} = -\frac{e^2}{2r} \tag{12}$$

Substituting the radius from Equation (11):

$$E = -\frac{1}{n^2} \cdot \frac{2\pi^2 \cdot e^4 \cdot m}{h^2} = \frac{1}{n^2} \cdot E_1 \tag{13}$$

where E_1, the energy of the ground state according to (13) is:

$$E_1 = -13.605\ eV = -1313.0\ kJ/mol \tag{14}$$

This is the amount of energy needed to separate the proton and electron. E_1 is the negative of the ionization energy, I, of the ground state. The calculated value agrees with the experimental ionization energy (1312.2 kJ/mol), and this agreement can be further improved when the rotation of nucleus and electron around a common center of gravity is introduced by using the reduced mass, μ, in Equations (11) and (13):

$$\mu = \frac{m \cdot M}{m + M} \tag{15}$$

m = electron mass
M = nuclear mass

For the hydrogen atom, $\mu = 0.9995\ m$, yielding a value of 1312.6 kJ for E_1.

The velocity, v, can also be calculated from the kinetic energy according to Equation (12) which gives $E_{kin} = -E$. For the ground state, $v = 2.2 \cdot 10^6$ m/s. At this velocity, the relativistic mass of the electron differs from rest mass, m_0, by less than 0.005%.

The circular path of the electron is an *orbital*, a term used in other atomic models to characterize the location of the electron, even if the path of the electron around the nucleus is no longer spherical. The *orbital energy* is the amount liberated when an electron is captured from a large distance into the corresponding orbital. Orbital energies, therefore, are always negative, and are a measure of the stability of the orbital.

For the hydrogen atom, the orbital energy is identical with the negative ionization energy of the corresponding state, and can be calculated directly from Equation (13). For states of principal quantum numbers 1 through 6, the orbital energies, E, and radii, r, are listed in Table 2, and depicted in Figures 1 and 2.

Bohr's concept can be applied without restriction to isoelectronic atoms and ions, such as, D, T, He^+, Li^{2+}, taking into consideration higher nuclear masses, M, and Z,

Table 2   Orbital Energies, E, and Radii, r, for various States of the Hydrogen atom according to Bohr (n = principal quantum number).

n	1	2	3	4	5	6	∞	
E	−1312	−328	−146	−82	−52	−36	0	[kJ/mol]
r	0.53	2.12	4.76	8.47	13.23	19.05	∞	[Å]

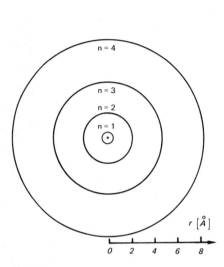

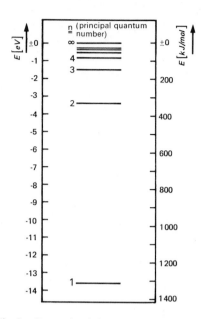

Fig. 1. Electron paths of the hydrogen atom according to Bohr in the ground state (n = 1) and the first three excited states.

Fig. 2. Energy level diagram of the hydrogen atom according to Bohr. E is the orbital energy of the respective state. States of n between 7 and infinity are not shown.

the number of charges of the nucleus (proton number). The reduced mass, μ, at increasing nuclear mass approaches the mass of the electron, m, since the sum of M + m hardly differs from M, and m can be used instead of μ in Equations (11) and (13), and Ze^2 for e^2 for heavy single electron atoms. Thus, e.g., the radius of He^+ in the ground state is calculated as 0.30 Å, with an ionization energy of 5251 kJ/mol.

The Bohr theory provided an explanation of the line spectra of atoms. The emission lines of the hydrogen atom on electrical discharge can be calculated exactly from the theory. The energy difference, ΔE, between two states, i and j, of the hydrogen atom is found from Equation (13) to be:

$$\Delta E = E_j - E_i = \frac{2\pi^2 \cdot e^4 \cdot \mu}{h^2} \cdot \left(\frac{1}{n_i^2} - \frac{1}{n_j^2} \right) \qquad n_j > n_i \qquad (16)$$

Transitions between states i and j are possible according to Bohr when photons of frequency v are absorbed or emitted. The condition:

$$\Delta E = h \cdot v \qquad (17)$$

must be fulfilled (Bohr's correspondence principle). The electron arrives at an orbital of greater or smaller quantum number by absorption or emission of a photon.

Spectroscopists characterize a spectral line by wavenumber instead of frequency, v, for which the symbol v is also used. Wavenumber is the reciprocal of wavelength, λ, expressed in cm, or the number of waves per centimeter, $v = 1/\lambda$. From Equations (16), (17) and (4), the hydrogen atom should possess spectral lines:

$$v\,[\mathrm{cm}^{-1}] = \frac{2\pi^2 \cdot e^4 \cdot \mu}{c \cdot h^3}\left(\frac{1}{n_i^2} - \frac{1}{n_j^2}\right) = R_H \cdot \left(\frac{1}{n_i^2} - \frac{1}{n_j^2}\right) \qquad (18)$$

The Rydberg constant, R_H, for the hydrogen atom is 109,677.58 cm^{-1}. The emission spectrum of the hydrogen atom (Figure 3) consists of series of lines. These series, named after their discoverers, are the result of electronic transitions from more energetic and unstable orbitals to less energetic and more stable orbitals (cf. Table 3). The positions of the individual lines agree with those calculated from Equation (18). Atomic spectroscopy thus experimentally confirms the existence of atomic states as well as the orbital energies.

Usefulness and Limitations of the Bohr Atomic Model

The Bohr theory postulates quantization of angular momentum and the radiationless rotation of the electron about the nucleus, without justifying these premises. The quantization of energy can be derived and orbital energies can be calculated. The quantitative agreement between the calculated and observed series of the line spectra of the hydrogen atom shows that the assumption of the existence of the electron in discrete stationary states is correct. Coulombic attraction is the only important force between the nucleus and electron. A great advantage of this model is its vividness.

However, the shortcomings of the Bohr theory become apparent in applying it to many-electron atoms and to problems related to bonding. Even some properties of the hydrogen atom remain unexplained or contradict the theory. Thus, the hydrogen atom should possess an angular momentum in all energy states resulting from the circular motion of the electron about the nucleus, which, owing to the charge of the electron, should give rise to a magnetic moment. This is not true for the ground state of the hydrogen atom, however. Also, the behavior of the hydrogen atom in a magnetic field, in which certain energy states are split, cannot be explained by the Bohr theory. Finally, the theory fails to rationalize the fine structure of the spectrum. For example, it can be shown with increased resolution that each line of the Balmer series consists of at least two lines, whose positions differ by very little.

A. Sommerfeld (1915) tried to overcome some of the imperfections of the Bohr theory by introducing elliptical orbitals of varying eccentricity. Sommerfeld introduced a second quantum number corresponding to the characterization of an ellipse by two parameters, the semi-major and semi-minor axes. This may be considered as the precursor of the azimuthal

Table 3.   Series of the Hydrogen Atom Spectra

	The Main Lines	n_i	n_j
Lyman Series	uv	1	2,3...
Balmer series	visible	2	3,4...
Ritz-Paschen series	ir	3	4,5...
Brackett series	ir	4	5,6...
Pfund series	ir	5	6,7...

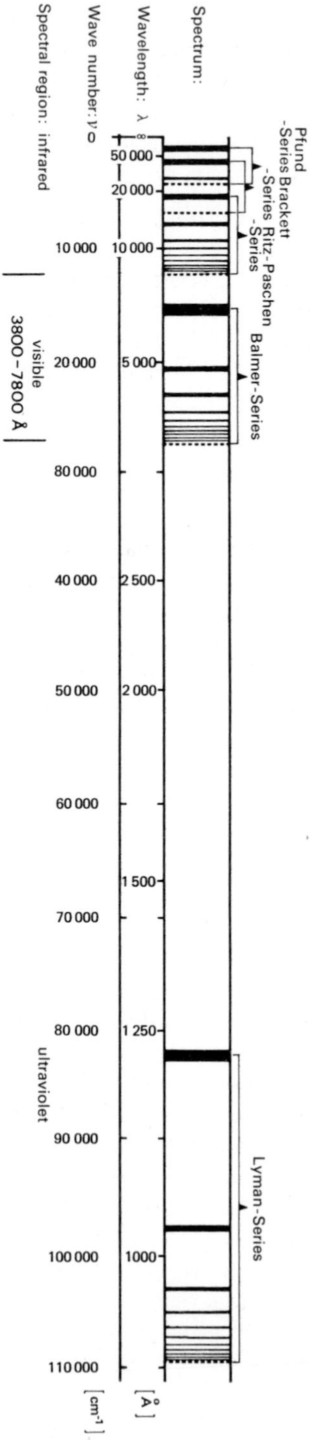

Fig. 3.   Schematic representation of the hydrogen spectrum. The line thickness corresponds to the intensities. Dotted lines denote the limit of the series which is sometimes followed by a continuum.

quantum number l, which arises in the wave-mechanical atomic model. The complex spectra of many-electron atoms, however, cannot be explained by the Bohr-Sommerfeld theory.

The Wave-mechanical Theory of the Hydrogen Atom

The wave-mechanical theory of the hydrogen atom, in contrast to the Bohr-Sommerfeld theory. is based upon two important principles of atomic physics, Heisenberg's Uncertainty Principle, and wave-particle dualism. The result is an alternative mathematical description of the hydrogen atom, which is in better agreement with the experimental data, although less graphically descriptive.

Heisenberg's Uncertainty Principle states that it is impossible in principle to determine simultaneously both the position and momentum of a particle exactly. Both parameters have an uncertainty Δx or Δp, respectively:

$$\Delta p \cdot \Delta x \geqslant \frac{h}{2\pi} \qquad (19)$$

The more uncertainty in the position coordinate, x, the sharper the determination of the momentum $p = m \cdot v$, can become and vice versa. Owing to the small value of Planck's constant, Equation (19) plays a role only at atomic dimensions and masses. The uncertainty relation can be plausibly explained in that the position of an atom or a particle is only determined by perturbing the system, which then brings about change in the momentum (velocity).

The Bohr theory contradicts the uncertainty relation which holds for rotatory motion in the form:

$$\Delta p_l \cdot \Delta \varphi \geqslant \frac{h}{2\pi} \qquad (20)$$

$$p_l = m\omega r^2 \text{ momentum}$$
$$\varphi \text{ angle}$$

For a given initial state (e.g., $\varphi = 0$) any subsequent position, φ, of the electron can be calculated since the radius and circumference of the orbital are known, and the velocity is also known. This, however, is in contradiction with Equation (20), since p_l is accurately defined by quantum condition (10), and hence φ cannot be determined accurately. Thus, only complete uncertainty of φ, i.e., spatial delocalization of the electron makes it possible to determine the momentum accurately. This can be accomplished by considering the electron mathematically as a *matter wave*. L. de Broglie showed in 1924 that a particle of mass, m, and velocity, v, has an associated wavelength, λ:

$$\lambda = \frac{h}{m \cdot v} \qquad (21)$$

which was derived using the results of the special theory of relativity in the form of Equations (1) and (2).

Table 4 summarizes the calculated wavelengths for some small particles. Electron beams, in passing through a crystal lattice, show similar diffraction and interference phenomena as do x-rays, and wavelength values obtained from interference experiments agree with those calculated from Equation (21). Employing suitable crystals (e.g., LiF) interferences can also be observed with heavier particles, e.g., atoms (He) or even molecules (H_2). The lattice constants are chosen for their similarity to the de Broglie wave length.

The matter waves of de Broglie have two properties important for atomic structure. Firstly, they are harmonic vibrations, and secondly, they are infinite.

Table 4   de Broglie Wave Length of some Particles

Particle	Mass [g]	Kinetic Energy [eV]	Velocity [m/s]	Wavelength [Å]
Electron	$0.91 \cdot 10^{-27}$	1	$5.9 \cdot 10^5$	12.3
Electron	$0.91 \cdot 10^{-27}$	100	$5.9 \cdot 10^6$	1.23
Electron	$0.91 \cdot 10^{-27}$	10 000	$5.9 \cdot 10^7$	0.12
Proton	$1.67 \cdot 10^{-24}$		$1.38 \cdot 10^5$	0.029
$H_2 (200 °C)$	$3.3 \cdot 10^{-24}$		$2.4 \cdot 10^3$	0.82

Electrons travelling in field-free space have zero potential energy. Such beams correspond to infinitely long propagating waves according to Equation (21). The wavelength can assume any value since the kinetic energy of the electrons is not quantized in this case. Electrons in a centrally symmetrical field of an atomic nucleus, however, undergo rotatory and oscillating movements about an equilibrium position which is the common center of gravity of the system. This is a standing matter wave.

In standing waves, the maximum amplitude is a function of the position coordinates, i.e., standing waves are described by a function $\psi(x, y, z)$. The amplitude function has certain nodes at which its value is zero. In a three-dimensional wave, the nodes may be planar or curved, and correspond to the nodal points of a one-dimensional wave, e.g., of a rod, and nodal lines of two-dimensional standing waves, e.g., of an oscillating circular section.

The amplitude of a wave is time- and position-dependent, and is described by a function $\Psi(x, y, z, t)$ which is the product of two functions, one depending on position coordinates and the other on time. A harmonic oscillation is described:

$$\Psi(x, y, z, t) = \psi(x, y, z) \cdot e^{-2\pi i v t} \tag{22}$$

where $\psi(x, y, z)$ is the maximum amplitude at position x, y, z. The electron is conceived in the wave-mechanical theory as a spherical standing wave. The spherical symmetry is a consequence of the nuclear field. The change in amplitude, Ψ, of a spherical wave, which expands in space with velocity, u, as function of time, t, and coordinates x, y, z, is, according to wave mechanics, described by the wave equation:

$$\frac{\partial^2 \Psi}{\partial x^2} + \frac{\partial^2 \Psi}{\partial y^2} + \frac{\partial^2 \Psi}{\partial z^2} - \frac{1}{u^2}\frac{\partial^2 \Psi}{\partial t^2} = 0 \qquad (23)$$

This differential equation can be easily modified for the hydrogen atom:
a) the partial differentials in Equation (23) are obtained by differentiation of Equation (22)
b) for each vibration, $u = \lambda \cdot v$ holds
c) $h/m \cdot v$ is substituted for λ from Equation (21)
d) the difference $E - U$ is substituted for $1/2\, mv^2$, the kinetic energy, from Equation (7).

The time-independent *Schrödinger equation* (1926) is obtained from Equation (23):

$$\frac{\partial^2 \psi}{\partial x^2} + \frac{\partial^2 \psi}{\partial y^2} + \frac{\partial^2 \psi}{\partial z^2} + \frac{8\pi^2 m}{h^2}(E - U)\,\psi = 0 \qquad (24)$$

The Schrödinger equation contains only the time-independent amplitude function $\psi(x, y, z)$ which is termed a wave function. Each function, ψ, which fulfills the conditions of the Schrödinger equation is one of its solutions.

An exact solution of the Schrödinger equation is only possible for one-electron atoms because the potential energy, U, is only exactly obtainable for the one-electron problem. Just as in the Bohr theory, U is expressed:

$$U = -\frac{e^2 \cdot Z}{r} \qquad (25)$$

where Z = nuclear charge number.

The wave-mechanical theory is thus based on Coulombic attraction between the nucleus and electron, and neglects other forces like gravity, which is many orders of magnitude smaller. The total energy, E, of the atom for all stationary states must be negative.

Of all solutions with negative values of E, only those in which the wave function $\psi(x, y, z)$ is single-valued, finite and continuous, and vanishes at infinity are physically meaningful. Solutions, termed eigenfunctions, ψ, exist under these limiting conditions for certain definite values of E, E_1, E_2, E_3 ... which are termed eigenvalues. They correspond to those stationary states of the atom in which matter waves are not extinguished by interference. Thus, the quantization of energy of the atom, within the framework of the wave-mechanical theory, arises as a necessary corollary of the wave properties of the electron. A further important result is the existence of degenerate electron states. If the eigenvalues are placed in an ascending series taking into account the signs, then there are for each value, E_n, exactly n^2 eigenfunctions, ψ, i.e., n^2 atomic states of like energy, which differ in the spatial structure of the standing electron waves, where n is the principal quantum number. The wave function, ψ, itself gives no description of the residence of the electron. According to M. Born (1926), however, the product

$$\psi^2 dV = \psi^2 dxdydz \tag{26}$$

is a measure of the probability, of encountering the electron at a certain time within the volume dV. The value ψ^2 is the probability density, and must be finite everywhere except at the nodes and at infinity, where it becomes zero. If the function is normalized, (see p. 34) the product $\psi^2 dV$ gives not only the relative, but the absolute residence probability.

According to W. Heisenberg, the energy, E, of a system at time, t, cannot be defined accurately because the uncertainty principle also holds for the simultaneous determination of E and t and has the form $\Delta E \cdot \Delta t \geq h/2\pi$. If the energy of an atom with eigenvalues is defined exactly (small ΔE), then the corresponding states can only be conceived as an average over a large time interval (large Δt), in order not to violate the uncertainty principle. At any given instant only statements of probability can be made concerning the atom. The eigenfunction, ψ, with its associated eigenvalue, E_n, therefore, is itself a probability parameter. On this basis the wave mechanical model portrays only the most probable structure of the atom.

However, the total energy, E, of a very large number of atoms, e.g., 1 mmole corresponding to ca. $6 \cdot 10^{20}$ particles, is so large, that even large absolute errors, ΔE, lead to vanishingly small errors in energy. Hence, energy in a macroscopic system is well defined and can be measured in a short time interval.

The Schrödinger equation can be written differently, giving the sum of the second partial differentials for the three coordinates the symbol ∇^2 which is the Laplace operator:

$$\frac{\partial^2}{\partial x^2} + \frac{\partial^2}{\partial y^2} + \frac{\partial^2}{\partial z^2} \equiv \nabla^2 \tag{27}$$

∇^2 thus is identical with the first part of the Schrödinger equation (24). Solving for $E \cdot \psi$:

$$E \cdot \psi = \left(-\frac{h^2}{8\pi^2 m}\nabla^2 + U\right)\psi \tag{28}$$

where the expression in the brackets is called the Hamiltonian operator, with symbol, H. Thus,

$$E \cdot \psi = H\psi \tag{29}$$

The symbol, H, then signifies a mathematical operation on the function ψ. $H\psi$ is, therefore, not a product as $E \cdot \psi$ on the left side of equation (29). This equation is the simplest form of the time-independent Schrödinger equation.

Electron States of the Hydrogen Atom

The Schrödinger equation provides n^2 degenerate electron states for each eigen-value, E_n. The corresponding eigenfunctions differ with respect to symmetry, arrangement of nodes, and in spatial orientation.

The electron-states (orbitals) derived from the Schrödinger equation are charac-terized by three quantum numbers. The *principal quantum number*, n, gives the number of nodes minus one of the wave function. All states having the same princi-pal quantum number are degenerate; n is an integer: 1, 2, 3 ... n.

The quantum number l, the *azimuthal quantum number*, gives the number of nodes passing through the center of the atom, and determines the angular momentum of the electron resulting from its motion about the nucleus. For each value of n, there are $(n-1)$ l-values, commencing with zero, thus: 0, 1, 2 ... $(n-1)$.

The *magnetic quantum number*, m, determines the spatial orientation of the orbitals of equal principal, n, and azimuthal, l, quantum numbers and is an axial quantum number. In a magnetic field, the energy of the orbitals depends on their orientation, i.e., on m. The quantum number m may take the values: 0, ± 1, ± 2 ... $\pm l$.

Figure 4 depicts the possible electron states of the hydrogen atom for n = 1 through

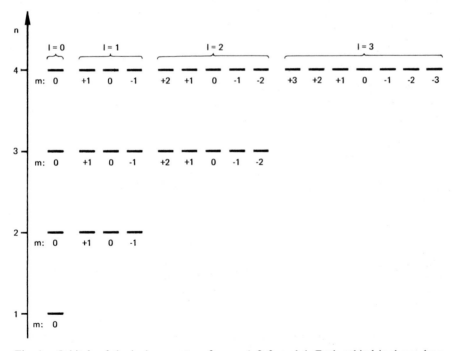

Fig. 4.   Orbitals of the hydrogen atom for n = 1, 2, 3, and 4. Each orbital is shown by a heavy horizontal bar.

4. States of like principal quantum number form a *shell*. In each shell the states (orbitals) of like azimuthal quantum number form a distinct level. The shells of increasing values of n are given the symbols K, L, M ...

Symbols are used instead of the quantum numbers to characterize the orbitals. The designation for each orbital is composed of the principal quantum number, n, a symbolic letter for the azimuthal quantum number, l, and an index which shows the spatial orientation in cartesian coordinates (x, y, z) with origin at the nucleus. Symbolic letters for l, and degree of degeneracy or the number of orbitals forming the level are:

Azimuthal quantum number	Level symbol	Degree of degeneracy
0	s	1
1	p	3
2	d	5
3	f	7

Of three coordinate axes, z is chosen to coincide with the molecular axis of linear molecules or with the direction of an applied field. Orbitals which are rotationally symmetrical with the z axis have a magnetic quantum number of zero. With these assumptions, the following designations are chosen:

Magnetic quantum number	Orbitals
0	s, p_z, d_{z^2}
± 1	p_x, p_y, d_{xz}, d_{yz}
± 2	$d_{xy}, d_{x^2-y^2}$

The indices are explained on p. 37.

The description in Figure 4 can now be extended using the orbital symbols as in Figure 5.

The energy of a one-electron atom depends only on the principal quantum number, n. Eigenvalues of energy, E_n, corresponding to these n states, are given by:

$$E_n = -\frac{1}{n^2} \cdot \frac{2\pi^2 e^4 Z^2 m}{h^2} = \frac{1}{n^2} \cdot E_1 \qquad (30)$$

Z = nuclear charge number

The Schrödinger equation can only be solved for values, E_n, which fulfill the conditions given on p. 29. Hence, the wavemechanical theory for the hydrogen atom leads to the same states (cf. p. 23) as the Bohr theory, despite the latter's contradiction of the Heisenberg Uncertainty Principle. The energy states which are obtained are those which have the greatest probability in the wave-mechanical theory. Therefore, the energy level diagram in Figure 2 (p. 24) is also valid for the wave-mechanical hydrogen atom model.

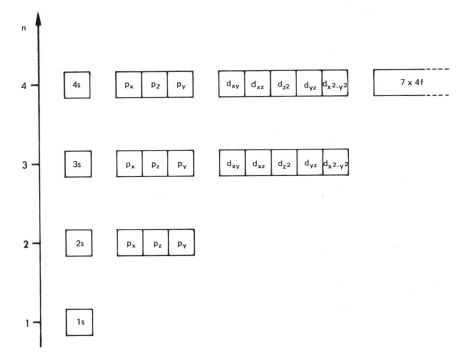

Fig. 5.   Orbitals of the hydrogen atom in the shells of principal quantum number 1, 2, 3, and 4.

The degeneracy of orbitals having the same principal quantum number presupposes a spherically symmetrical field for the motion of the electron. This degeneracy, therefore, only holds for one-electron atoms and only then in the case of a non-relativistic treatment of electronic motion. A second electron would travel in a field resulting from the superposition of the nuclear field and that of the first electron. Degeneracy is lifted even in one-electron atoms by the application of an external electric or magnetic field. Fields of this kind may arise, e.g., from neighboring atoms or ions (ligand and crystal fields). In such cases orbital energy is dependent upon all the quantum numbers.

Graphical representations of wave function ψ (orbitals) and residence probabilities $\psi^2 dV$:
The wave function, ψ, depends only on the coordinates x, y, z (see p. 28). In order to visualize the spatial extent of the electronic distribution or orbital geometry, the origin of the coordinate system is placed at the nucleus. As the electric field of the nucleus is centrally symmetrical, wave functions are generally given as functions of the polar coordinates r, φ, ϑ. Polar coordinates of a given point can be obtained from the cartesian coordinates by a simple transform, following the three equations which result from Figure 6.

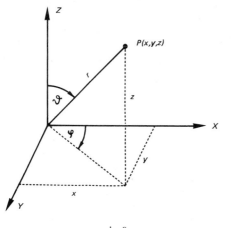

Fig. 6.   Correlation between Cartesian (x, y, z) and polar coordinates (r, φ, ϑ) at a point.

$$x = r \cdot \sin \vartheta \cdot \cos \varphi \qquad y = r \cdot \sin \vartheta \cdot \sin \varphi \qquad z = r \cdot \cos \vartheta$$

Some eigenfunctions of the hydrogen atom in polar coordinates are listed for principal quantum numbers 1 through 3 in Table 5, using the atomic unit of length, a_0 (cf. p. 20). In one-electron atoms containing more than one proton, the nuclear charge number, Z, must also be introduced.

The wave functions, ψ, consist of several parts: a constant (normalizing factor), a radius-dependent portion, and parts depending on the angles φ and ϑ. The total probability of encountering the electron anywhere in space must be unity, hence:

$$\int \psi^2 \, dV = \int\int\int_{-\infty}^{+\infty} \psi^2 \, dx \, dy \, dz = 1 \tag{31}$$

Wave functions satisfying Equation (31) are said to be normalized.

Table 5   Some Eigenfunctions of the Hydrogen Atom.

n	l	m	Orbital	Wave functions in polar coordinates
1	0	0	1s	$\psi = \dfrac{1}{\sqrt{\pi a_0^3}} \cdot e^{-\frac{r}{a_0}}$
2	0	0	2s	$\psi = \dfrac{1}{4\sqrt{2\pi a_0^3}} \cdot \left(2 - \dfrac{r}{a_0}\right) \cdot e^{-\frac{r}{2a_0}}$
2	1	+1	$2p_x$	$\psi = \dfrac{1}{4\sqrt{2\pi a_0^3}} \cdot \dfrac{r}{a_0} \cdot e^{-\frac{r}{2a_0}} \cdot \sin \vartheta \cdot \cos \varphi$
3	2	+1	$3d_{xz}$	$\psi = \dfrac{1}{81\sqrt{2\pi a_0^3}} \cdot \dfrac{r^2}{a_0^2} \cdot e^{-\frac{r}{3a_0}} \cdot \sin \vartheta \cos \vartheta \cos \varphi$

It is useful to consider $\psi(r, \varphi, \vartheta)$ as a product of three separate functions, each depending only on one coordinate:

$$\psi(r, \vartheta, \varphi) = R(r) \cdot \theta(\vartheta) \cdot \Phi(\varphi) \tag{32}$$

If the angular dependence of ψ is neglected, then R gives the change of wave function with radius, r, measured from the nucleus. R is thus termed a radial function. The angular function $\theta\Phi$ is the factor with which R is multiplied with respect to the directions ϑ and φ, in order to obtain the correct values of ψ. This scale factor is independent of r. The product $\theta \cdot \Phi$ gives the change of ψ on the surface of a sphere with radius, r, as in Equation (32). In discussing the wave function and its symmetry properties, it is appropriate to consider radial function R and the angular function $\theta \cdot \Phi$ separately.

The normalized radial functions of the hydrogen eigenfunctions for principal quantum numbers 1, 2, and 3 are shown in Figure 7. The function R(r) in general consists of the product of two parts as can be seen from the last column in Table 5. One is an exponential function, the other consists of exponents of r/a_0. Hence, for small or medium values of r, R(r) may assume positive or negative values (cf. Figure 7). For very large r values, R approaches zero exponentially. Zero values of R(r) correspond to nodes of the standing wave in three-dimensional space.

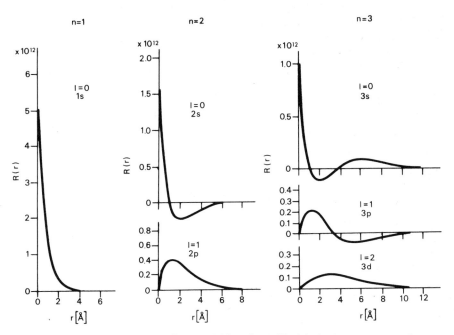

Fig. 7.   Dependence of the normalized radial function R(r) of the hydrogen atom on distance from the atomic center, shown for various quantum numbers.

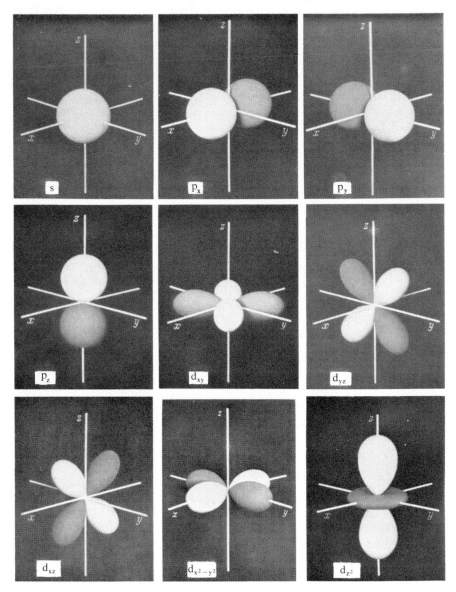

Fig. 8.   Polar diagrams of the angular function $\theta\Phi$ for s-, p-, d-orbitals. The surfaces on which $\theta\Phi > 0$ are lightly shaded; those where $\theta\Phi < 0$ are more darkly shaded. The scale is selected to give maximal projections of s:p:d_{xy}, d_{xz}, d_{yz}, $d_{x^2-y^2}$:d_{z^2} as $1:1.732:1.936:2.236$. (L.E. Sutton, Chemische Bindung und Molekülstruktur, Springer Verlag, Berlin, 1961).

R(r) depends only on the quantum numbers n and l; thus orbitals which differ only in the magnetic quantum number, m, that is, in their orientation in space, have the same radial function.

Neither the function ψ, nor R(r), is capable of precise visual interpretation (see p. 30). However, since $\psi^2 dV$ expresses the residence probability of the electron in the volume element dV, $R^2 dV$ must have a corresponding meaning. R depends only on the coordinate r; thus $R^2 dV$ is proportional to the probability of finding the electron in volume dV along this coordinate, neglecting the angular dependence. From Figure 7 can be derived how R^2 varies with r. (Negative R values give positive R^2 values.)

Polar diagrams are useful for visualization of the angular function. These are spatial models (surfaces) as shown in Figure 8. Construction of these polar diagrams is carried out by plotting $\theta \cdot \Phi$ in each direction of ϑ and φ in the form of vectors, beginning at the origin. The vector has direction ϑ, φ; its length corresponds to the value of the angular function and its end is a point on the surface of the model in Figure 8.

Polar diagrams of individual angular functions differ in form and orientation. The angular functions of certain s-, p-, or d-states are identical for all values of the principal quantum numbers (e. g., for all s, p_x, d_{z^2} orbitals, etc.). The angular function may be positive or negative in different directions, and thus some surfaces in Figure 8 are lightly ($\theta \Phi > 0$) or darkly ($\theta \Phi < 0$) shaded. If the functions, taking the sign into account, have an inversion center they are called gerade, or if not, ungerade. This classification holds also for the corresponding complete wave functions. Hence, s- and d-functions are gerade, while p- functions are ungerade.

The eigenfunction of the hydrogen atom associated with the 1*s-state* does not depend upon ϑ and φ (cf. Table 5). Therefore, $\theta \cdot \Phi$ is constant for all angles, and the polar diagram is a spherical surface. This is true of all s-orbitals. The number of the nodes is $n - 1$ (where n = principal quantum number). The 1s orbital is, therefore, a spherical cloud, in which the values of ψ diminish exponentially from the center as shown in Figure 7.

The hydrogen atom itself in the ground state is a spherically symmetrical system, whose probability density, ψ^2, also decreases exponentially from the center. A sec-

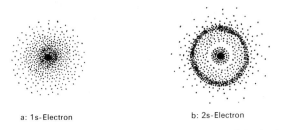

a: 1s-Electron   b: 2s-Electron

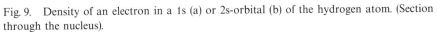

Fig. 9.   Density of an electron in a 1s (a) or 2s-orbital (b) of the hydrogen atom. (Section through the nucleus).

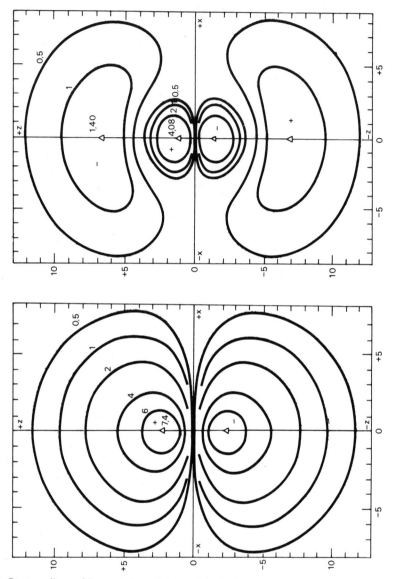

Fig. 10.   Contour lines of $2p_z$ and $3p_z$ orbitals of the hydrogen atom, showing the values of the normalized wave function multiplied one hundredfold. Maximum positive and negative values are shown by triangles. The distance from the nucleus, multiplied by $2/n$ (where n = principal quantum number), is shown on the axes in units of $a_0 = 0.53$ Å.

tion through the hydrogen atom gives an electron density distribution as shown in Figure 9a.

The 2s-*orbital* has a nodal surface in the form of a sphere at which the wave function changes sign. The electron density, ψ^2, forms a distribution like that shown in Figure 9b. Inside the spherical nodal surface the total residence probability is only 5.4%, while outside it corresponds to 94.6%. For the 3s-*orbital* which has two nodal planes the probabilities are, within the first nodal sphere, 1.5%; between nodal spheres, 9.5%; outside, 89.0%.

The angular functions of the three *p-orbitals* are symmetrical with respect to rotation about a defined axis. The product $\theta \cdot \Phi$ is a maximum (positive or negative) along this axis; and zero in all directions perpendicular to the axis. Therefore, xy is a nodal plane for the p_z orbital. The index z is assigned to the axis along which the angular function has maximum values.

Sometimes polar diagrams of angular functions, as shown in Figure 8, are termed orbitals, while strictly this term should apply only to the total wave function.

The form of p- and d-orbitals can best be seen in *contour diagrams* which are formed from a section through an orbital, including the nucleus, in which all points having the same ψ values are connected. In the case of the spherically symmetrical s-orbitals, the section can have any orientation with respect to the three axes x, y and z. The contour lines of s-orbitals are concentric circles. Those lines which connect points with $\psi = 0$ are the spherical nodes. The section for p-orbitals is selected so that the axis of rotation lies in the plane. The p_z orbital contour diagram for principal quantum numbers 2 and 3 is shown in Figure 10. *Boundary lines*, taken from the contour diagram, are used for a simple graphical representation of an orbital. These are selected so that the residence probability within such an area has a high value. The dimensions of the orbital representation depend on this value. Boundary lines are connections of points where the wave function ψ is invariant.

Figure 11 depicts a 2p-orbital of the hydrogen atom where the probability of encountering an electron within the boundary lines is 99%. The three-dimensional orbital is formed for s- and p-orbitals from the contour or boundary-line diagram by rotation around a diagonal (s-orbital) or around the preferred axis (p-orbital). Such three-dimensional models are shown schematically in Figure 12.

Aside from s- and p-orbitals, there are five d-orbitals which are important in chemical bonding. Among the d-orbitals, which are energetically equal in the hydrogen

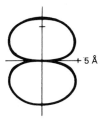

Fig. 11.   Representation of a 2p-orbital of the hydrogen atom with boundary lines. Rotation around the vertical axis yields the three-dimensional orbital, with a probability of encountering the electron within it of 99%.

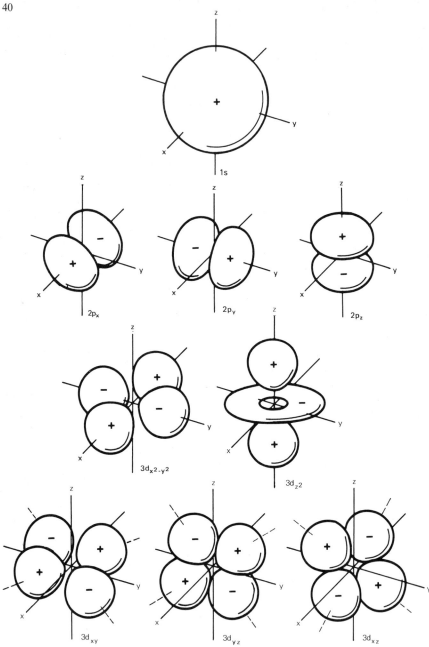

Fig. 12. Schematic representation of hydrogen atom orbitals with the signs of the wave function, ψ_1, shown. The s- and d-orbitals are symmetrical with respect to the origin; the p-orbitals are asymmetrical. The scales used for s-, p-, and d-orbitals are different.

atom, only the d_{z^2} has rotational symmetry. The other four each have three two-fold axes of rotation, and differ from each other in their spatial orientation in the coordinate system. The relative orientations are indicated by indices (cf. Figure 8). Contour diagrams of the representative orbitals $3d_{z^2}$ and $3d_{xz}$ are shown in Figure 13. The orbitals $3d_{xz}$, $3d_{yz}$, $3d_{xy}$ and $3d_{x^2-y^2}$ each possess two perpendicular nodal planes which pass through the nucleus. In the $3d_{z^2}$-orbital, the values of $\psi = 0$ are found on 2 conical surfaces, which are rotationally symmetrical with respect to the z-axis, and whose apices come into contact at the nucleus. The angle at the apices of the cones is $110°$. A spatial representation of the five 3d-orbitals is shown in Figure 12; however, it should be noted that this is only one possible set of d-orbitals; for example, the x-, y- and z-axis labels could be switched in any order and equally valid sets of orbitals would result.

The sign of the wave functions of the individual orbital lobes is shown in Figure 12. The function ψ itself has no physical reality, but the atom, characterized by ψ^2, exists. Two wave functions belong to each atomic state ψ^2, corresponding to the two roots $\sqrt{\psi^2} = \pm\psi$. The Schrödinger equation is independent of the sign of the wave function. Functions $+\psi$ and $-\psi$ thus describe identical states. The + sign is selected in the direction of the positive x, y, and z axes. The sign is then fixed in the remaining lobes, since it must change in passing through a nodal plane. For a 1s-orbital having no nodal plane, ψ may be either positive or negative everywhere. The wave mechanical theory of the hydrogen atom does not allow a precise definition of the position of the electron at any given time. Rather, it describes the electron as a cloud around the nucleus. The residence probability is large close to the nucleus; and even far from the nucleus it is not completely zero. The electron thus is not near the nucleus at all times. Wave mechanics treats the electron as a point charge and point mass, but the position coordinates are only given as probability statements. Owing to this uncertainty regarding the distance of the electron from the nucleus, the velocity has an associated uncertainty as well, and can only be described by a probability statement. For the ground state of the hydrogen atom, the most probable velocity, v, is $1.2 \cdot 10^6$ m/s.

The uncertainty in the velocity is reflected in the momentum, $p = m \cdot v$, of the electron. In contrast, owing to the complete uncertainty of its position, the angular momentum, $p_1 = m\omega r^2$, can be exactly defined (cf. p. 27):

$$p_1 = \sqrt{l(l+1)} \cdot \frac{h}{2\pi} \qquad (33)$$

l = azimuthal quantum number

It follows from Equation (33) that an s-orbital electron ($l = 0$) moves so that, on a time average, it has no angular momentum. This is a principal difference between the wave mechanical and the Bohr theories. In the latter the circular electron motion produces an angular momentum, as demanded by the quantum condition [Equation (10)] for all atomic states; but in the former, p_1 depends only on l.

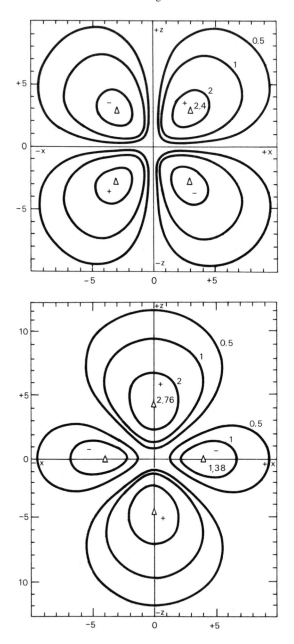

Fig. 13.   Contour diagrams of both $3d_{xz}$ and $3d_{z^2}$ orbitals of the hydrogen atom. The numerical values have the same significance as in Figure 10. Of the five different d-orbitals, only the d_{z^2} is rotationally symmetrical.

Owing to the negative charge of the electron, the orbital angular momentum, p_l, gives rise to a *magnetic moment*, μ_l:

$$\mu_l = \frac{e}{2\,mc} \cdot p_l = \frac{eh}{4\pi mc} \cdot \sqrt{l(l+1)} = \sqrt{l(l+1)} \cdot \mu_B \tag{34}$$

The constant factor, $\mu_B = eh/4\pi mc$ (where c = the velocity of light), is called the Bohr magneton, abbreviated B.M. This magnetic, or orbital, moment depends only on l as well, and for the s-, p-, and d-electrons is zero, $\sqrt{2}$ B.M. and $\sqrt{6}$ B.M., respectively. Electrons in orbitals of the same azimuthal quantum number have numerically identical orbital momenta, p_l, and magnetic moments, μ_l. Since p_l and μ_l are vectors, their direction is dependent of the spatial orientation of the orbital, and, therefore, dependent on the magnetic quantum number, as well. This fact is important for the structure of many-electron atoms, and, therefore, will be discussed in greater detail.

Consider an excited, one-electron atom with its electron occupying a d-orbital (l = 2). The orbital momentum vector can lie in one of five directions, depending upon which of the five d-orbitals is occupied. The orientation of this vector in space is completely random and undefined in the absence of an external field. However, in a magnetic field, the electron will tend to orient and align its angular momentum and magnetic moment. This results in precession, which, being a periodic motion, is quantized. The angular momentum vector precesses about the vector of the applied field. Only certain values of the angle between field and angular momentum vectors can exist for each orbital owing to the quantization of energy. This is the directional quantization of the angular momentum and its associated magnetic moment. Figure 14 shows the *directional quantization* for the case of l = 2 in a magnetic field, where the field vector coincides with the z-axis of the coordinates.

The angular momentum vectors of a electron in each of the five d-orbitals (m: $+2 \ldots -2$), and the precession of these vectors around the axis of the field are shown. While the length of the vectors for all d-orbitals is the same, the angular momentum component in the direction of the field, i.e., the projection of the angular momentum vector on the z-axis, depends on the magnetic quantum number, m, and has the value $m \cdot h/2\pi$. The magnetic quantum number thus specifies the angular momentum component of an electron in the direction of the z-(field) axis in units of $h/2\pi$. The same holds for the magnetic orbital moment, whose component in the z-direction is $m \cdot \mu_B$.

The number of possible values of m is $2l+1$, corresponding to the degree of degeneracy which is removed, however, by application of a field, resulting in $2l+1$ separate energy levels.

So far, the wave-mechanical treatment of the hydrogen atom, based upon the Schrödinger equation, has necessitated three quantum numbers: n, l, and m to distinguish the relative orbitals according to energy (n), orbital momentum (l), and orientation (m). The relativistic extension of wave mechanics by P.A.M. Dirac

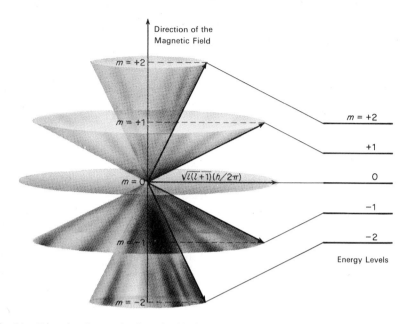

Fig. 14.   Directional quantization of orbital momentum in a magnetic field for the case l = 2 (l = azimuthal, m = magnetic quantum number), after W.J. Moore, Physical Chemistry, 5th ed., Prentice Hall Inc., Englewood Cliffs, New Jersey, 1972.

(1928) brought about two further developments: a quantized angular momentum of the electron, and a slight dependence of orbital energy on the azimuthal quantum number. This latter tiny energy difference between orbitals of identical principal quantum number, but differing azimuthal quantum number, led to the previously mentioned fine structure of the Balmer lines in the atomic spectrum of hydrogen (p. 25). It is of no chemical importance, and will be neglected.

The angular momentum of the electron is also quantized, and an additional quantum number, called the *spin quantum number*, with symbol s, must be introduced. The spin angular momentum, p_s, is:

$$p_s = \sqrt{s(s + 1)} \cdot \frac{h}{2\pi} \tag{35}$$

The spin quantum number, s, can take only the values $+\frac{1}{2}$ or $-\frac{1}{2}$, and can be visualized by assigning a rotatory motion to the electron about an axis going through its center. Since this motion can occur in only two directions, the spin quantum number is restricted to two values. The electron spin, and its associated angular momentum were demonstrated in 1925 by G. Uhlenbeck and S. Goudsmit from the emission spectra of the alkali metals, lithium, sodium, and potassium.

The spin angular momentum, like the orbital angular momentum, has an associated magnetic moment, the *spin moment:*

$$\mu_s = 2\sqrt{s(s+1)} \cdot \mu_B \tag{36}$$

The direction of both vectors p_s and μ_s coincides.

The introduction of the two spin states, $s = \pm\frac{1}{2}$, results in doubling of the number of electron states of the hydrogen atom shown in Figure 4. Each principal quantum number now has $2n^2$ different states, each characterized by four quantum numbers. The two spin states $+\frac{1}{2}$ and $-\frac{1}{2}$ are conventionally symbolized by vertical arrows ($\uparrow$ and $\downarrow$), which are fed into the "orbital boxes" in Figure 5.

Electron states which differ only in spin quantum number, s, are degenerate. Hence, the degree of degeneracy of s-, p-, and d-levels, taking into account the spin, is 2-, 6- and 10-fold, respectively. The energy of an atom in an electrical or magnetic field depends not only on the values of the quantum numbers, n, l, and m, but also on s, which gives, by analogy with the directional quantization of the orbital angular momentum, the component of the spin angular momentum in the direction of the field, in units of $h/2\pi$.

In the technique of *electron spin resonance* (ESR) a substance with unpaired electrons is placed in a strong, homogeneous magnetic field. The two electron spins $+\frac{1}{2}$ and $-\frac{1}{2}$ correspond to two different energy levels between which transitions are possible. Interaction of the energy difference, ΔE, with microwaves ($\Delta E = h\nu$) results in spin reversal of the electron, i.e., in a transition to the higher energy state. The absorption of radiation $h\nu$ is an indication of the presence of unpaired electrons. The fine structure of the absorption signal can be interpreted in terms of the localization of the electron in a particular atom, which gives rise to splitting owing to electron spin-nuclear spin coupling.

One-Electron Atoms with Z > 1:

In a one-electron atom with a nuclear charge > 1, the electron, according to Equation (30), has a lower potential energy than in the hydrogen atom. The wave functions are dependent upon nuclear charge, and the term $1/a_0$ in the hydrogen atom function is replaced by Z/a_0:

$$\psi(1s) = \sqrt{\frac{Z^3}{\pi \cdot a_0^3}} \cdot e^{-\frac{Zr}{a_0}} \tag{37}$$

$$\psi(2p_z) = \frac{1}{4}\sqrt{\frac{Z^3}{2\pi a_0^3}} \cdot \frac{Zr}{a_0} \cdot e^{-\frac{Zr}{2a_0}} \cdot \cos\vartheta \tag{38}$$

$$\psi(3d_{xz}) = \frac{1}{81}\sqrt{\frac{Z^3}{2\pi a_0^3}} \cdot \frac{Z^2 r^2}{a_0^2} \cdot e^{-\frac{Zr}{3a_0}} \cdot \sin\vartheta \cos\vartheta \cos\varphi \tag{39}$$

Partitioning these functions in R(r) and $\theta(\vartheta) \cdot \Phi(\varphi)$ parts reveals the dependence of the radial portions on Z. The radial functions of the orbitals contract strongly with increasing Z, lowering the orbital energy. The residence probability of an electron in the neighborhood of the nucleus thus increases with Z. The graphical contour and boundary line diagrams change scale with increase in Z, but the geometry and symmetry remain the same.

4.1.2. Many-Electron Atoms

A realistic view of many-electron atoms can only be provided by the wave-mechanical theory. Although the Bohr-Sommerfeld and other models for one-electron atoms furnish useful results for the hydrogen atom, they fail completely for many-electron atoms in that they no longer explain experimental results, even qualitatively. The following holds for neutral, many-electron atoms or monoatomic ions formed by ionization or electron capture.

In the wave-mechanical theory, the number of atomic electron states ($2n^2$) is independent of the nuclear charge number. Thus, the orbitals derived for the hydrogen atom are also present in many-electron atoms, and are occupied with increasing orbital energy until all the electrons are accommodated. The *Pauli principle* (1925), which postulates that no two electrons with all four quantum numbers identical may co-exist in an atom, holds for this filling process. Thus a state defined with discrete values of n, l, m, and s can only by occupied by a single electron.

The potential energy of an electron in a many-electron atom is determined by the electrostatic attraction of the nucleus and electron and the repulsion among the remaining electrons; i.e., the energy of each electron depends upon the coordinates of all electrons. However, since electrons do not move in a determinable manner, their potential energy cannot be calculated exactly, and hence, nor can the Schrödinger equation be solved exactly. However, approximate methods which give a satisfactory description of the electronic states in many-electron atoms are used for the solution of this problem. One method was devised by J.C. Slater.

The Slater treatment begins with the movement of electrons in a field which is considered to be the superposition of the nuclear and electron fields. Any one electron is thus not exposed to the total nuclear charge, Z, but to a smaller *effective nuclear charge*, Z*, which has been reduced by partial screening arising from the other electrons. The difference, $\sigma = Z - Z^*$, is called the screening constant. A centrally symmetric field is assumed to result on a time average. The state of an electron in this field can be described by functions similar to those given for single-electron atoms on p. 45. The radial portion now contains the effective nuclear charge, Z*.

Slater (1930) developed certain empirical rules for the determination of the sreening constants, σ, taking into account the more powerful screening effect of nearby electrons, and the dependence of Z* on the type of electron under consideration; Z* is always smallest for a valence electron, and largest for a 1s electron. The Slater rules are used by first placing the electrons into groups:

1s/2s, 2p/3s, 2p/3s, 3p/3d/4s, 4p/4d/4f/5s, 5p etc.

The screening constant, σ, for an electron is obtained by summation of the individual contributions:

a) Electrons in groups higher than the electron in question do not contribute.
b) Each electron in the same group contributes 0.35; if the electron in question is in a 1s orbital, the other electrons of the group contribute 0.30.
c) Electrons in the next lower-lying group contribute 0.85 each, if the electron for which σ is to be calculated is an s- or a p- electron. Electrons in still lower groups contribute 1.00. For a d- or f-electron, however, all electrons in lower groups contribute 1.00.

The treatment applies even if the orbitals of lower groups are only partially occupied, as is the case in excited states. The use of Slater rules can be illustrated with reference to two examples, a 3p-electron of a phosphorus atom in the ground state, and for the 3d-electron of the same atom in an excited state. The Z* values are calculated as follows:

$$P(1s^2\ 2s^2p^6\ 3s^2p^3)\text{: }\sigma\text{(for a 3p-electron)} \quad = 2\cdot1.0 + 8\cdot0.85 + 4\cdot0.35 = 10.2$$
$$Z^* = 15 - 10.2 = 4.8$$

$$P(1s^2\ 2s^2p^6\ 3sp^3d)\text{: }\sigma\text{(for the 3d-electron)} = 14\cdot1.0 = 14.0$$
$$Z^* = 15 - 14 = 1.0$$

Improved rules for calculation of Z* have been published by E. Clementi and D. L. Raimondi (J. Chem. Physics **38**, 2686 (1963)).

The simple Slater approximation provides only a crude picture of the electron configuration of many-electron atoms. More precise (and complicated) calculations use the *self-consistent field theory* of D.R. Hartree (1928) in which each electron is again assumed to move in a time-averaged, centrally symmetrical field of the nucleus and other electrons. For an atom with n electrons the average field of n − 1 electrons is calculated from their assumed wave functions. The wave function of the first electron is then calculated. This function and postulated functions for the remaining n − 2 electrons are used to obtain the wave function for the second electron, and the calculation is repeated for all n electrons. This iterative process is continued until the wave equations, or atomic orbitals become "self-consistent", i.e., until continued iterative calculations do not result in further refinement of the orbitals. This so-called SCF method was further improved by V. Fock.

Hartree-Fock calculations furnish the best descriptions of electron distributions in many-electron atoms at the present time. Experimental tests of the SCF method are afforded by the data on radial electron density distributions from scattering experiments. For example, good agreement is found for the argon atom (Z = 18). Orbitals calculated with the Hartree-Fock method differ from hydrogen atom

orbitals only in the radial function R(r). The geometry of many-electron orbitals is hydrogen-like, and the separation into s-, p-, d-, and f-levels is maintained.
The degeneracy of orbitals of same principal quantum number is, however, removed in part by mutual repulsion of the electrons. The orbital energy now depends on both quantum numbers n and l; and increases with increasing values of n or l.
Orbital energies, E (in units of E_H), are plotted in Figure 15 on a double logarithmic scale vs. the electron or nuclear charge number. The larger the value E/E_H, the more stable is the corresponding level. Orbitals of a given subshell are still degenerate.

The orbital energies in a many-electron atom are not identical with the negative ionization energies of the respective electrons. When, for example, an electron is removed from the 2p orbital of a fluorine atom, all the electron-electron repulsions in the resulting F^+ ion are reduced, and hence its orbitals become more stable. Thus, stabilization energy is gained

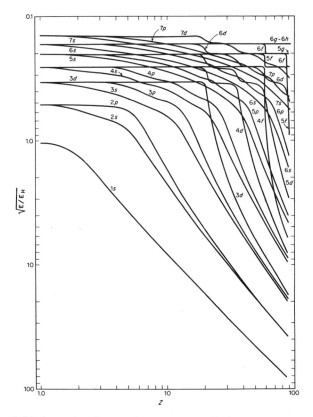

Fig. 15.   Orbital energies of many-electron atoms calculated as a function of Z. The energy is plotted in units of $E_H = -13.6$ eV, where E_H is the energy of the hydrogen atom in the ground state (after W.J. Moore, Physical Chemistry, 4th Ed., Prentice Hall Inc., Englewood Cliffs, New Jersey, 1972).

through ionization. Since better methods for the determination of orbital energies are not available currently, the negative ionization energy obtained from atomic, or in the case of bonded atoms, from photoelectron spectra, is used as a "pseudo-orbital energy".
In photoelectron or ESCA spectroscopy a compound is ionized by UV or x-rays of known wavelength and energy, E. The difference, $E - E_{kin}$, the kinetic energy of the emitted electrons, gives the energy required for ionization. Beside of the sign this is the orbital energy of the atomic or molecular orbital from which the electron originated.

Figure 16 shows how the energies of the valence orbitals of the first row elements are related to their position, i.e., how the energies depend upon the effective nuclear charge or number of electrons in their respective levels.

4.1.3. The Aufbau Principle and the Periodic System

Figure 15 shows that there is no simple progression of orbital levels with increasing nuclear charge, Z, which is valid for all atoms. While for the hydrogen atom, for example, $E(3d) < E(4s)$, this relationship is reversed at increasing atomic number (near $Z = 7$); at further increases in Z, $E(4s)$ is again larger than $E(3d)$.
Similar situations are found for other orbitals or levels. The formal buildup of the elements by stepwise addition of electrons outside as well as protons and neutrons within the hydrogen nucleus results in a sequence of levels of increasing energy which differs considerably from that found in the hydrogen atom. The observed sequence as shown in Figure 17 of increasing energy (decreasing stability) is:

$$1s\ 2s\ 2p\ 3s\ 3p\ 4s\ 3d\ 4p\ 5s\ 4d\ 5p\ 6s\ 4f \approx 5d\ 6p\ 7s\ 6d \approx 5f$$

The stepwise occupation of orbitals by electrons is also indicated, but is only approximately correct, since levels decrease in energy with increasing occupation by electrons, and can fall below the preceding orbital. The energy of a p-level, e.g., decreases while 1 to 6 electrons are added with a simultaneous increase in nuclear charge number, as shown in Figure 16.

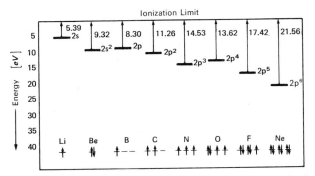

Fig. 16.   Relative stabilities of the highest occupied orbitals of atoms in the first period. The length of the vertical arrows corresponds to the respective ionization energies.

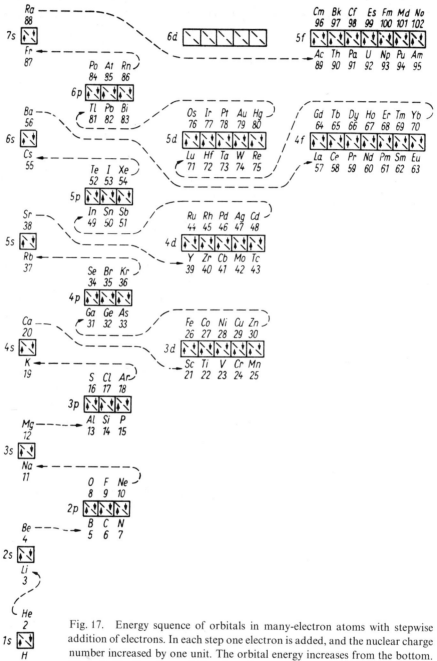

Fig. 17. Energy squence of orbitals in many-electron atoms with stepwise addition of electrons. In each step one electron is added, and the nuclear charge number increased by one unit. The orbital energy increases from the bottom. (From L. Pauling, General Chemistry, 3rd Ed., Freeman, San Francisco, 1970).

The electron configuration of the atoms underlies the organization of the periodic system, where elements with analogous valence electron configurations form vertical groups, whereas in the horizontal rows (periods) the number of electrons increases stepwise. Main group elements in the ground state have empty or completely filled d- and f-levels. The valence electrons of the main group elements in the ground state are s- or p-electrons.

The electron configuration of atoms is abbreviated using the following symbolism; for example, sulfur in the ground state:

$$S(KL3s^23p^4) \quad \text{or} \quad S[Ne]3s^23p^4$$

Shells K and L are completed; in shell M ($n = 3$) there are two electrons in the s-, and four in p-level (one p-orbital is doubly-occupied, the other two singly); all other levels are unoccupied. An excited state of the sulfur atom is written

$$S(KL3s3p^33d^2) \quad \text{or} \quad S[Ne]3s3p^33d^2$$

Here one s- and one p-electron are promoted to the d-level, requiring ca. 2390 kJ/mol. The atom now contains 6 unpaired electrons.

It can be seen that these symbols only refer to the principal and azimuthal quantum numbers of the electrons, but do not specify the values of m and s, and, hence, do not uniquely define the state of the atom. Term symbols will be introduced later to accomplish this.

To this point only the electrostatic repulsion of the electrons has been considered. However, electron magnetic orbital and spin moments, which give rise to weaker interactions, will now be discussed, with spin moments first. Consider an atom with two electrons in different orbitals. The two electron magnetic spin moments are not independent in their direction, but are mutually coupled since each electron moves in the magnetic field of the other. This results in an interaction in which both vectors tend to become parallel to achieve energy minimization. Owing to the directional quantization of spin, only two possible values, $+\frac{1}{2}$ and $-\frac{1}{2}$, are available to each electron. At the minimum of energy, both electrons have the same spin quantum number. As they rotate a force is exerted which results in magnetic coupling and a precession of the rotational axis. This in turn gives rise to a precession of the magnetic moment and angular momentum vectors.

The two angular momentum vectors, which have a value of $s \cdot h/2\pi$ in the z-direction, add vectorially to give a total angular momentum, characterized by the total spin quantum number, S:

$$\sqrt{S(S+1)} \cdot \frac{h}{2\pi} \tag{40}$$

In many-electron atoms with i electrons, S is the sum of all spin quantum numbers:

$$S = \sum_i s_i \tag{41}$$

The total spin quantum number, S, may take whole or half integer values, and is used without a sign. Since the Pauli principle stipulates that orbitals are filled by two electrons of opposing spin, S is equal to zero for all filled orbitals, levels, and shells. Thus, only partially-occupied orbitals are taken into account in calculating S. The magnitude $2S + 1$ is called the *spin multiplicity* of a state. In contrast to S, the multiplicity is always a whole integer. States with $2S + 1 = 1, 2, 3, \ldots$ are called singlet, doublet, triplet, etc., states.

These considerations are the basis for *Hund's first rule:*

An atom in the ground state assumes the highest spin multiplicity possible with the electrons in the most stable orbitals.

According to this experimentally confirmed rule, the p-, d-, and f-electrons start filling their levels with spins parallel. With two electrons in a p-level, of the two configurations

S = 0, 2S + 1 = 1 (Singlet)   S = 1, 2S + 1 = 3 (Triplet)

the triplet (ground) state has a lower energy than the singlet (excited) state. Only after the orbitals are occupied singly do the electrons of anti-parallel spin begin to fill.

Two electrons of antiparallel spin occupying the same atomic orbital, and whose angular momentum and magnetic spin moments hence are compensated, are called an electron pair. A schematic representation using box orbitals is found here and in Figure 5. The conventional representation should not be interpreted to imply that the two electrons are close together. On the contrary, owing to their like charge they assume a maximum distance from each other. The effect of electrostatic repulsion is referred to as charge correlation.

Corresponding to the magnetic coupling of the spin moments, there is a similar interaction of the orbital magnetic moments of the individual electrons. The magnetic orbital moment of an electron, with the value $\sqrt{l(l+1)}\,\mu_B$, interacts with a second electron so that directional quantization occurs. Both vectors can assume only certain discrete values, so that the vector component of the angular momentum along the z-axis is $mh/2\pi$ (cf. Figure 18). The angular momentum vectors add to give a total orbital angular momentum, defining the z direction and specified by quantum number L:

$$\sqrt{L(L+1)} \cdot \frac{h}{2\pi} \tag{42}$$

L is the sum of the magnetic quantum numbers of all electrons:

$$L = \sum_i m_i \tag{43}$$

L is always an integer and, like S, has no sign. Formally, L gives the z-component of the total orbital angular momentum in units of $h/2\pi$. For values of L = 0, 1, 2, 3, ..., capital letters S, P, D, F, ... are used. The total orbital angular momentum of a completed shell or a filled level is always zero, thus only the partially occupied levels are used to calculate L; L is also zero for a half occupied level (one electron per orbital). Formally, L gives the z-component of the total orbital angular momentum in units of $h/2\pi$.

If two electrons occupy a p-level, the Pauli principle and Hund's first rule demand that they reside in different orbitals: e.g.:

a: | ↑ | ↑ | | L = 1   b: | ↑ | | ↑ | L = 0
    m: +1   0   −1   m: +1   0   −1

The two states (a) and (b) do not have the same energy. In the energy minimum the two magnetic orbital moments tend to be as close to parallel as possible to give the largest total orbital angular momentum. Therefore, (a) (P-state) has lower energy than (b) (S-state). This is accounted for in *Hund's second rule* which states: If there are several atomic states possible with highest spin multiplicity, the ground state is the one with the largest L-value.

The coupling of angular momenta to give a total angular momentum, S, and of orbital angular momenta to give a total orbital angular momentum, L, is referred

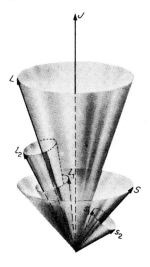

Fig. 18.   An example of Russel-Saunders coupling. Vectors l_1 and l_2 add to give the total orbital angular momentum vector, L. In the same fashion, angular momentum vectors s_1 and s_2 add to form S. L and S then precess around total angular momentum vector J (after W.J. Moore, Physical Chemistry, 5th Ed., Prentice Hall Inc., Englewood Cliffs, New Jersey, 1972).

to as Russel-Saunders-, or L, S-coupling. Although spin and orbital moments of individual electrons do not mutually interact, magnetic coupling of the total spin momentum with the total orbital momentum does take place. Vector addition produces the total angular momentum and total magnetic moment of the atom (cf. Figure 18).

The total angular momentum, J, has the magnitude:

$$\sqrt{J(J+1)} \cdot \frac{h}{2\pi} \tag{44}$$

If $J \neq 0$, the atom has a permanent magnetic moment.

Russel-Saunders-coupling holds only for lighter atoms, with ≤ 55 electrons (Cs), embracing most of the non-metals. For given L and S values, J can assume values from $|L + S|$ to $|L - S|$. Thus, e.g., J for $L = 1$ and $S = 1/2$ can assume values of $3/2$ and $1/2$. If L is larger than S, the number of J-values is $2S + 1$; if S is larger than L, $2L + 1$ J-values can exist. If L and S are both zero (completed shells or levels), J must also be zero.

Hund's third rule is invoked for the J-value of an atom in its ground state:

The energy of a state increases with J, in the case of a level less than half-occupied. Otherwise the energy decreases with J.

For the example just cited ($L = 1$, $S = 1/2$) the ground state $J = 1/2$. Based on this vector model, term symbols can now be introduced of the form:

$$^{2S+1}L_J$$

which, in conjunction with the symbols introduced earlier (p. 51) permit an exact description of the electronic state of an atom. The upper left exponent of L, $2S + 1$, is the spin multiplicity, the lower right subscript the quantum number, J. The term symbol for the hydrogen atom is thus $^2S_{1/2}$, since $s = 1/2$, hence $S = 1/2$; in addition, $l = 0$, hence $L = 0$ (symbol S) and correspondingly $J = 1/2$.

Quantum numbers S, L, and J for completed levels are zero, and are not used in deriving the term symbol, where only partially-filled levels must be considered. In the main groups of the periodic system, elements with analogous valence electron configuration are arranged vertically and have the same term symbols as the lightest member of the group:

Li	Be	B	C	N	O	F	He
$^2S_{1/2}$	$^1S_0$	$^2P_{1/2}$	$^3P_0$	$^4S_{3/2}$	$^3P_2$	$^2P_{3/2}$	$^1S_0$

Finally, as an example, in the photolysis of NO_2, oxygen atoms in the ^1D state are formed because of spin conservation:

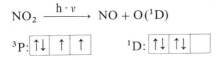

According to Hund's first rule, the ^{1}D-state of the oxygen atom is of higher energy than its ^{3}P ground state, with an energy difference of 201 kJ/mol.

In summary, the energy of each electron in a many-electron atom depends upon all four of its quantum numbers n, l, m, and s, because of the mutual interaction of the electrons based upon electronic charge and on magnetic orbital and spin momenta.

4.2. The Structure and Properties of Atomic Nuclei

Nuclei consist principally of protons and neutrons. A nuclide characterized by certain numbers of protons and neutrons is written in symbols such as $^{10}_5$B or $^{35}_{17}$Cl, where the left exponent is the mass number (the sum of the proton and neutron numbers) and the subscript the proton number (= atomic number, Z). Nuclides which differ only in the number of neutrons are isotopes.

Charge number can be obtained, e.g., from analysis of the x-ray spectra. According to H. Moseley (1913), the wave number of the K(α) line of the x-ray emission spectrum of an atom is:

$$\nu[\text{cm}^{-1}] = \frac{3R}{4} \cdot (Z - 1)^2 \tag{45}$$

R = Rydberg constant
Z = Atomic charge number

The K(α) line is the shortest major line of the atomic spectrum, and results from the decay of an electron from the 2p to a vacancy in the 1s level arising from the ejection of an electron by high energy irradiation. The Z of Moseley's law, as was discovered later, corresponds to the nuclear charge and atomic number of the element in the periodic system. The law can be derived from the Rydberg series [Equation (18)] by making the following substitutions.

$n_i = 1$, $n_j = 2$; for $e^4 : e^4 Z^{*2}$; and for $Z^* : Z - 1$ (screening of nuclear charge by the second 1s electron).

The masses of nuclides can be determined by high-resolution mass spectrometers with great accuracy. The mass of the atom is the sum of the nuclear and electronic masses. The nuclear mass, however, is found to be less than the sum of the individual neutron and proton masses by a quantity called the mass defect. Considerable binding energy is liberated when nucleons combine, and this energy corresponds to a loss of mass through the mass-energy relationship [Equation (1)]. Binding energies per nucleon for nuclei with mass numbers >15 are ca. 8 MeV (1 MeV: 1 Megaelectron volt $= 10^6$ eV); and are less for lighter nuclei. Orbital energies of valence electrons, and chemical bond energies which range between 1 and 100 eV, are by comparison vanishingly small and do not show a mass defect. Table 6 lists some exact nuclide masses which allow such nearly identical molecules as $^{14}N_2\,^1H_4$ (32.037) and $^{16}O_2$ (31.990) to be distinguished. Mass numbers are rounded-off nuclide masses.

Table 6.   Exact Masses of Selected Nuclides (in Atomic Units).

^1H	1.007825	^{19}F	18.998405
^2H	2.014102	^{28}Si	27.976929
^{12}C	12.000000	^{31}P	30.973765
^{13}C	13.003354	^{32}S	31.972073
^{14}N	14.003074	^{35}Cl	34.968851
^{16}O	15.994915	^{79}Br	78.918329
^{18}O	17.999160	^{127}I	126.904470

Atomic nuclei are only stable with certain proton/neutron ratios. For the lighter nuclides, the ratio is nearly unity; for heavier nuclides, it decreases steadily. Nuclides with large deviations from these ratios undergo radioactive disintegration, the final products of which are stable nuclides.

The atomic weight of an element (atomic mass) is an average of the masses of its naturally occurring nuclides, in their relative abundances.

Nuclear radii were shown by Rutherford to be ca. 10^{-13} cm. The precise value can be obtained by scattering of α-particles or neutrons. These methods yield the relationship:

$$r[\text{cm}] = 1.3 \cdot 10^{-13} \cdot \sqrt[3]{M} \qquad (46)$$

r = nuclear radius
M = mass number

The average density of a nucleus can then be calculated from Equation (46) to be ca. $2 \cdot 10^{14}$ g/cm^3; i.e., 1 cm^3 of nuclear matter would weigh 200 million tons.

Nuclei, like atoms, have no sharp boundaries; their radii are determined by the gradual decrease of nuclear forces. The extent of these forces, however, is much smaller than the Coulombic forces responsible for the gradual decrease in density of the electrons in the atom. Nuclei can be ellipsoidal (oblate or prolate), but deviations from spherical are small.

Protons and neutrons have a spin connected with their magnetic moments (cf. Table 1), indicating that these particles are also structured with charge separation between their center and periphery. In addition to spin, these nucleons also display angular momentum. Like electrons in the atom, both momentum components add vectorially to give a total nuclear angular momentum which is termed *nuclear spin* and symbolized by I. This quantum number may take positive whole or half integer values. The nuclear magnetic moment is written as:

$$\mu = g \cdot \sqrt{I(I+1)} \cdot \mu_K \qquad (47)$$

where g is the magnetogyric ratio; μ_K is the nuclear magneton and corresponds to the atomic Bohr magneton as its electronic counterpart:

$$\mu_K = \frac{e \cdot h}{4\,Mc} \qquad (48)$$

$M = $ mass of protons
$c\ = $ light velocity

Nuclear spins are directionally quantized. In a magnetic field, only $2I+1$ discrete orientations of the angular momentum vector to the field axis are permitted, each corresponding to different energies with the particular energy differences dependent on the magnetic field (cf. Figure 19).

In an atom with a non-zero angular momentum of the electron cloud, nuclear spin and electron angular momentum can take only certain directions to each other. This leads to the hyperfine structure in atomic spectral lines from which nuclear spin, I, can be determined. Values of I for selected nuclides are listed in Table 7.

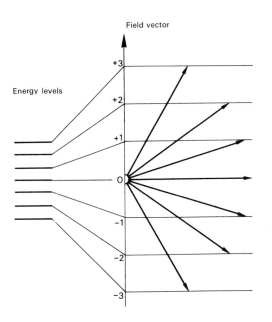

Fig. 19.   Directional quantization of nuclear spin in a magnetic field for the case $I = 3$.

Table 7.   Nuclear Spin, I, and Nuclear Quadrupole Moment, eQ, of Selected Nuclides

Nuclide	I	Q in multiples of $e \cdot 10^{-24}$ cm^2
^{1}H	1/2	0
^{2}H	1	0.00277
^{10}B	3	0.111
^{11}B	3/2	0.0355
^{19}F	1/2	0
^{31}P	1/2	0
^{35}Cl	3/2	-0.0797
^{37}Cl	3/2	-0.0168

Nuclei with even mass numbers have even-numbered nuclear spins in the ground state, while odd mass number nuclei have half integer values of I.

In the nuclear magnetic resonance (NMR) method, the nuclear spin is directionally quantized by an external magnetic field. An additional energy in the form of an electromagnetic, radio frequency field (E = hν) brings about transitions between different energy levels. Electron screening in the atom modifies the magnetic field in the neighborhood of the nucleus based on the bonds made to neighboring atoms. Therefore, the resonance of a given nucleus, i.e., the transitions from ground to excited states, is dependent on chemical environment. The frequency is usually held constant while the magnetic field strength is varied until resonance occurs. Nuclei with spin I = 1/2 are particularly useful in NMR.

A model of nuclear angular momentum as spin is depicted in Figure 20. Nuclei with I ≧ 1, in addition to nuclear spin, also have an electric quadrupole moment e · Q. In terms of Figure 20, the quadrupole moment arises from the flattening or extension of the nucleus in the direction of the axis of rotation. The magnitude of e · Q is a measure of the deviation from sphericity; nuclei with e · Q = 0 are spherically symmetrical, whereas quadrupolar nuclei behave as if their electrical charge is distributed over a rotating ellipsoid. A spherical nucleus of protons and neutrons (or their components) distributed in a non-statistical array would also have a non-spherical charge distribution. Nuclear quadrupole resonance spectroscopy is based on the electric quadrupole moment.

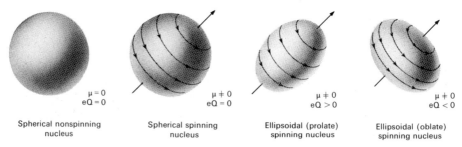

μ = 0 eQ = 0	μ ≠ 0 eQ = 0	μ ≠ 0 eQ > 0	μ ≠ 0 eQ < 0
Spherical nonspinning nucleus	Spherical spinning nucleus	Ellipsoidal (prolate) spinning nucleus	Ellipsoidal (oblate) spinning nucleus

Fig. 20.   Models for atomic nuclei with various dipole and quadrupole moments (after W.J. Moore, Physical Chemistry, 4th ed., Prentice Hall, Inc., Englewood Cliffs, 1972).

5. Theory of the Chemical Bond

A satisfactory theory of the chemical bond should be able to provide answers to the following fundamental questions:

> Why do atoms combine to form molecules?
> Why do they do so in certain ratios, and often in several ratios, as, e.g., NO, NO_2?
> Why do molecules and crystals have discrete structures?

In order to answer these and related questions, it is appropriate to consider certain distinct bond types:

> ionic bonds
> covalent (and related coordinate) bonds
> van der Waals bonds.

There is a continuous transition between bond types, and in most compounds more than one of these idealized types is present.

5.1. The Ionic Bond

5.1.1. Introduction

A large number of compounds crystallize in ionic lattices consisting of periodic, regular, three-dimensional arrays of anions and cations. The ions may be elemental or complex (derived from compounds), as:

Li^+ and H^- in LiH	NO^+ and HSO_4^- in $NOHSO_4$
Ca^{2+} and F^- in CaF_2	H_3O^+ and ClO_4^- in H_3OClO_4
Al^{3+} and O^{2-} in Al_2O_3	NH_4^+ and BF_4^- in NH_4BF_4

Stoichiometric ratios of anions and cations achieve electrical neutrality in the crystal. Crystal type, i.e., geometry and symmetry of the lattice, is dictated by the relative sizes of the ions and their charges. Ions are generated from atoms by supplying the ionization energy, I, or by liberating the electron affinity, A.

5.1.2. Ionization Energy

The first ionization of a gaseous atom, X:

$$X(g.) + I_1 \rightarrow X^+(g.) + e^- \tag{49}$$

requires the first ionization energy, which is positive, i.e., energy must be supplied to the system. The value of I_1 is influenced by the position of X in the periodic system. Metal atoms ionize most easily and noble gas atoms with most difficulty. Ionization energies range from ca. 4 to 25 eV, or 400 to 2400 kJ/mol. Figure 21 plots the first ionization energy as a function of atomic number.

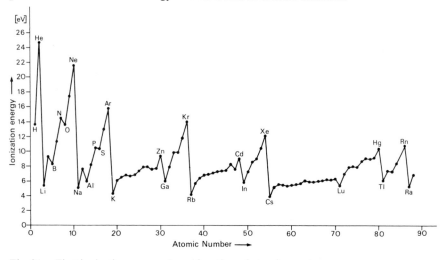

Fig. 21.   First ionization energy, I_1, as function of atomic number.

The second ionization energy, I_2, follows from:

$$X^+(g.) + I_2 \rightarrow X^{2+}(g.) + e^- \tag{50}$$

and is always considerably larger than I_1 because the electron must now be removed from a particle already bearing a positive charge. The value of I_2 changes periodically with atomic number, like I_1, and maximizes at the ions having the rare gas configuration (e.g., Na^+, K^+, etc.). Figure 21 would be thus shifted to the right by one atomic number for I_2, and once again for the third ionization energy, I_3.
The markedly high ionization energy of rare gas atoms and rare gas-like ions is of great importance and can be traced to the high effective nuclear charge, Z^*, acting on the valence electrons in this configuration, which can be calculated by the rules on p. 47. For the light elements of the first period, the following values are obtained:

Li	Be	B	C	N	O	F	Ne
1.30	1.95	2.60	3.25	3.90	4.55	5.20	5.85

5. 1. 3. Electron Affinity

Many non-metallic atoms capture electrons in the gas phase in an exothermic reaction:

$$Y(g.) + e^- \rightarrow Y^-(g.) + A_1 \tag{51}$$

The electron affinity, A_1, is by definition positive if energy is released by the system. For most non-metallic atoms, values of A_1 range from ca. -0.6 to 3.6 eV, or -58 to $+350$ kJ/mol. Addition of a second or third electron is always strongly endothermic, i.e., A_2 and A_3 are always negative. The numerical values shown in Table 8 are from experiment and theory.

Table 8.   Electron Affinities of Non-Metals at 0 K in eV; values of A_1 and (in brackets) of A_2.

H: 0.755					He: -0.22
B: 0.2–0.3	C: 1.12	N: -0.2 (-8.3)	O: 1.47 (-8.1)	F: 3.45	Ne: -0.30
	Si: 1.4	P: 0.78	S: 2.07 (-6.1)	Cl: 3.61	Ar: -0.37
	Ge: 1.4	As: 0.6–0.7	Se: 2.1–2.2 (-4.4)	Br: 3.36	Kr: -0.42
		Sb: 0.6	Te: 2.0–2.1	I: 3.06	Xe: -0.45

The magnitudes of the electron affinities also illustrate the special stability of the rare gas-like atoms and ions. The largest A_1 values are found for the halogens whose anions have spherically symmetrical, rare gas configurations.

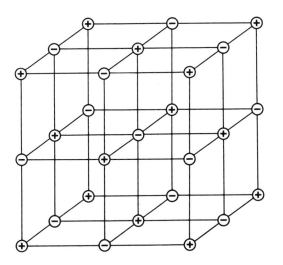

Fig. 22.   Ion positions in the rock salt (NaCl) lattice.

5.1.4. Ion Lattices and Ionic Radii

Rock salt will be discussed as an example of an ionic lattice. Precise internuclear distances can be dermined from X-ray diffraction. Sodium chloride has a cubic, face-centered unit cell. The unit cell is the smallest array containing all the symmetry elements of the crystal which can be formed from it by periodic repetition.

The location of the ions in a rock salt crystal is shown in Figure 22.

From precise x-ray diffraction data it is possible to obtain, in addition to ion location, the total electron density distribution in a crystal. For rock salt, the result is shown in Figure 23 in the form of a contour diagram for a segment of the (110) plane. The electron density decreases from the center of the atoms to the outside;

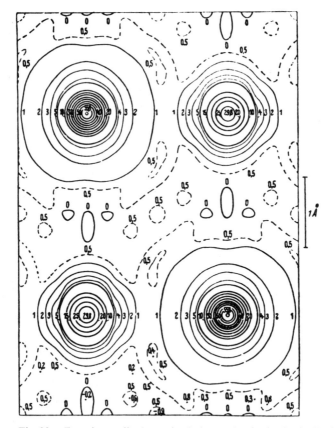

Fig. 23. Experimentally determined charge density in the (110) plane of a NaCl crystal in $e^-/Å^2$. The larger Cl^- ions have a maximum charge density of 55.8, the Na^+ ions, 29.8 $e^-/Å^2$. The (110) plane is so positioned that like ions are vertically stacked (cf. Figure 22). (After H.A. Stuart, Molekülstruktur, Springer-Verlag, Berlin, 3rd Ed. 1967.)

rapidly, at first, then more slowly, as in the profile in Figure 24. Along the cation-anion axis the charge falls to <0.2 electron/$Å^3$. This point can be considered as the boundary of the two contacting ions of opposite charge, with the ionic radius defined as the distance from the nucleus to this point. Integration of the electron density over the spherical volumes limited by the ionic radii yields 10.05 electrons for Na^+ and 17.70 electrons for the Cl^- ion, compared to 10 and 18 for the pseudo-Ne und Ar configurations. The missing 0.25 electrons are found in the interstices in the spherical packing which were not considered in the integration (cf. Figure 23). Thus the lattice consists of ions and not atoms.

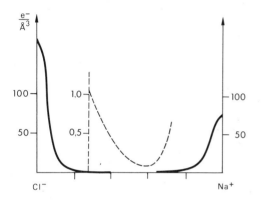

Fig. 24. Experimentally determined electron density along the Na—Cl internuclear axis in the NaCl lattice. Two scales are indicated.

Ionic radii derived from electron density diagrams are not constant, but depend on the particular lattice in which the ion finds itself. Dependence of the ionic radius on the number of electron shells and the charge is illustrated below:

	Cation radius [Å]	Anion radius [Å]
NaCl	1.15	1.66
KCl	1.45	1.70
LiF	0.92	1.09
CaF_2	1.26	1.10
MgO	1.02	1.09

Ionic radii can be defined in various ways, but all increase with increasing principal quantum number of the valence electrons, and decrease with increasing positive charge because of the change in effective nuclear charge and electron-electron repulsion. The radii of the isoelectronic ions shown below increase from cations to anions:

$$Ca^{+2} < K^+ < Cl^- < S^{-2}$$

5.1.5. Lattice Energy

The stability and the properties of ionic compounds derive from the lattice energy, U_0, which is liberated when molar equivalent amounts of gaseous cations and anions form a single crystal:

$$X^+(g.) + Y^-(g.) \rightarrow XY(s.) + U_0 \tag{52}$$

U_0, as energy liberated by the system, always carries a negative sign.

Table 9.   Components of Lattice Energy (in kJ/mol)

Energy	LiF	NaCl	CsI
Coulomb Interaction	−1200	−860	−618
Repulsion	+180	+100	+61
Van der Waals Attraction	−16	−16	−46
Zero Point Energy	+16	+8	+29

Lattice energy consists of several component parts which are summarized for three alkali halide crystals in Table 9. The greatest contribution arises from electrostatic attractions and repulsions. In addition, van der Waals attraction acts independently of charge (see Section 5.3). At equilibrium attractive forces are balanced by repulsion arising from mutual interpenetration of the electronic clouds of neighboring ions. Like atoms, ions have no fixed boundaries, and in NaCl crystal formation the electron clouds of cations and anions mutually penetrate to reach an internuclear distance of 2.814 Å, resulting in some contraction of the ions. This increases the potential energy of the ions, and lowers the lattice energy as shown in the second line of Table 9.

The fourth component of lattice energy, the zero point energy, is the vibrational energy of the ions, which exists even at 0 K. This energy can be calculated from the lattice vibration frequencies obtained from infrared spectra. The zero point energy has a negligible effect on the lattice energy; it is generally neglected, together with the van der Waals force contribution, since both are small, have opposite signs, and thus tend to compensate.

5.1.6. Determination of Lattice Energy

Lattice energy values are obtained by indirect means. For simple salts, U_0 can be calculated easily from Coulombic interaction, but if thermodynamic properties are known, the Born-Haber cycle is used to calculate U_0. The electrostatic equation treats the ions of a salt, MY, having an NaCl structure, as point charges. Each ion, M^+ and Y^-, in this salt is surrounded by six oppositely-charged neighboring ions at a distance, d. At distance $\sqrt{2}d$ lie 12 like-charged ions; farther, at distance $\sqrt{3}d$, lie 8 ions of opposite charge, etc. Taking into account at first only Coulombic attraction and repulsion forces to calculate the potential energy U:

$$U = - \frac{Z_C \cdot Z_A \cdot e^2 \cdot N_L \cdot A}{d} \qquad (53)$$

e   = elementary charge
$Z_{C,A}$ = ionic charge number
N_L = Avogadro's number
A   = Madelung constant
d   = smallest distance between oppositely charged ions

The Madelung constant, A, takes into account the successive shells of neighboring ions, and is the result of a convergent series depending only on the lattice type[1]. Values of A range between 1.6 and 4.0, increasing with the coordination number of the ions [NaCl (rock salt): 1.75; CsCl: 1.76; CaF_2 (fluorite): 2.52; TiO_2 (rutile): 2.39; Al_2O_3 (corundum): 4.04].
The repulsive energy from the electron clouds of neighboring ions is determined using the Born Landé equation. The value for the potential energy, U (Equation 53), is multiplied by the factor $(1 - 1/n)$:

$$U_o = - \frac{Z_C \cdot Z_A \cdot e^2 \cdot N_L \cdot A}{d} \cdot \left(1 - \frac{1}{n}\right) \qquad (54)$$

The number, n, can be derived from salt compressibilities, and lies in the range 7 to 10. The repulsion of neighboring ions thus decreases the lattice energy by 10 to 15% (cf. Table 9).
The smaller the internuclear distance, (e.g., LiF vs. CsI) or the larger the ionic charges (e.g., Al_2O_3, cf. Table 10), the larger the lattice energy. A change in n of one or two units, by contrast, has little effect on U_0.

Table 10.   Lattice Energies of Some Salts (in kJ/mol).

	F^-	Cl^-	Br^-	I^-	O^{2-}
Li^+	−1033	−846	−798	−740	
Na^+	−959	−778	−740	−693	
K^+	−814	−709	−680	−641	
Rb^+	−778	−687	−659	−623	
Cs^+	−748	−653	−633	−602	
Mg^{2+}	−2939		−2399	−2290	−3932
Ca^{2+}	−2609	−2227	−2131	−2039	−3479
Al^{3+}					−15111
Ag^+		−874			

1 Definitions of the Madelung constant vary. For example, the charge number, Z, is sometimes included in A. The reader should check on the definition of these constants before using tabulated values [cf. D. Quane, J. Chem. Educ. **47**, 396 (1970)].

Born-Haber Cycle:
According to Hess' law of constant heat summation, the energy difference between two states is independent of the route taken from initial to final state. In order to arrive, e.g., at the lattice energy of NaCl one can traverse a hypothetical cyclic scheme (M. Born and F. Haber, 1919):

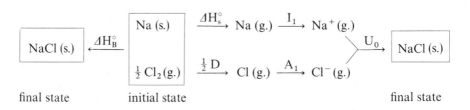

final state   initial state   final state

The individual energy and enthalpy parameters in this cycle are:

U_0 lattice energy of NaCl
ΔH_B° standard enthalpy of formation of crystalline NaCl
ΔH_S° standard enthalpy of sublimation of Na
I_1 first ionization energy of the Na atom
D dissociation energy of the Cl_2 molecule
A_1 electron affinity of the Cl atom

The contributions are exclusively enthalpy values, ΔH°, since the determinations are carried out at constant pressure (at 25 °C). Thus the Born-Haber cycle method gives lattice enthalpies, differing from lattice energies by the volume work, $p\Delta V$:

$$\Delta H = \Delta U + p\Delta V \tag{55}$$

$$(p = const.)$$

For NaCl, $p\Delta V$ at 25 °C amounts to 5.0 kJ/mol which is within the limits of error of the measured lattice energies, where accuracy is seldom $<2\%$. Hence, the difference between the lattice energy [as derived from Equation (54)] and lattice enthalpy may be neglected in most cases.

The Born-Haber cycle gives the relationship of U_0 and other parameters.

$$U_0 = \Delta H_B^\circ + A_1 - \tfrac{1}{2}D - I_1 - \Delta H_S^\circ \tag{56}$$

Lattice energies so determined agree with those derived from Equation (54) to within a few kJ/mol, consistent with the fundamental ionic crystal model.
Larger deviations occur only when, as with e.g., CuBr, the system is no longer purely ionic.

5.1.7. Importance of the Lattice Energy

Lattice energy reflects the strength of the bonds between ions in a crystal. The chemical and physical properties of salts are related to their lattice energies, but direct comparisons can be made only for salts of the same lattice type.

Lattice energies correlate directly with melting and boiling points and hardness, and inversely with coefficients of thermal expansion and compressibility. The high lattice energy of Al_2O_3 makes the hard corundum a suitable grinding agent, but the high melting point of the oxide (2045 °C) necessitates an eutectic mixture with Na_3AlF_6 for the electrolysis carried out in winning aluminum metal.

The solubility of a salt is greatly influenced by the lattice energy. The Coulombic attraction, F, of oppositely-charged ions in a solvent of dielectric constant, ε, is reduced the greater the value of ε:

$$F = \frac{e^2}{\varepsilon \cdot d^2} \tag{57}$$

d = interionic distance

To dissolve a salt, its lattice energy must be overcome in a process furnishing energy. This process is the solvation of the ions. Solvation energy is defined as the energy (or more precisely, enthalpy) liberated when one mole of gaseous ions is placed in an infinite amount of solvent:

$$X^+_{(g)} + H_2O_{(l)} \rightarrow X^+_{(aq)} + \Delta H_{solv} \tag{58}$$

As an energy furnished by the system, ΔH_{solv} is negative. In practice, solvation energies are derived from a thermochemical cycle.

The solvation energy can be written, according to Born:

$$\Delta H_{solv} = \frac{Z^2}{2r} \cdot (1 - \varepsilon) \tag{59}$$

Z = charge number of the ion
r = ionic radius

From Equations (57) and (59) the solubility of an ionic substance is directly related to the dielectric constant of the solvent.

Water and water-like substances such as liquid NH_3, SO_2, and HF, as well as anhydrous H_2SO_4 and HSO_3F are good solvents for salts, as are strongly polar organic solvents, such as dimethylsulfoxide (DMSO), tetramethylenesulfone, nitromethane, nitrobenzene, acetonitrile, tetramethylurea, dimethylformamide (DMF), tetrahydrofuran (THF), dioxane, etc. Salt melts can also serve as solvents and reaction media.

The most important solvent for salts is water, and enthalpy values for the hydration of selected ions are listed in Table 11. The values indicate that ΔH_{hydr}, consistent with Equation (59), is more negative with smaller, more highly charged ions.

Table 11.   Hydration Enthalpies of Selected Ions in kJ/mol (± 10 kJ/mol).

H^+	-1092	Ca^{2+}	-1577	OH^-	-641
Li^+	-519	Sr^{2+}	-1443	F^-	-515
Na^+	-406	Ba^{2+}	-1305	Cl^-	-381
K^+	-322	Al^{3+}	-4665	Br^-	-347
Rb^+	-293			I^-	-305
Cs^+	-264	Ag^+	-470		

Halogen Exchange Reactions:
Many non-metallic chlorides, E—Cl (E = B, C, P, S, etc.) react with alkali metal halides in halide exchange reactions. For example, the mixed halide SOBrCl is obtained from thionyl chloride and a salt MBr:

$$O{=}S\overset{\diagup Cl}{\underset{\diagdown Cl}{}} + MBr \rightleftharpoons O{=}S\overset{\diagup Br}{\underset{\diagdown Cl}{}} + MCl$$

The lattice energies tell which bromide, MBr, should engage in the most facile exchange. The exchange equilibrium will be displaced to the right the smaller the lattice energy of MBr vs. MCl. The equilibrium constant, K, is written:

$$\ln K_p = -\frac{\Delta G^\circ}{RT} \quad \text{or at } 25^\circ C: \quad \log K_p = -3.065 \cdot \Delta G^\circ (\text{in kJ/mol}) \quad (60)$$

The change in free energy, ΔG°, can be approximated by the enthalpy change, ΔH°. Both parameters are connected in the second law of thermodynamics:

$$\Delta G^\circ = \Delta H^\circ - T\Delta S^\circ \tag{61}$$

and in general, the entropy change, ΔS°, and the product $T \cdot \Delta S^\circ$ are small at room temperature vis-a-vis ΔH°.
The data in Table 10 indicate that the difference in lattice energy, ΔU_0, between the alkali metal chlorides and bromides decreases slightly from lithium to cesium. Hence, from a thermodynamic point of view, LiBr is best suited for the Cl/Br exchange reaction, while CsCl should be employed for the reverse reaction.
Similar considerations hold for other halogen exchanges, for example, in

$$\geqslant C - Cl + MF \rightarrow \geqslant C - F + MCl$$

finding experimentally that LiF is worst- and CsF best-suited for replacement of Cl by F agrees with the theoretical prediction.

Complex Formation of Metal Halides:
Many metal halides form complexes with halide compounds Y, the stability of

which is a function of cationic size, for example:

$$MF + HF \quad \rightarrow \quad M[HF_2]$$
$$MCl + ICl_3 \quad \rightarrow \quad M[ICl_4]$$
$$MF + SF_4 \quad \rightarrow \quad M[SF_5]$$

Reactions of this type can be considered as a thermochemical cycle:

$$MX(s.) + Y \xrightarrow{\text{I}} M^+(g.) + X^-(g.) + Y$$
$$\Big\uparrow IV \qquad\qquad\qquad \Big\downarrow II$$
$$M[XY](s.) \xleftarrow{\text{III}} M^+(g.) + [XY]^-(g.)$$

The enthalpy of step IV determines whether decomposition of the complex into its components is favorable:

$$\Delta H(IV) = -[\Delta H(I) + \Delta H(II) + \Delta H(III)] =$$
$$= -U_o(MX) - \Delta H(II) + U_o(M[XY]) \tag{62}$$

The dissociation equilibrium IV will tend to shift to the side of the complex salt the larger is $\Delta H(IV)$. Since $\Delta H(II)$ is cation-independent, $\Delta H(IV)$, depends only on the difference in lattice energies between the halide, MX, and complex salt, M[XY] (steps I and II, respectively). If Y is neutral, M[XY] will have a smaller lattice energy than MX, owing to the larger size of the anion. However, differences in lattice energies also depend on cation size, and are smallest for the largest size cations (cf. Table 10). Hence, complex salts, M[XY], containing large cations are the most stable with respect to decomposition to the starting materials. For this reason, in preparative chemistry, large species such as tetraalkyl ammonium, phosphonium, or arsonium cations, are often employed to stabilize complex anions as salts.

5.1.8. Polarization of Anions by Cations

Simple anions such as S^{2-} or Br^- have larger radii than the isoelectronic cations. Their voluminous electron cloud can be deformed when exposed to strong, unidirectional forces.

Only a small degree of polarization can occur in a lattice with regular arrangement of cations and anions, since the ions are mutually surrounded in nearly spherical symmetry. If, however, the symmetry is distorted as, e.g., in a melt, or by dissolving or vaporizing the crystal, ion pairs and larger aggregates are formed in which the bonds are no longer purely ionic.

Diatomic molecules of NaCl exist in the vapor of NaCl, and can be isolated at low temperatures in a solid matrix of an inert gas at liquid helium temperatures. The method of matrix isolation permits NaCl molecules to be studied spectroscopically. The bonds in ion pairs of this type will be discussed for the case of LiH, for which quantum mechanical calculations of the electron density distribution have been carried out. Electron distributions can be calculated for diatomic molecules with the aid of the spin-free, one-electron density function, $P(1)$, which is defined for any given n-electron wave function, $\psi(1, 2, \ldots, n)$, as:

$$P(1) = n \cdot \int |\psi(1, 2, \ldots, n)|^2 \, ds_1 \, d\tau_2 \ldots d\tau_n \qquad (63)$$

The integral, $\int ds_1$, is taken over the spin coordinates of electron 1 and $\int d\tau_2 \ldots d\tau_n$ over the position and spin coordinates of all the remaining electrons. The value of ψ is determined using the Hartree-Fock method where $P(1)dV_1$ is the probability of encountering an electron of either spin, independent of position and spin of all remaining electrons, in the volume element, dV_1. The density function thus derived can pictorialize the bonds by contour diagrams, such as that shown in Figure 25 for the molecule LiH. This molecule consists of the isoelectronic ions Li^+ and H^-, both with the spherically symmetrical helium configuration. Figure 25, however, shows the ions as only approximately spherical. The smaller cation partially penetrates the electronic cloud of the anion, and thus disturbs its charge distribution. Some of the contour lines surround the entire molecule, but between the nuclei the electron density is 0.2 $e^-/Å^3$, the same value found between the Na^- and Cl^- in

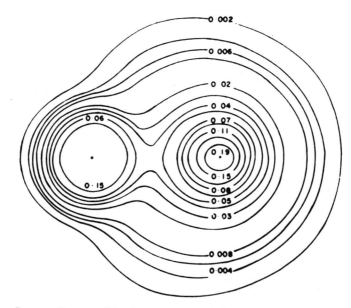

Fig. 25.   Contour diagram of the electron density of the LiH molecule. The lithium nucleus is to the left. The numbers stand for electrons per a_0^3 ($a_0 = 0.53$ Å; 1 $e^-/a_0^3 \approx 5$ $e^-/Å^3$). The measured internuclear distance is 1.60 Å.

rock salt in which there is markedly greater electron density in the neighborhood of the nuclei. In the solid state LiH crystallizes in an ionic lattice of the rock salt type.

The dipole moments of the ion pairs confirm their partially covalent character. If, for example, the molecules KCl, KBr, and CsCl consisted of two spherically symmetrical ions at distance, d, the dipole moments would in theory be $\mu = e \cdot d$ (e = elementary charge) or 13.1 to 14.7 D. The experimental dipole moments, however, are 8 to 10 D.

The relative stabilities of the ion pair and ionic lattice can be examined on enthalpy grounds. The enthalpy of formation for molecular NaCl is the ΔH_f° for:

$$Na(g.) + Cl(g.) \rightarrow NaCl(g.)$$

which is -441 kJ/mol. The internuclear distance in the molecule is 2.36 Å; the value in rock salt is 2.81 Å. The *ca.* 16% lower value in the molecule reflects the polarization of the anion by the cation, but while some energy is gained, and the molecule stabilized, NaCl forms an ionic lattice.

The enthalpy of formation of rock salt is the ΔH_f° for:

$$Na(g.) + Cl(g.) \rightarrow NaCl(s.)$$

which is (see Section 5.1.6.):

$$\Delta H_f^\circ = \Delta H_f^\circ - \Delta H_{subl}^\circ(Na) - \tfrac{1}{2}D(Cl_2) = -411 - 109 - 121 = -641 \text{ kJ/mol}$$

The polymerization of one mole of NaCl molecules liberates $-641 + 411 = -230$ kJ/mol at 25 °C., explaining why NaCl does not exist as molecules but as an ionic lattice.

Polarization and the concommitant transition of the purely ionic to the polar bond occurs not only in ion pairs, but also in crystals, when, e.g., a large anion such as I^- or S^{2-} is in contact with a small cation like Ag^+ or Li^+, and when polarization is favored by a low coordination number or by coordination of low symmetry. The proton, which owing to its small size has the greatest field strength and can pentrate the electron clouds of anions deeply, has the greatest polarization capability, and, therefore, does not form ionic lattices even with anions of the most electronegative non-metals (cf. HF and H_2O).

5.2. The Covalent Bond

A purely non-polar bond is present in homonuclear diatomic molecules such as H_2, Cl_2, N_2, etc. The theoretical treatment of these molecules starts with the isolated hydrogen, chlorine and nitrogen atoms, whose orbitals are employed in the formation of covalent bonds.

Exact solutions are available for the hydrogen atom, whereas the description of many-electron atoms is based on assumptions and approximations.

The H_2 molecule forms spontaneously from two hydrogen atoms, liberating 432 kJ/mol (4.48 eV), which must be expended as the dissociation energy in the cleavage of the H_2 molecule. The internuclear distance in the H_2 molecule is 0.742 Å, or 1.40 a_0. A theory of the covalent bond should be capable, starting with the hydrogen atom, of explaining molecule formation and of calculating the internuclear distance and dissociation energy. The solution will be sought through the two models of the hydrogen atom, the first based on the Bohr atomic model, which will permit the development of concepts of importance in the wave mechanical explanation of molecule formation.

5.2.1. The Bohr Atomic Model and Molecule Formation

When two hydrogen atoms approach each other, the energy of the system changes as a function of internuclear distance, r, as is shown in Figure 26. The energy of the completely separated atoms is zero by definition. The plot was calculated from vibrational levels obtained from the ultraviolet spectrum of the H_2 molecule. The energy minimum lies at the equilibrium distance, d = 0.742 Å; the corresponding energy is the bond energy, B.E., which is positive and larger than the dissociation energy by the zero point energy, $\frac{1}{2}h\nu$:

$$\text{B.E.} = D_0 + \tfrac{1}{2}h\nu \tag{64}$$

where ν is the frequency of the ground state vibration. The zero point energy is the vibrational energy which the molecule retains even at 0 K, and is expecially large in H_2 (ca. 25 kJ); D_0 is 432 kJ/mol so that B.E. = 457 kJ/mol (4.74 eV).

The approach of two hydrogen atoms, according to Bohr, also leads to a state of negative energy similar to the curve in Figure 26, when the two atoms assume the position shown in Figure 27. In this model the electrons rotate synchronously, but with a phase shift of 180° in two parallel planes normal to the internuclear axis.

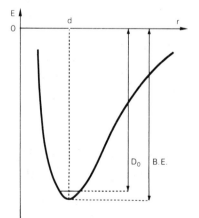

Fig. 26.   Energy, E, of the molecule H_2 as a function of internuclear distance, r; d is the equilibrium distance. B.E. denotes bond energy and D_0 the dissociation energy.

The electron-electron and nuclear-nuclear repulsions are overcome by the attraction of each nucleus for both electrons.

This model reproduces the experimental energy curve of Figure 26, if certain perturbations of the two approaching hydrogen atoms are taken into consideration. Each electron in the molecule moves in the field of both nuclei, i.e., the effective nuclear charge is >1.

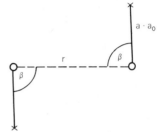

Fig. 27. Model of the hydrogen molecule using Bohr-type hydrogen atoms. The electrons (x) rotate synchronously on circular orbits which are perpendicular to the plane of the paper and parallel to each other.

The orbital radius, a_0, at $r = \infty$ is, consequently, diminished to a value $a \cdot a_0$ ($a < 1$) where the factor a varies linearly with energy of formation of the molecule, namely from $a = 1$ at $r = \infty$ to $a = 0.873$ at equilibrium distance, d (the energy minimum). The angle, β, between the internuclear and nuclear-electron axes also varies slightly, from $90°$ at $r = \infty$, to $88.8°$ at $r = d$.

With these assumptions, the energy minimum and the corresponding bond energy correspond exactly to the experimentally obtained values, and the change in energy with distance agrees exactly with the experimental curve. This means that the molecule is formed solely by Coulombic interaction of four elementary charges, but the model has been accomodated to the experimental parameters, and only empirical arguments for the values a and β can be furnished. The main advantage of the model lies in its vividness.

The covalent bond present in the H_2 molecule is often called the electron pair bond, a concept originating with G.N. Lewis (1916), and sometimes interpreted as if two electrons resided between two nuclei and thus removed the repulsion between them. Figure 27 shows, however, that the electrons neither reside between the nuclei, nor form a spatially localizable pair. The idea is only suggested by formalisms such as H:H, which is best replaced by a line representation, H—H.

The wave mechanical model of the hydrogen molecule can be built by two methods which, while differing in their mathematical formula and in their interpretation, mostly agree in their results; the method developed by W. Heitler and F. London (1927) (called the valence bond or VB method), and the molecular orbital (MO) method developed by R.S. Mulliken and F. Hund (1928).

5.2.2. The VB Theory of the Covalent Bond

The VB treatment of the H_2 molecule starts with the wave mechanical description of the hydrogen atom. The one-center atomic function (ψ_{1s}) of the hydrogen atom is utilized in the formulation of the wave function of the hydrogen molecule.
In two hydrogen atoms, H_A and H_B, in the ground state, the electrons are designated (1) and (2), respectively, and the 1s wave functions ψ_A and ψ_B describe the atomic states. The symbols $\psi_A(1)$ and $\psi_B(2)$ are employed to apportion the electrons.
At infinite internuclear distance, the wave function of the system is given by:

$$\psi_1 = \psi_A(1) \cdot \psi_B(2) \tag{65}$$

The function ψ_1 fulfills the conditions of the time-independent Schrödinger equation, assuming that ψ_A is invariant in the ambit of ψ_B, and vice versa. The potential energy, U, follows from the mutual Coulombic interaction of the two electrons and two nuclei with each other giving 4 attraction and 2 repulsion terms[2].

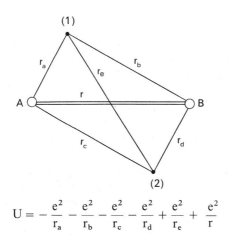

$$U = -\frac{e^2}{r_a} - \frac{e^2}{r_b} - \frac{e^2}{r_c} - \frac{e^2}{r_d} + \frac{e^2}{r_e} + \frac{e^2}{r} \tag{66}$$

This value of U is now used in the Schrödinger equation to obtain the dependence of E on all 6 distances, and the function $E = f(r)$ is minimized.
At infinity, E is simply $2E_H$, i.e., the sum of the energies of the two hydrogen atoms, and the B.E. is, consequently, zero. The energy $2E_H$ is subtracted from the total energy of the molecule to give the B.E. Its dependence on the internuclear distance, r, is calculated from Equation (65), and is depicted as curve (a) in Figure 28, which goes through a minimum. The process of molecule formation can thus be explained solely from the properties of the wave function, $\psi(1s)$, of the unperturbed hydrogen

---

2 Van der Waals interaction of the two hydrogen atoms will be neglected here, as well as subsequently.

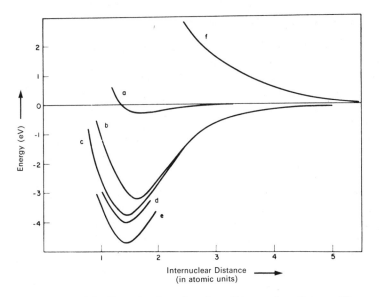

Fig. 28.   Bond energy of the H_2 molecule as function of internuclear distance. The curves a through d correspond to the several steps of the quantum mechanical calculation. Curve e shows the experimentally determined energy course.

atoms. The calculated energy minimum, however, corresponds to neither the experimental bond energy, nor to the measured internuclear distance; the experimentally determined function is shown by curve (e).

For better agreement between theory and experiment, the mathematical formulation and corresponding molecular model must be considerably modified. Allotting each electron to its own nucleus is justifiable at very large inter-nuclear distances, but not at smaller values of r, and particularly not at the equilibrium distance, since the location of an electron from the Heisenberg's Uncertainty Principle cannot be precisely determined. Since the two electrons cannot be distinguished, Equation (67)

$$\psi_2 = \psi_A(2) \cdot \psi_B(1) \tag{67}$$

must be as acceptable as (65).

It is an important property of the Schrödinger equation that if two functions ψ_a and ψ_b are solutions of the equation, then a so-called linear combination

$$\psi = a \cdot \psi_a + b \cdot \psi_b \tag{68}$$

is also a solution. The coefficients (a) and (b) are the mixing ratio of functions ψ_a and ψ_b which must be such that ψ is normalized.

If the wave functions ψ_1 and ψ_2 are equally valid equations representing certain properties of the molecules, then their linear combination:

$$\psi_3 = c_1 \cdot \psi_1 + c_2 \cdot \psi_2 \qquad (69)$$

must be a better solution than either of the two functions ψ_1 or ψ_2 alone. On symmetry grounds, the two functions must have equal weights in the linear combination. Since the weight is proportional to the square of the coefficients, then $c_1^2 = c_2^2$. Thus two linear combinations, corresponding to the two roots of the square: $c_1 = c_2$ (symmetrical state) and $c_1 = -c_2$ (antisymmetrical state) must exist:

$$\psi_{cov} = c_1(\psi_1 + \psi_2) \qquad (70)$$

$$\psi_{cov}^* = c_1(\psi_1 - \psi_2) \qquad (71)$$

where the subscript cov is used to indicate the covalent nature of the bond described by these functions.

The curves (b) and (f) shown in Figure 28 correspond to the states described by the wave functions ψ_{cov} and ψ_{cov}^*. The first is a stable state ($c_1 = c_2$) with an energy minimum at -3.14 eV (the experimentally obtained bond energy is 4.747 eV). The second is a state ($c_1 = -c_2$) which has for all finite internuclear distance a higher energy than the isolated atoms, corresponding to repulsion of the two atoms. Such states are called antibonding, and the corresponding wave functions are marked with an asterisk.

Assuming delocalization of the electrons over both atomic orbitals, ψ_{cov} gives an internuclear distance of 0.87 Å and a bond energy of 303 kJ/mol for the H_2 molecule, just $\frac{2}{3}$ of the experimental bond energy. The electron density, calculated from this wave function by W. Heitler and F. London, is shown in Figure 29(b). In a covalent bond the electron density along the internuclear axis does not fall to nearly zero, as in heteropolar bonds, but remains high, so that the repulsion of the two atomic nuclei is compensated. In the unstable, antibonding state described by function ψ_{cov}^*, the electron density is different [cf. Figure 29(a)] in having a nodal plane which bisects the internuclear axis and is normal to it and in which the electron density is zero. The reason for the instability of this state is the resulting internuclear repulsion.

An improvement over the Heitler-London model was proposed by S. C. Wang in 1928, who recognized that the electrons are exposed to an effective nuclear charge, $Z^* > 1$, which leads to a contraction of the 1s orbitals of both hydrogen atoms. Therefore, the correct linear combination should employ by analogy with He^+ and Li^{2+} (cf. p. 45) a higher charge number, Z^*, which like the corresponding perturbation factor (a) in the Bohr treatment of the H_2 molecule, is a function of internuclear distance, e.g., for $r = \infty$, $Z^* = Z = 1$; for the equilibrium state a Z^* value of 1.666 is used to obtain a bond energy of 363 kJ/mol or almost 80% of the experimental value [curve (c) in Figure 28].

With further improvements, internuclear distances and bond energies which agree with the experimental values can be obtained.

Polarization can be taken into account. The 1s orbitals cannot be assumed to remain

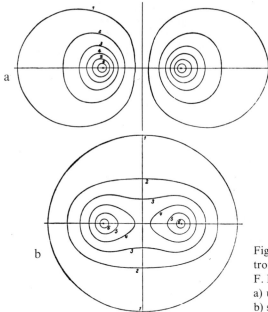

a

b

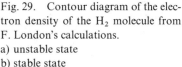

Fig. 29. Contour diagram of the electron density of the H_2 molecule from F. London's calculations.
a) unstable state
b) stable state

spherically symmetrical in the molecule, and thus modified wave functions ψ_A and ψ_B are used which account for the polarization in the direction of the internuclear axis. This raises the calculated bond energy to 85% of the experimental.

A second improvement considers possible ionic structures. There is a small probability that both electrons will find themselves simultaneously at the same nucleus giving an ion pair consisting of a proton and a hydride ion,

$$H^- \, H^+ \quad \text{or} \quad H^+ \, H^-$$

to which the wave functions $\psi_A(1) \cdot \psi_A(2)$, and $\psi_B(1) \cdot \psi_B(2)$ correspond. Linear combination of these two products gives a function, ψ_{ion}, which accounts for the ionic contribution:

$$\psi_{ion} = c_3 \cdot \psi_A(1) \cdot \psi_A(2) + c_4 \cdot \psi_B(1) \cdot \psi_B(2) \tag{72}$$

This yields the total function for the H_2 molecule:

$$\psi = N(\psi_{cov} + \lambda \cdot \psi_{ion}) \tag{73}$$

N = normalizing factor
λ = mixing coefficient

The value λ is obtained by minimizing the energy. This energy minimum lies at 396 kJ/mol which is 87% of the experimental value [curve (d) of Figure 28]. This negligible improvement shows the insignificance of ionic structures for the H_2 molecule (λ is ca. $\frac{1}{4}$), however, in bonds between different atoms the contribution may be considerably larger leading to a continuous transition to ionic bonds.

Wave functions like those in Equation (73) are often described as the resonance of a covalent structure H—H, described by ψ_{cov}, with two ionic structures,

$$H\text{---}H \leftrightarrow H^+ H^- \leftrightarrow H^- H^+$$

However, these limiting structures have no physical reality themselves, but only contribute with appropriate weight to the ground state of the molecule. Limiting structures are never isolated, nor are they individually responsible for properties of the molecule. Limiting resonance structures should not differ in the number of unpaired electrons, nor in their internuclear distances. The contribution of a structure to the ground state of a molecule depends inversely on its energy content: the larger the energy, the smaller its contribution. The more energetic ionic limiting structures contribute only about 2% to the B.E. of the H_2 molecule, and are generaally neglected as well as for other homonuclear, diatomic molecules which are considered to posses purely covalent bonds.

Conditions for Formation of Bonds:

The VB description requires that unpaired electrons be present in the atoms which form an electron pair bond. Therefore, two He atoms do not form a He_2 molecule, but two Li atoms can form Li_2.

The presence of unpaired electrons is a necessary, but not sufficient condition for bond formation. Electrons forming a pair must be in orbitals of suitable symmetry, as will be discussed later.

A vivid picture of the covalent bond is derived from the wave function in which the product $\psi_A \cdot \psi_B$ is contained. If this product and, therefore, ψ_{cov}, are not to fall to zero, giving a zero B.E., the value of ψ_A in the neighborhood of ψ_B must be non-zero and vice versa. That implies that wave functions or atomic orbitals must mutually overlap. The stronger this overlap, the larger is ψ_A in the area of ψ_B, and vice versa, and the larger the B.E. becomes. A more precise mathematical treatment leads to the overlap integral, S:

$$S = \int \psi_A \cdot \psi_B \, dV \tag{74}$$

The product $\psi_A \cdot \psi_B$ is integrated over all volume elements, dV, and S, a general measure of the strength of the bond, is obtained. On the other hand the bond strength can be estimated from the experimental B.E. or from the valence force constants (cf. Section 5.5). The overlap integral, S, depends on the signs and values of ψ_A and ψ_B, and can be positive, zero, or negative, and can assume all values between zero and 1. Thus:

$$S > 0 = \text{bonding (attraction)}$$
$$S = 0 = \text{no bonding}$$
$$S < 0 = \text{repulsion.}$$

Approximate numerical values of S will be discussed later for suitable examples for which comparisons can be made.

The principle of wave function, or orbital overlap is illustrated by the overlap of the 1s functions of the hydrogen atoms in the H_2 molecule shown schematically in Figure 30. Since both wave functions are positive, the product $\psi_A \cdot \psi_B$ is also positive and S is, therefore, everywhere > 0.

Lithium atoms have three electrons, one of which is unpaired in a 2s orbital, and lithium exists in the vapor phase as Li_2 molecules like the other alkali metals. The overlap relations in Li_2 are shown in Figure 31, where the total overlap integral, S, is the sum of positive and negative values of the product $\psi_A(2s) \cdot \psi_B(2s)$. With closer approach, the larger the overlap becomes in those regions where the functions have opposite signs. The maximum B.E. of 109 kJ/mol, much smaller than for H_2, is thus realized at the relatively large internuclear distance of 2.67 Å.

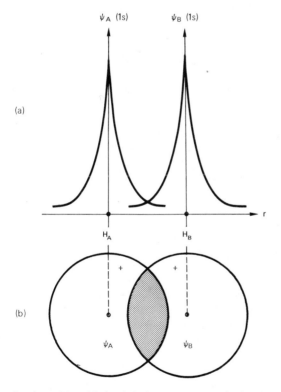

Fig. 30.   Overlap of 1s orbitals of the hydrogen atoms in the H_2 molecule.
(a) Profile of the wave functions   (b) Boundary line diagram

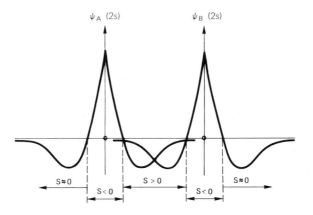

Fig. 31.   Overlap of the 2s-orbitals of the lithium atoms in the Li_2 molecule.

Other examples of orbital overlap are shown in Figure 32. The overlap of orbitals with σ-symmetry leads to σ-bonds, which can be distinguished from π overlap leading to π-bonds. σ-Orbitals are rotationally symmetrical about the internuclear axis. The principle of maximum overlap states that the greatest bond energy is realized in a σ-bond if the constituent orbitals are combined so that their axes of symmetry coincide. Thus the geometry of molecules is mainly determined by the geometry of their σ-orbitals. Strictly speaking, maximum positive and minimum negative overlap should be referred to. π-Bonds, on the other hand, are characterized by a nodal plane which contains the internuclear axis. In this case the overlap does not take place along the symmetry axes of the orbital lobes.

Orbitals are drawn using contour or boundary lines. However, while the orbitals differ in their quantum numbers n, l, and m, the inner nodal planes of the orbitals of higher quantum number can be disregarded, since they have no effect on the overlap in the peripheral regions. For this reason, s, p, and d-orbitals are sometimes drawn as projections through the radial function (see Figure 8) and independent of principal quantum number, as in Figure 32, but the overlap, in particular of π-bonds, is portrayed imprecisely that way, since calculations have demonstrated that the magnitude of S is considerably influenced by the radial function R (r). Contour diagrams should, therefore, be employed.

Large positive values of the overlap integral are obtained only when the orbitals to be combined possess suitable symmetry, and are of similar energies. Figure 33 shows the overlap of orbitals of orthogoal symmetry for which S is zero, although considerable overlap takes place.

Atoms containing more than one unpaired electron can enter into several covalent bonds with similar atoms, for example the nitrogen atom with configuration $K2s^2p^3$ and unpaired electrons in the $p_x, p_y,$ and p_z orbitals which can overlap with the corresponding orbitals of a second atom leading to one σ- (p_z/p_z) and two π-bonds

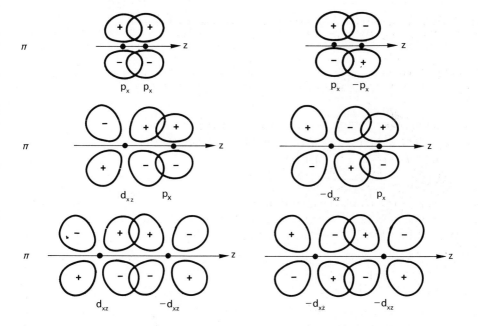

Fig. 32.   Overlap of atomic orbitals leading to positive or negative overlap integral, S.

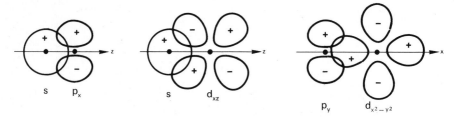

Fig. 33.   Overlap of orthogonal atomic orbitals. The overlap integral, S, is zero on the grounds of symmetry.

$(p_x/p_x$ and $p_y/p_y)$. In addition, each atom has one 2s electron pair which does not participate in bonding. This is written $:N{\equiv}N:$

The electron density distribution in the N_2 molecule is shown in Figure 34 in the form of a contour diagram calculated with the aid of the one-electron density function, P(1) (see p. 70). The spherical symmetry of the two separated nitrogen atoms is no longer recognizable. The charge redistribution is best illustrated as in Figure 35 by a density difference diagram which is constructed by subtracting the electron density distributions in two superimposed nitrogen atoms held at the internuclear distance from the distribution in the molecule. The solid contour lines connect points of equal positive values of electron density difference (in e^-/a_0^3); in these areas density has increased in the formation of the molecule. The regions enclosed by the dotted contour lines indicate a corresponding decrease in density.

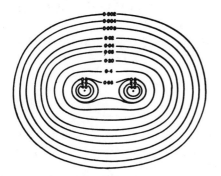

Fig. 34.   Contour diagram of the charge distribution in the N_2 molecule.

Figure 35 shows that the originally spherical electron density about the two atoms has now moved predictably into the area between the two nuclei where the effective nuclear charge is highest. Surprisingly, however, a portion of the density is found in the distal regions, opposite the internuclear axis, owing to sp-hybridization of the σ-orbitals of the nitrogen atoms, which will be discussed later.

A problem unsolved within the framework of VB theory concerns the molecule O_2. Oxygen atoms, with electron configuration $K2s^2p^4$, possess two unpaired electrons

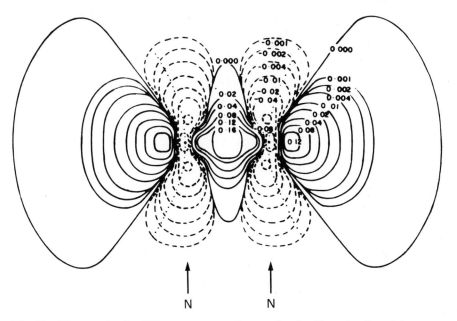

Fig. 35.   Electron density difference contour diagram for the N_2 molecule and two non-bonded nitrogen atoms held at the molecular internuclear distance.

and hence can form a molecule with a double bond: O=O with a σ- and a π-bond accounting for the high bond energy of 494 kJ/mol for O_2, which lies in the range of other double bonds. However, oxygen is paramagnetic and contains two unpaired electrons, based on its magnetic susceptibility, to give a ground state triplet. The VB theory offers no explanation for this discrepancy.

The Promotion of Electrons:
Even atoms having no unpaired electrons in the ground state, such as Be, Kr, and Xe, can from covalent bonds. Other non-metallic atoms form more bonds than would be expected from their number of unpaired electrons, e.g., SiH_4, SF_6. Promotion of electrons from doubly-occupied to vacant orbitals is assumed, giving an excited state which requires considerable energy.
The boron atom in the ground state, for example, with electron configuration $K2s^2p$, could yield the molecule BF. The stable boron trifluoride, BF_3, however, is formed when a 2s electron is promoted to a 2p orbital, allowing three covalent bonds to be formed:

Excited states with a larger number of unpaired electrons can be populated spectroscopically with the energy furnished at high temperatures or by electrical discharge. Such spectroscopic states have physical reality, but do not appear during bond formation as intermediates. Rather, they serve as hypothetical steps, into which a reaction is conceptually separated. A cyclic process of this kind is given below (p. 92).

Valence electron promotion also explains why the valence states of the non-metals always differ by two units. Sulfur, for example, forms the fluorides SF_2, SF_4, and SF_6, derived from the following valence configurations:

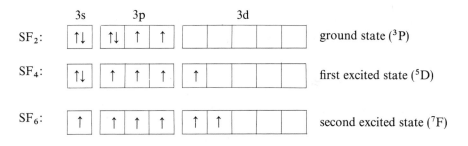

The sulfur fluorides SF, SF_3, or SF_5 are unknown, or exist as radicals, i.e., unstable molecules with unpaired electrons.

Electrons in non-metals are promoted within one shell only, so that the required energy can be recovered in bond formation. Hence, the maximum number of bonds is limited to the number of realizable, singly-occupied atomic orbitals in the valence electron shell. However, coordinate-covalent bonds may be formed in addition (cf. Section 5.2.4.).

The Hybridization of Valence Orbitals:

The molecule silane, SiH_4, is derived from the following valence electron configuration of the silicon atom:

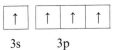

It might be expected that three of the four Si-H bonds would be different from the fourth which is formed by overlap of a 3s-orbital with the 1s orbital of hydrogen. According to the geometry of the silicon orbitals, the first three bonds should be orthogonal like the p_x, p_y and p_z orbitals; no statements can be made about the direction of the fourth bond because of the spherical symmetry of the 3s-orbital. The SiH_4 molecule is, by contrast, tetrahedral, with all Si-H distances equal and valence angles of $109°\,28'$.

To overcome this discrepancy L. Pauling, in 1931, introduced the concept of hybridization, which is a mathematical mixing of various orbitals to give new orbitals of different geometry. Linear combinations of pure atomic orbitals, e.g., ψ_1 and ψ_2, give hybridized orbitals:

$$\psi = N(\psi_1 + \lambda \cdot \psi_2) \tag{75}$$

where N is a normalizing factor, and λ is a mixing coefficient. As shown on p. 76, each hybridized function, ψ, has a conjugate function:

$$\psi' = N(\psi_1 - \lambda \cdot \psi_2) \tag{76}$$

By hybridization of two, or in general n atomic functions, two or n hybridized functions result with different shapes.

sp-Hybridization:

The beryllium atom with ground state electron configuration $K2s^2$, can be excited to configuration $K2sp$. The singly occupied 2s and $2p_z$ orbitals can be mixed partially ($\lambda < 1$) or completely ($\lambda = 1$); complete hybridization gives:

$$\psi = \sqrt{1/2}(2s + 2p_z) \quad \text{and} \quad \psi' = \sqrt{1/2}(2s - 2p_z) \tag{77}$$

These orbitals are shown schematically in Figure 36 in which the hybrid orbitals, in contrast to the atomic orbitals, are no longer symmetrical or antisymmetrical with respect to the nucleus, but are more directional, permitting better overlap of the bonding orbitals which are located along the symmetry axes of the hybrid orbitals. The 180° axis in the sp-hybrid should, based on the principle of maximum overlap, give a linear structure for covalent BeX_2 compounds and this is found for $BeCl_2$ in the gas phase (in condensed phases, however, most beryllium compounds are polymers). The carbon atoms in acetylene, C_2H_2, and nitrogen in N_2 are likewise sp-hybridized.

sp-Hybridization is assumed only between orbitals of the same shell. With 3s and 3p orbitals, the inner nodes must be taken into account, however, neither the peripheral shape of the hybridized orbitals nor their direction is affected by the nodes. Therefore, Figure 36a is approximately independent of the principal quantum number on the premise that the differences in the radial portions of the s- and p-functions at the periphery of an atom may be neglected. This can be shown using only the angular portions of functions ψ_s, and ψ_p, which are independent of the principal quantum number (Figure 36b). A corresponding argument is used for the sp^2 and sp^3 hybrid orbitals discussed later.

The sp-hybridized orbitals are orthogonal. Strong overlap with the ligand orbitals is possible along the axis of rotation, but small deviations from this direction cause only insignificant diminishing of the overlap integral (this also holds for pure atomic and other hybrid orbitals, although to a lesser degree). The angle-retaining forces in a molecule are thus always much smaller than the distance-maintaining forces.

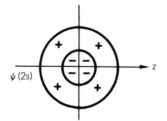
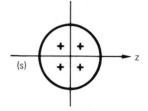
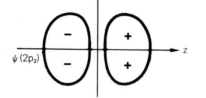
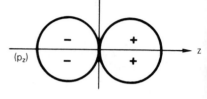
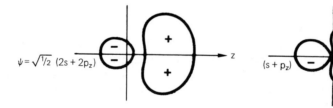
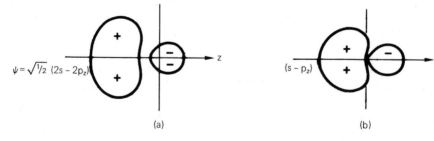

(a)   (b)

Fig. 36.   Schematic representation of the hybridization of the 2s and $2p_z$ atomic orbitals to form sp-hybridized orbitals (a). The lines connect points of equal ψ-value. The nucleus is at the origin of the coordinate system. In (b) the overlap of the angular portions, which are independent of the principal quantum number, is shown by polar diagrams.

This is important in reaction mechanisms and transition states. The thermal energy of a molecule at room temperature, for example, produces angular changes of ca. 5°, and thus molecules are more flexible than is often assumed.

sp²-Hybridization:

The three unpaired electrons which are present in an excited boron atom in the $K2sp^2$ state occupy one 2s and two 2p orbitals. Complete hybridization, as shown in Figure 37, leads to three equivalent hybrid orbitals, whose axes lie in a plane at 120° angles. Hence, compounds of the type BX_3 are trigonal-planar.

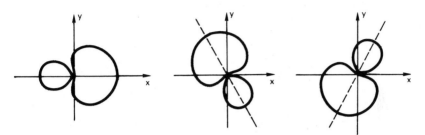

Fig. 37.   Polar diagrams of the sp^2 hybrid orbitals formed by complete hybridization of atomic orbitals s, p_x and p_y.

sp³-Hybridization:

Complete hybridization of one s- with three p-orbitals leads to four hybrid orbitals, whose axes point toward the corners of a tetrahedron. Compounds of the type AX_4 which are derived from an sp^3 configuration are consequently tetrahedral, e.g., CCl_4, SiH_4, NH_4^+, BF_4^-, etc. The contour diagram of an sp^3 hybrid orbital shown in Figure 38 strongly resembles the sp or sp^2 hybrids (cf. Figure 36b).

The lobes of the hybrids shown in Figures 36 to 38 are as voluminous as those of atomic orbitals and the contours which are near the nucleus encompass a large portion of that space. The lobes of a hybrid orbital are rotationally symmetrical like the p-orbitals, and maximum overlap occurs with a σ-bond orbital along the axis of rotation, determining the geometry of the molecule.

Hybrids from d-Orbitals:

Hybrids may also be formed using d-orbitals, for example in the linear pd hybrid formed by combination of a p_z with a d_{z^2} orbital, singly occupied in the $KLMN5s^2p^5d$ excited state of xenon. In the VB theory the pd hybrid accounts for the shape of the molecule XeF_2.[3]

3 Xenon may also be considered as sp^3d-hybridized, which does not change the result.

Table 12   Hybridization of Atomic Orbitals and Resulting Molecular Geometry

Number of Substituents	Hybridization of Center Atom	Arrangement of Substituents	Examples	Strong π-Orbitals
2	$sp(sp_z)$ $p^2(p_xp_y)$ $pd(p_zd_{z^2})$	linear angular linear	CO_2, N_3^- H_2Se XeF_2	$p^2d^2(p_xp_yd_{xz}d_{yz})$ d_{xz} und p_z oder d_{yz} $p^2d^2(p_xp_yd_{xz}d_{yz})$
3	$sp^2(sp_xp_y)$ p^3	trigonal-planar trigonal-pyramidal	BF_3, SO_3, CO_3^{2-} AsH_3	$pd^2(p_zd_{xz}d_{yz})$ —
4	sp^3	tetrahedral	SiH_4, SO_4^{2-}	$d^2(d_{z^2}d_{x^2-y^2})$
5	$sp^3d_{z^2}$ $sp^3d_{x^2-y^2}$	trigonal-bipyramidal tetragonal-pyramidal	PF_5, SOF_4 $Sb(C_6H_5)_5$	$d^2(d_{xz}d_{yz})$
6	$sp^3d^2(d_{z^2}d_{x^2-y^2})$	octahedral	SF_6, SiF_6^{2-}, $XeOF_4$	$d^3(d_{xz}d_{yz}d_{xy})$
7	sp^3d^3	pentagonal-bipyramidal	IF_7	—

Orbitals available for strong π-bonding are listed in column 5. Weaker π-orbitals which are capable of only small positive overlap with ligand orbitals often exist in addition.

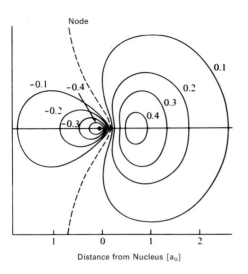

Fig. 38.   Contour diagram of an sp^3 hybrid orbital of carbon. The three-dimensional orbital is obtained by rotation around the horizontal axis. The node does not bisect the nucleus of the atom.

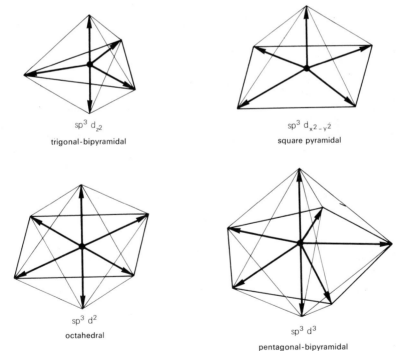

Fig. 39.   Preferential orientation of orbital lobes in some s-p-d-hybridizations.

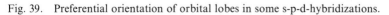

The orbital axes in other hybrid states are shown in Figure 39. Some examples for the most important non-metallic structures are listed in Table 12.

The hybrid orbitals are for identical ligands equivalent and usually differ only in spatial orientation, but three exceptions are shown in Table 12.

The trigonal bipyramidal $sp^3d_{z^2}$ hybrid consists of three equatorial and two axial orbitals, rather than five equivalent orbitals. Whereas in SiH_4 (sp^3) and SF_6 (sp^3d^2), all internuclear distances are identical[4], there are two different P-F distances in PF_5, with the smaller distances (and hence stronger bonds) in the equatorial plane. Likewise, the rarer hybrid $sp^3d_{x^2-y^2}$ directs five nonequivalent orbitals toward the four identical base corners of a square pyramid and to the unique apical position, giving two internuclear distances, as in $Sb(C_6H_5)_5$.

Two sets of hybrids are again encountered in the sp^3d^3 hybrid with five equivalent equatorial bonds and two shorter and stronger axial ones. The IF_7 molecule is somewhat deformed, however, with the five equatorial fluorine atoms non-planar and with $\angle\, FIF = 171°$ between the axial fluorine atoms.

Hybridization Energy:
Hybridization leads to orbitals which differ not only in geometry, but also in energy from atomic orbitals. The energy of the hybrids is the mean of the atomic orbital energies (cf. Fig. 40).

After promotion of an s-electron in each case, the valence orbitals in beryllium, boron and silicon atoms prior to hybridization are singly occupied, and the total energy of these excited atoms would not change with hybridization because the

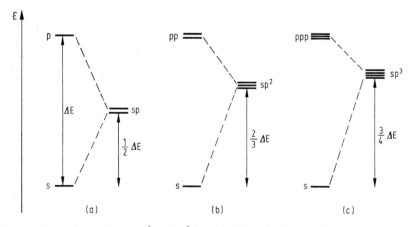

Fig. 40.   Energy levels for sp, sp^2 and sp^3 hybrids (E = orbital energy).

---

4 This holds strictly only for the gaseous state, since in the condensed phase, intermolecular interactions may lead to lowered symmetry, and differing internuclear distances.

destabilization of the remaining s-electron is compensated by the stabilization of the p-electrons. For example, in the silicon atom (KL3sp³), the destabilization is about $3/4\,\Delta E$ and the stabilization $3\cdot 1/4\,\Delta E$. All the atomic orbitals to be hybridized in the three cases cited are singly occupied, and the total electron density remains unchanged during hybridization. This means that the excited atomic or hybridized orbitals describe the same state of the atom.

For the nitrogen atom, which e.g., in NH_3 is approximately sp³ hybridized, the 2s level is doubly, and the three p-orbitals singly occupied. Additional energy is, therefore, required to promote the s-electrons to the sp³ level amounting to $2\cdot 3/4\,\Delta E - 3\cdot 1/4\,\Delta E = 3/4\,\Delta E$. Likewise, the oxygen atom in H_2O regains the hybridization energy in bond formation since hybrid orbitals lead to much stronger bonds (greater overlap integrals, cf. Section 5.5), than pure atomic orbitals. This gain in B.E. is the justification for hybridization.

The Valence State:

Covalent bond formation requires atoms with unpaired electrons of spins oriented with equal populations of $s = +\frac{1}{2}$ and $s = -\frac{1}{2}$. For example, if silicon atoms hybridized in the excited sp³ state are to react with ground state hydrogen, in which the electrons are equipartitioned between the spin states $s = \pm\frac{1}{2}$, the same statistical occupancy must hold for silicon atoms if all the hydrogen atoms are to react. The statistically distributed spins are symbolized by a line:

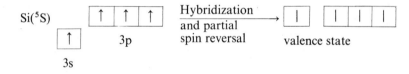

$Si(^5S)$ 3p   Hybridization and partial spin reversal   valence state   3s

The electron distribution of the valence state is similar to that of the atom in the molecule. Valence states are formed by separating the atoms without spin reversal, so that each bond is homolytically cleaved, resulting in a statistical spin distribution. Thus the valence state is neither a stationary nor a nonstationary (unstable) atomic state, but a statistical mean of stationary states of many atoms with random spin distributions. These spin distributions do not conform to the ground state, and the valence state is richer in energy, as can be calculated from quantum mechanics.

The hypothetical steps of molecule formation are shown for SF_6 in an energy level diagram in Figure 41. The starting materials are rhombic sulfur (S_8) and gaseous fluorine at standard conditions (25 °C, 1 bar). Firstly, $\frac{1}{8}$ mole of S_8 is transformed into 1 mol of sulfur (^3P) atoms, requiring 276 kJ/mol. Secondly, 3 mol of F_2 molecules are dissociated into atoms, requiring 465 kJ/mol. Promoting two valence electrons of sulfur to the 3d level requires 2360 kJ/mol and gives a configuration with six unpaired electrons. Finally, the sulfur atom must be transformed into the octahedral valence state with sp³d² hybridization and statistical spin distribution. The reaction with fluorine liberates sufficient energy so that the net enthalpy is

−1210 kJ/mol. The hypothetical scheme in Figure 41 for the spontaneous ignition of sulfur in a stream of fluorine, and subsequent burning to SF_6, is not meant to imply that the promoted and hybridized atomic states are reaction intermediates. The real reaction traverses other, largely unknown intermediate steps.

Hybrid orbitals only exist in bonded atoms. The geometry of molecules can be rationalized with their aid as regular polyhedra. Hybridization is assigned from experimentally determined molecular structure, and not the other way around. Non-metallic atoms make use of hybridized orbitals to maximize bond energy. Hybridization improves overlap markedly, since hybrid orbitals are more directed. The hybridized state of an atom can be continuously varied using quantum mechanical methods, to find the minimum state energy. Hybrids will be asymmetrical when the

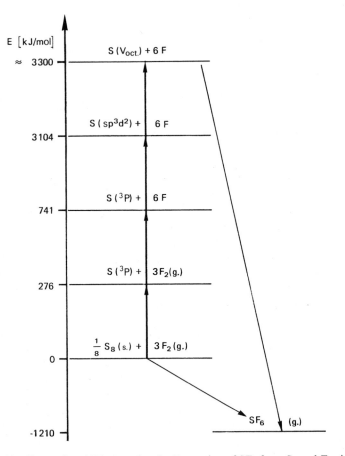

Fig. 41.   Energy Level Diagram for the Formation of SF_6 from S_8 and F_2 via hypothetical intermediate steps.

coordination of the atom under consideration is asymmetrical, i.e., when there are different bond partners (as in SiH_3Cl), or free electron pairs (as in OF_2) present.
In SiH_3Cl, for example, the $\angle HSiH$, despite the small spacial requirement of the hydrogen atoms, is $110.2°$ larger than $\angle HSiCl$ of $108.7°$.
The valence angles increase with increasing s-content in the hybrid orbitals (cf. Table 13). The hybridization of the silicon atom in SiH_3Cl is asymmetrical in that the s-content in the orbitals directed toward the hydrogen atoms is $>25\%$, and in the remaining orbital $<25\%$. The same is true for CH_3Cl ($\angle HCH = 110.5°$).
The s-content can be calculated from the angle, α, between equivalent sp-hybridized orbitals, from the formula:

$$\cos \alpha = \frac{s}{s-1}$$

(s in fractions of 1)

Table 13.   Valence Angles in Various sp-Hybrids.

Hybrid orbitals	p^2 or p^3	sp^3	sp^2	sp
s-content	0%	25%	33%	50%
Angle between the axes of the orbitals	$90°$	$109.5°$	$120°$	$180°$

The s-content of hybrid orbitals is, excluding steric effects and d-orbital participation, smaller the more electronegative the ligand atom and larger the more electronegative the central atom. The valence angles listed in Table 14 can be interpreted in that manner.
Free electron pairs in molecules such as NH_3 and H_2O can occupy hybrid orbitals, just as bonding pairs do. They reside in regions of the molecule in which no binding takes place (cf. Figure 42 and Table 14). Only electrons in s-orbitals or in completed shells are spherically symmetrical about the nucleus.

Table 14   Valence Angles of Some Simple Compounds.

OH_2	$104.5°$		H_2O	$104.5°$
OF_2	$103.3°$		H_2S	$92.2°$
			H_2Se	$91.0°$
NH_3	$107°$		H_2Te	$89.5°$
NF_3	$102.5°$			
			NH_3	$107°$
CF_4	$109.5°$		PH_3	$93.8°$
HCF_3	$108.8° (FCF)$		AsH_3	$91.8°$
H_2CF_2	$107° (FCF)$		SbH_3	$91.5°$

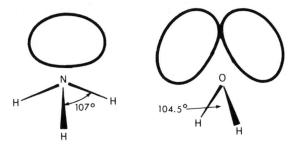

Fig. 42.   Mutual orientation of bonding and free electron pairs in the NH_3 and H_2O molecules.

Hybridization in molecules with multiple bonds:

The three bonds in the N_2 molecule can be described in two ways in the framework of VB theory. According to the σ-π-model, one σ- and two π-bonds are present (cf. p. 80) with the π-bonds arising from overlap of p_x and p_y orbitals, but it cannot be decided whether two pure p_z, or sp-hybrids formed from the filled 2s and singly-occupied $2p_z$ overlap in the σ-bond:

N:

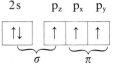

The resulting distribution would be similar to the one in acetylene, $HC\equiv CH$, in which the σ-bonds originate from sp-hybridized orbitals of the carbon atoms.

If N_2 is sp-hybridized, the two free electron pairs would be located on the distal sides of the molecule. The density difference diagram in Figure 35 shows that this is clearly the case, and the free electron pair is involved in the hybridization of the σ-bond framework in almost all compounds of three-valent nitrogen. The magnitude of this effect cannot be determined in the diatomic N_2 molecule which lacks valence angles.

A lesser-known model with individual arc bonds will be discussed for ethylene, C_2H_4. According to the σ-π-model, the carbon atoms of C_2H_4 have trigonal-planar, sp^2 hybridization with valence angles of $120°$. A π-bond lying above and below the carbon-carbon σ-bond has no influence upon the symmetry, but hinders free rotation around the bond axis. The energy barrier is a function of the strength of the π-bond.

The arc model starts with the carbon atoms as sp^3 hybrids, two orbitals of which bind the hydrogen atoms, and two of which engage in double bond formation (cf. Figure 43). Maximum overlap is assumed to take place along arc lines which connect the carbon atoms. The bond length is thus larger than the internuclear distance, and the model yields $\angle\,HCH = 109.5°$ and $\angle\,HCC = 125.6°$, and predicts restricted rotation.

Despite the differences in predicted valence angles, it is difficult to decide between the two models. The valence angles of C_2H_4 correspond well to the σ-π-model, but the angles of most molecules of the type $X_2A{=}AX_2$ lie between the predicted values, inferring that each model has its justification.

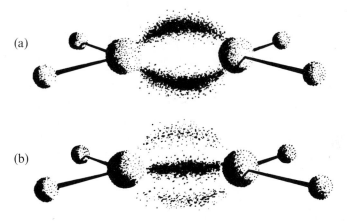

Fig. 43.   Two models for the C_2H_4 molecule
a) The Arc-bond Model   b) The σ-π-Model

Both models predict a linear arrangement for the acetylenes. The arc model is also in agreement with the density difference diagram for the N_2 molecule in Figure 35. Experimental evidence is available for arc bonds in certain molecules. The maximum electron density in a normal σ-bond lies on the internuclear axis and diminishes perpendicular to it as shown in the contour charts for H_2, N_2 and LiH (Figures 25, 29, and 34). The arc model predicts an asymmetrical charge distribution about the internuclear axis as is found for cyclopropane, C_3H_6, and for P_4, the constituent of white phosphorus. The angle between ring forming atoms in both molecules is 60°, and since no atomic orbitals embrace such a small angle, the overlap problem is of interest. Information from X-ray diffraction and quantum mechanical calculations on cyclopropane derivatives demonstrate that the maximum electron density does not lie along the internuclear axes, but outside of the triangle formed by the carbon centers. The s-contribution in the hybrid orbitals which bond the hydrogen atoms is calculated from the $\angle\,HCH = 118°$ as 32%, while the arc bonds are formed by orbitals with 18% s-character, which lie at a $103°\angle$ to each other. The hybridization cannot be calculated from the valence angles for P_4, but should correspond to those in similar phosphorus compounds. Angles at the phosphorus atom range from 99° to 103° in PCl_3 and in the PS_3 group present in the phosphorus sulfides P_4S_3, P_4S_5, and P_4S_7, corresponding to an s-contribution to the hybrid orbitals of 14–18%. Only arc bonds are possible between the orbitals of P_4, and the assumption that the d-orbitals of phosphorus do not participate is borne out by the photoelectron spectrum. The free electron pairs of the phosphorus atoms are located in hybrid orbitals with 50% s-content, which are oriented outward from the center of the P_4 tetrahedron (cf. Figure 44).

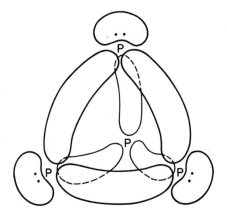

Fig. 44.   Schematic representation of the four arc bonds in the P_4 molecule (after R.J. Gillespie, Molecular Geometry, van Nostrand-Reinhold, New York, 1972).

This model agrees with the quantum mechanical calculations using 3s and 3p orbitals of phosphorus, where density maxima are found outside the internuclear axes, and the electron pairs are localized outside the tetrahedron.

The internuclear distances in cyclopropane and in the P_3 rings of the molecules P_4 and P_4S_3 correspond to typical single bond values, confirming that the arc bonds are not weak.

The Hybridization of π-orbitals:
Hybrid atomic orbitals are used in the construction of orbitals of π-symmetry, just as with σ-. The orbitals available for π-hybridization have been listed in Table 12. Hybridization of three π-orbitals is assumed in the SO_3 molecule which is trigonal planar with three S=O bonds

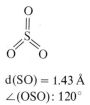

$d(SO) = 1.43 \text{ Å}$
$\angle(OSO): 120°$

The bonds in SO_3 can be constructed by starting with the excited state of the sulfur atom with configuration $KL3sp^3d^2$.

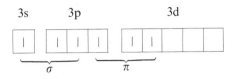

The σ-bond framework starts with a sulfur sp^2-hybrid (orbitals s, p_x, p_y) with the remaining three unpaired electrons occupying the orbitals p_z, d_{xz}, and d_{yz} singly. Since all three S=O distances are equal, the three π-bonds must be equivalent. This can only be attained by hybridizing the π-orbitals.

5.2.3. The Molecular Orbital Theory of the Covalent Bond

The molecular orbital (MO) theory assumes that the electrons of a molecule are delocalized over the molecular framework. They are no longer in the atomic orbitals (AO's) of particular atoms, but in molecular orbitals (MO's) which extend over the whole molecule.

Starting with the H_2^+ ion which consists of two protons and one electron, the model can be extended to H_2 and other many-electron molecules, assuming that the orbitals derived for H_2^+ are also present in other homonuclear diatomic species and can be successively occupied with electrons following the Pauli principle. The molecular ion, H_2^+, thus has a similar significance in MO theory as the hydrogen atom in atomic theory.

The molecule H_2^+ is formed in the ionization of H_2, in a gaseous discharge, or in the ion source of a mass spectrometer. The spectroscopically determined value for internuclear distance is $d = 1.06$ Å $(= 2.00\ a_0)$. The B.E. (269 kJ/mol) is markedly smaller than for H_2.

The electron in H_2^+ moves in a field of two nuclei, and the MO's are two-centered. The problem in MO theory is finding ψ-functions which describe such electronic states. Suitable poly-centered wave functions are constructed by a linear combination of single-center, atomic functions, called the LCAO (from linear combination of atomic orbitals) approximation.

The LCAO Approximation:

The nuclei in the hydrogen molecule ion, designated A and B, are first located at the experimentally determined distance. An electron moves with greatest probability near the nuclei, and when close to nucleus A will be little influenced by nucleus B and vice versa. In the absence of nucleus B, the electron would be in the 1s AO of A, so it can be assumed that the MO near A simulates the 1s AO of A, and similar considerations hold for B. The following linear combination neglects the influence of nucleus B on A and vice versa

$$\psi = c_1 \cdot \psi_A(1s) + c_2 \cdot \psi_B(1s) \tag{79}$$

This equation is generally used in the form:

$$\psi = N[\psi_A(1s) + \lambda \cdot \psi_B(1s)] \tag{80}$$

N: normalizing factor
λ: mixing coefficient

The mixing coefficient, λ, assigns relative weight to the two AO's and reflects the polarity of the MO. The coefficient can assume values between $-\infty$ and $+\infty$, but for H_2^+ the weights of both AO's are the same on symmetry considerations. The weights are proportional to the squares of the coefficients, c_1 and c_2, and the squares are equal: $c_1^2 = c_2^2$; thus $\lambda^2 = 1$, and $\lambda = \pm 1$. It follows that the linear combination in Equation (80) describes two MO's ψ and ψ^*:

$$\psi = N(\psi_A + \psi_B) \tag{81}$$
$$\psi^* = N^*(\psi_A - \psi_B) \tag{82}$$

The two-center functions are used to determine the B.E. of the ion H_2^+ as a function of internuclear distance. Inserting ψ and ψ^* into the Schrödinger equation yields the total energy, E', of the electron, if the potential energy, U, is known:

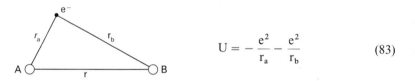

$$U = -\frac{e^2}{r_a} - \frac{e^2}{r_b} \tag{83}$$

The bond energy, E, of H_2^+ from its dissociation products H and H^+ is:

$$E = E' + \frac{e^2}{r} - E_H \qquad\qquad E_H = -13.6\,eV \tag{84}$$

The energy of a hydrogen atom is subtracted from the energy E' of the electron, and the repulsion energy of the nuclei has been taken into account.

The bond energy of the H_2^+, calculated from functions ψ and ψ^*, is shown as a function of internuclear distance in Figure 45 where function ψ describes a stable state. The function ψ^* represents an unstable state whose energy at all finite internuclear distances is greater than that of the dissociation products. The electron of H_2^+ can reside either in the ψ (bonding) or ψ^* (antibonding) orbital with the molecule ion either stable or unstable. The calculated electron density distributions for the two states are shown as a section through the ion in Figure 46. The electron density in the antibonding state falls to zero along the internuclear axis, and thus has a nodal plane perpendicular to this axis. The electron density between the nuclei

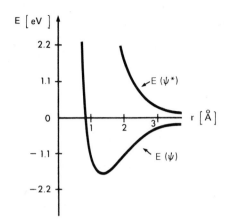

Fig. 45.   Bond energy of the ion H_2^+ as a function of internuclear distance, r, calculated for the bonding and antibonding states. The minimum of the curve $E(\psi)$ lies at 1.3 Å, and -1.76 eV.

is large in the bonding state, however, and the repulsion of the two nuclei is thus compensated by the attraction of the electron for the nuclei, resulting in a stable state.

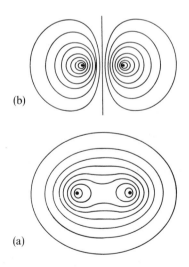

Fig. 46.   Contour diagrams of the electron density of the molecule ion H_2^+ for (a) the bonding and (b) antibonding orbital.

The energy of the electron in each MO is:

$$E(\psi) = \frac{E_A + \beta}{1 + S} \qquad\qquad E(\psi^*) = \frac{E_B - \beta}{1 - S} \tag{85}$$

E_A denotes the energy of the electron in orbital ψ_A of atom A, having a distance d from atom B. A is a hydrogen atom, and the 1s-function is used for ψ_A. E_A is not identical, however, with the energy of an electron in an isolated hydrogen atom, since the potential energy of the electron at nucleus A is influenced by B, and the Equation 83 is substituted for U into the Schrödinger equation $E_A \cdot \psi_A = H\psi_A$, giving a Hamiltonian operator of the form:

$$H = -\frac{h^2}{8\pi^2 m}\nabla^2 - \frac{e^2}{r_a} - \frac{e^2}{r_b} \tag{86}$$

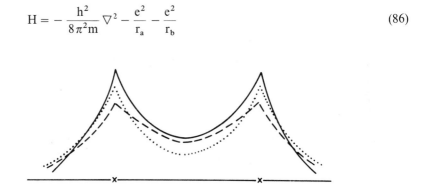

Fig. 47.   Change in the electron density, ψ^2, along the internuclear axis in the molecular ion, H_2^+

······ calculated neglecting bonding ($\psi^2 = \frac{1}{2}\psi^2{}_A + \frac{1}{2}\psi^2{}_B$)
– – – calculated from Equation (81)
——— calculated considering the effective nuclear charge (see p. 102)

As long as the electron is in the AO, ψ_A, it is closer to nucleus A, so that $r_a < r_b$ and the potential energy and E_A should be similar to that in the isolated hydrogen atom. The orbital energy of the isolated hydrogen atom is the negative of the ionization energy (cf. p. 23).

The overlap integral, S, can be neglected in the numerator of the energy equation since $S < 1^5$ to give:

$$E(\psi) = E_A + \beta \qquad\qquad E(\psi^*) = E_B - \beta \tag{87}$$

The parameter β is the resonance or exchange integral:

$$\beta = \int \psi_A H \psi_B \, dV \tag{88}$$

5   In H_2^+, $S = 0.59$, but generally $S = 0.2$ to 0.3.

This integral, which according to R.S. Mulliken is proportional to the overlap integral, S, but carries an opposite sign, reflects the stabilization or destabilization of an electron in an MO. In the bonding MO the electron has a lower energy than in the AO, ψ_A, because it now has available the space near both nuclei and can reside in an area of high effective nuclear charge between the nuclei. Delocalization of electrons in mutually overlapping AO's always leads to stabilization provided the MO has no nodal planes between the nuclei. The stabilization is larger the more the AO's overlap. These results are shown in a qualitative MO energy level diagram which is symmetrical as in Figure 48 only if the approximation $S = 0$, i.e., $E = E_A \pm \beta$ is employed, otherwise the destabilization of the antibonding MO is somewhat greater than the stabilization of the bonding MO.

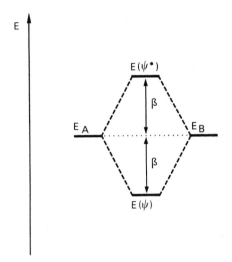

Fig. 48.   Energy level diagram for the formation of molecular orbitals by overlap of identical atomic orbitals to form a homonuclear diatomic molecule.

The MO's for H_2^+ are also present in H_2 and He_2^+ in which overlap of two identical 1s-orbitals takes place as well. Electrons are added in order of increasing orbital energy in conformity with the Pauli principle (Aufbau). In H_2^+ the single electron occupies the bonding orbital described by ψ to form a partial covalent bond assigned a bond order of one half.

In the H_2 molecule two electrons occupy the bonding MO, and the bond is correspondingly stronger. This is shown, e.g., in the B.E. values and internuclear distances in Table 15.

The molecule ion, He_2^+, found in gaseous discharges, contains three valence electrons, two of which lie in a bonding, and one in an antibonding MO. The attraction of the two bonding electrons more than compensates for the antibonding one and a stable ion results, which is similar to H_2^+ in B.E. and internuclear distance. However, if He_2^+ is neutralized by an additional electron, dissociation into two helium atoms takes place, since two antibonding electrons have a stronger repulsion than the attraction energy furnished by two bonding electrons. Consequently, the He_2 molecule is unstable.

The term bond order, b(MO), can be defined so that each bonding electron corresponds to one-half order. If bonding and antibonding electrons are present, those in the antibonding MO's are subtracted from the total number of bonding electrons, and the result divided by two. Non-bonding electrons are not counted. Increasing bond orders, as listed in Table 15, are reflected in bond energy increases and decreasing internuclear distance. Bond order, as used here, only holds for two-center bonds.

Table 15   Bond Properties of some Diatomic Molecules.

Molecule	Valence Electrons	Bond energy kJ/mol	Internuclear Distance, Å	Bond Order b(MO)
H_2^+	1	269	1.06	0.5
H_2	2	458	0.74	1.0
He_2^+	3	≈ 300	1.08	0.5
$[He_2]$	4	0	–	0

The LCAO approximation discussed so far leads to a calculated B.E. which is too small, and an internuclear distance of 2.5 a_0 for H_2^+ which is too large. The molecular function ψ needs to be improved by modifying functions ψ_A and ψ_B by introducing the effective nuclear charge number, Z^*, into the radial portion of the wave function. At large values of r, $Z^* = 1$; at 2.0 a_0 energy minimization gives $Z^* = 1.24$ which corresponds to a higher B.E. and smaller internuclear distance. A further improvement results from elongation of the spherically symmetrical AO's in the direction of the neighboring nucleus to improve the overlap. The calculated B.E. and internuclear distance, after appropriate modification of the atomic functions, differs only slightly from the experimental values.

Molecular Orbitals from p- and d-Atomic Orbitals:

MO's are formed when the overlap integral, S, for the AO's differs from zero. If $S > 0$, the MO is bonding, if $S < 0$, antibonding, if $S = 0$, non-bonding in character. If other orbitals aside from the 1s are involved, overlap relationships of all orbital combinations must be tested.

Overlap of two orthogonal orbitals is zero based on symmetry considerations, and non-zero S values only result in the overlap of orbitals of suitable symmetry. From orbitals with σ-symmetry with respect to the internuclear axis, σ- and σ^*-MO's result, and from the overlap of π-AO's π- and π^*-MO's result as shown schematically in Figure 49. Combinations with d-orbitals may also be of importance. The

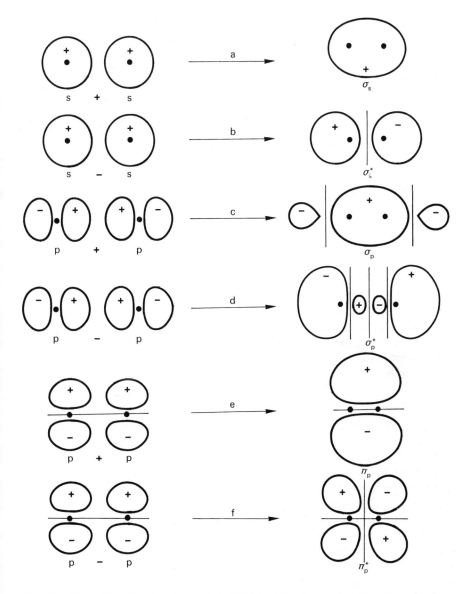

Fig. 49.   Formation of molecular orbitals (MO's) of bonding and antibonding character from atomic orbitals (AO's) of different symmetry. (a–d): σ-orbitals; (e–f): π-orbitals

splitting of two AO's into two MO's is greater with better overlap, since the resonance integral, β, is proportional to the overlap integral, S. At equal internuclear distance and orbital energy, the overlap of two σ-orbitals is usually stronger than that of two π-orbitals, for reasons of symmetry. Symmetry considerations aside, better interaction is achieved with smaller differences in the orbital energies.

The energy level diagrams of the homonuclear diatomic molecules of the first period will be discussed. Fluorine has nine electrons occupying all orbitals of the K- and L-shells, 1s, 2s, $2p_x$, $2p_y$, and $2p_z$. In the formation of F_2 only those orbitals combine which correspond in symmetry and energy as shown in Figure 50.

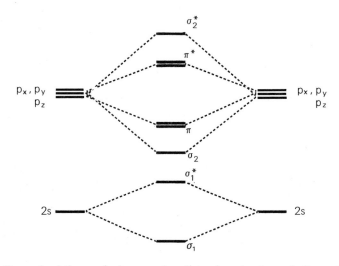

Fig. 50. Energy level diagram for homonuclear diatomic molecules neglecting sp-interactions.

The high effective nuclear charge draws the 1s electrons close about the nuclei, and, consequently, overlap with other orbitals does not take place. The 2s-orbitals, however, combine to form MO's σ_1 and σ_1^*. An interaction of 2s and $2p_z$ orbitals, which have the same symmetry is neglected because the orbital energies differ by more than 20 eV. The $2p_z$ orbitals, therefore, combine only with each other to form MO's σ_2 and σ_2^*. Owing to symmetry, the π-orbitals can only combine pairwise, p_x with p_x, and p_y with p_y. Because of the degeneracy of the π-orbitals and the equal overlap of both pairs, two degenerate π- and two π^*-MO's result. Unoccupied orbitals with principal quantum number >2 need not be taken into account since they have a much larger orbital energy. The 14 valence electrons of F_2 now occupy the 7 MO's of lowest energy, four bonding and three antibonding, as shown in Figure 50, resulting in $b(MO) = 1$.

Two additional valence electrons would be available in the hypothetical molecule Ne_2 which could only occupy the σ_2^*-MO, resulting in an equal number of bonding and antibonding electrons so that b(MO) and B.E. would be zero. The Ne_2 molecule does not exist.

The energy difference between the 2s and 2p levels decreases markedly in the series, $Ne > O_2 > N_2 > C_2 > B_2$, from >25 eV in the Ne atom to ca. 3 eV in the boron atom. The smaller this difference, the more likely is the interaction of the 2s- and $2p_z$-orbitals, leading to mixing (hybridization) and bringing about a different sequence in the energy of the MO's as shown in Figure 51. The σ_2-MO now lies higher than the two degenerate π-MO's. The σ- and σ*-MO's no longer have pure s- or p-character, but are sp-hybrids.

The properties of B_2 and C_2 reflect the change in the energy sequence of the MO's, and that the two π-MO's in these molecules are more stable than the σ_2-MO's. This also holds for N_2, since the unpaired electron in the N_2^+ ion has the properties of a σ-electron. Some sp-hybridization is also indicated for O_2 and F_2, although in these cases the sequence of MO's follows the diagram shown in Figure 50.

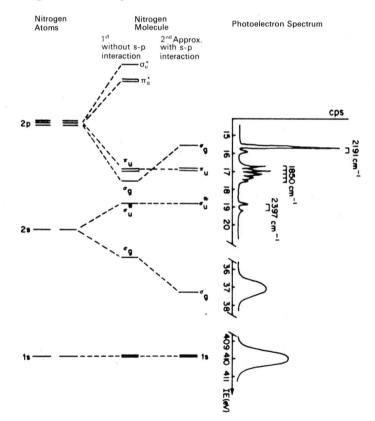

Fig. 51.   Comparison of the qualitative MO scheme for the nitrogen molecule with the photoelectron spectrum. [After H. Bock, *J. Chem. Educ., 51*, 506 (1974)].

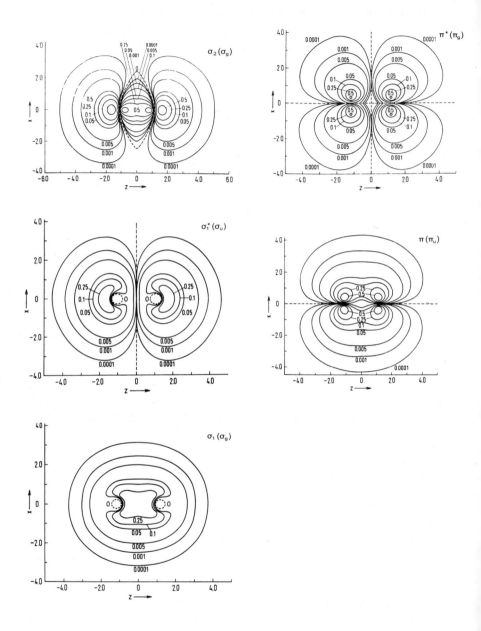

Fig. 52.   Contour diagrams of the electron density of various molecular orbitals of the O_2 molecule. These are the occupied orbitals of the valence shell (cf. Figure 50). [After R. S. Mulliken, *Angew. Chem.* **79**, 541 (1967).]

Table 16   Bond Properties of some Diatomic Molecules.

Molecule	Valence Electrons	Bond Order	Dissociation Energy, kJ/mol	Internuclear Distance, Å	Comments
$[Ne_2]$	16	0	0	–	unknown
F_2	14	1	151	1.42	singlet
O_2	12	2	494	1.21	triplet
N_2	10	3	945	1.09	singlet
C_2	8	2	628	1.31	singlet
B_2	6	1	289	1.59	triplet
$[Be_2]$	4	0	0	–	unknown
Li_2	2	1	109	2.67	singlet

Table 16 summarizes some properties of the molecules discussed here. In O_2, the 12 valence electrons fill the MO's σ_1, σ_1^*, σ_2 and π, but occupy the π^*-MO's with only two electrons with parallel spins which exist as far apart as possible. Each of these two degenerate orbitals is singly occupied, and O_2 is in the triplet state, a fact not explained by the VB theory. The b(MO) of 2 reflects the high dissociation energy. The ionization energy of 12.2 eV is smaller than that of the oxygen atom (13.6 eV), since the antibonding MO's are richer in energy than the AO's.
The electron density diagrams for several O_2 MO's are shown in Figure 52.
Both the MO- and the VB theories give a bond order of 3.0 for the N_2 molecule. The ionization energy of N_2 of 15.56 eV is larger than that of the nitrogen atom, (14.5 eV), since the highest occupied MO is bonding.
The molecule C_2 which occurs in carbon vapor has a double bond, and like N_2 is diamagnetic. This is explained by Figure 51, since Figure 50 would predict a triplet ground state. The molecule B_2 with 6 valence electrons is paramagnetic, and like O_2 has two unpaired electrons, as in the scheme in Figure 51, i.e., the unpaired electrons are in degenerate π-MO's.
The Li atom has no p-electrons. Both valence electrons of Li_2 are in a σ_1-MO; if two additional were added, as for the hypothetical Be_2, they could only occupy the σ_1^*-MO which would dissociate the atoms (Fig. 50).

Heteronuclear Diatomic Molecules:
The energy level diagrams of heteronuclear diatomic molecules, AB, present complications. The MO treatment of polyatomic molecules will be reserved for discussion in the part II.
The MO description of heteronuclear molecules requires knowledge of the symmetries and energies of all valence orbitals to tell which AO's combine, if σ-AO's hybridize, and whether strong overlap occurs. To determine the highest occupied AO's, the negative ionization energies are used as orbital energies. The AO energies used in the MO-diagrams are for atoms in the valence state which are in the proximity of the bond partners, but not bonded as yet.

The molecules HF and CO illustrate these relations. The construction of the MO's for HF in Figure 53 requires the ionization energies, $I_H = 13.6$ eV, and $I_F = 17.4$ eV. Of the valence orbitals of the fluorine atom only $2p_z$ has the proper energy and symmetry to interact with the hydrogen 1s orbital, while $2p_x$ and $2p_y$ are of π-symmetry with regard to the intermolecular axis and 2s is too low in energy. But since the energy gap between H1s and F$2p_z$ orbitals is quite large, the σ-MO which accepts the bonding electron pair is largely localized near the fluorine atom, and its energy and geometry are similar to the p_z-hybrid AO at fluorine, but the antibonding σ*-MO is similar to the 1s-AO of hydrogen. The strongly polar bond in HF, which results from the considerable difference in electronegativity between H and F, is reflected in the MO diagram along with the high bond energy arising mainly from the stabilization of the hydrogen electron to the level of the fluorine $2p_z$ orbital.

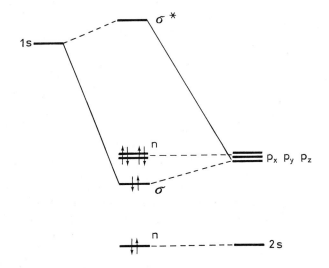

Fig. 53.   Energy level diagram for the bond in the HF molecule.

The MO scheme for CO, where sp-interaction of the σ-orbitals is taken into account, is shown in Figure 54. The diagram is complex even though the molecule is simple. The oxygen AO's are more stable owing to its higher effective nuclear charge than the corresponding carbon AO's. The experimental bond order is 2.8 from valence force constants. The CO molecule is an isostere of N_2, i.e., both molecules have the same:

a) number of atoms   c) arrangement of electrons
b) number of electrons   d) sum of nuclear charge numbers

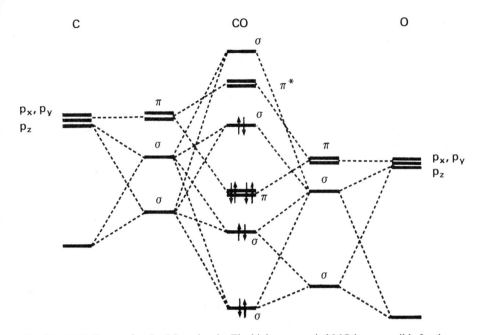

Fig. 54.   MO diagram for the CO molecule. The highest occupied MO is responsible for the Lewis-base properties of CO.

The MO scheme for N_2 can serve as an approximation for CO in that CO can be considered as a perturbed nitrogen molecule. Similar treatment is also accorded other isosteric or isoelectronic molecules, for example CN^- and NO^+ are also isoelectronic with CO and N_2, and the first three of the above conditions are fulfilled.[6]

Concluding remarks on MO theory:
The MO energy level diagrams are based on many simplifications and approximations, and hence are only useful in qualitative and comparative discussions. This holds particularly for b(MO) which apparently varies discontinuously, contrary to experimental evidence from bond energies, internuclear distances, and valence force constants which suggest that bond strength is smoothly variable as when substitution changes the electrons distribution, and orbital overlap. In contrast, b(MO) = 1 in H_2 as long as the two 1s-orbitals overlap to any degree, i.e., even at internuclear distances >0.74 Å. This shortcoming can only be remedied by a more realistic definition of b(MO).

6 Such compounds, which correspond in number of atoms and number and arrangement of valence electrons, but not in the total number of electrons, are also usually referred to as isoelectronic, cf. HF *vs.* HCl; SO_4^{2-} *vs.* BrO_4^-.

5.2.4. The Coordinate Bond

A covalent single bond is the result of two electrons of opposite spin occupying an orbital produced by overlap of two pure or hybridized AO's.
In the treatment of the covalent bond by either MO or VB theory, the origin of the two indistinguishable electrons is of no consequence. Thus either atom can contribute one electron each, as when two singly occupied AO's overlap; or both electrons can originate from one atom, as when a doubly-occuped orbital overlaps an unoccupied one:

$$A\cdot + \cdot B \rightarrow A - B \qquad\qquad A + :B \rightarrow \overset{\ominus}{A} - \overset{\oplus}{B}$$

The coordinate bond is portrayed by the second case in which an electron charge is transferred from one atom to the other, and formal charges are created. The donor atom formally loses negative charge which the acceptor atom gains. However, whole charges are not localized at each atom. The bonding electron pair is unsymmetrically distributed between both atoms, so that it is encountered with greater probability near the more electronegative partner. The atomic sizes also play a role in determining the polarity of the bond (cf. Section 5.5.5.) and partial instead of full charges, denoted by $\delta(+)$ and $\delta(-)$, are used:

$$\overset{\delta(-)\ \delta(+)}{A - B}$$

It is even possible, owing to other effects, for the actual polarity of a coordinate bond to be the reverse of that denoted by the formal charges.

Coordinate bonds, accordingly, cannot be distinguished from covalent bonds which often are polar because of the different electronegativities and sizes of the atoms, likewise symbolized by partial charges at both atoms. The coordinate bond thus only differs from the normal covalent bond with respect to the starting materials. Typical examples are:

$$H^+ + :NH_3 \rightarrow NH_4^+$$
$$SF_4 + F^- \rightarrow SF_5^-$$
$$SiF_4 + 2F^- \rightarrow SiF_6^{2-}$$
$$BF_3 + :NR_3 \rightarrow F_3\overset{\ominus}{B} - \overset{\oplus}{N}R_3$$

Empty d-orbitals of SF_4 and SiF_4 are employed in coordinate bonds, but the resulting bonds in the product ions SF_5^- and SiF_6^{2-} are indistinguishable. The hybridization of the central atom must change correspondingly, e.g., from sp^3 in SiF_4 to sp^3d^2 in SiF_6^{2-} which is octahedral.
Likewise, the planar BF_3 molecule (sp^2 hybrid at boron) changes into the pyramidal $-BF_3$ group (sp^3 hybrid) when the lone pair of, e.g., an amine, R_3N, or an ether,

R_2O, is utilized. According to the theory propounded by G. N. Lewis (1925), the formation of a coordinate bond may be regarded as salt formation between an acid (electron acceptor) and a base (electron donor). Lewis acids then are compounds with unoccupied orbitals in the valence electron shell, such as H^+, BF_3, SiF_4, PF_3, PF_5, BrF_3, SO_2, SO_3, etc. Lewis bases are compounds with free electron pairs in the valence shell, e.g., ammonia and amines, phosphines, water, alcohols, ethers, derivatives of hydrogen sulfide, halide ions, in addition to CO, N_2, NO, CN^- and other non-metallic compounds such as SO_2, $SOCl_2$, S_4N_4, NOCl, $POCl_3$, etc. The transition metal complexes with nonmetallic ligands are all based on coordinate bonds.

Other compounds which possess no unoccupied orbitals that can be utilized to form σ-bonds also function as Lewis acids. For example, in SO_3 all valence orbitals are occupied (sp^3d^2, cf. p. 97) and the three unoccupied d-orbitals are unsuited for bonding. The sulfur atom is saturated with 12 valence electrons, but SO_3 forms very stable donor-acceptor complexes of type $R_3N \cdot SO_3$ with nitrogenous Lewis bases since one of the sulfur π-orbitals can be cleared and then can accept the electrons of the Lewis base. The best suited orbital is the p-π-orbital:

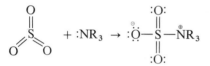

The sulfur atom changes its hybridization from sp^2 to sp^3 during this reaction and only two π-bonds remain, a reorganization which changes the geometry from trigonal planar to a deformed tetrahedron. Concomitantly, the SO bond order dedreases. The three SO-bonds in the reaction product are identical on the basis of resonance:

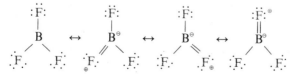

Boron trifluoride and many other non-metallic fluorides are similar. The BF bonds in BF_3 are polar and concentrate positive charge on the boron atom, enhancing its propensity as an electron-acceptor which is based on the unoccupied 2p-π-orbital, and allowing further stabilization by delocalizing the non-bonding electrons of the fluorine atoms into the 2p-π-boron orbital:

This diminishes the partial positive charge at boron and decreases its Lewis acidity. Owing to the planarity of the molecule, the delocalization can only occur if the corresponding π-orbitals overlap. These are four 2p-orbitals of the three fluorine and one boron atom which are perpendicular to the molecular plane. Linear combination of these four orbitals by the MO-method leads to one bonding, two non-bonding, and one anti-bonding π-MO's occupied by three electron pairs, i.e., just as in the VB method, one delocalized π-bond, and two lone electron pairs which remain at the fluorine atoms are obtained. The formal charges associated with this coordinate, π-bond, display an opposite polarity to that of the σ-bond, and enhance the strength of the BF bond. On adduct formation, the fluorine electrons are displaced from the 2p-π-orbital of the boron atom by the electrons of the incoming Lewis base.

Intramolecular π-bonds, as in BF_3, are frequently found in non-metallic compounds when an atom with unoccupied and energetically favorable p- or d-orbitals bonds another with a lone electron pair.

5.3. The Van der Waals Bond

All atomic and molecular gases (Ar, HCl, SiH_4, etc.) can be liquified and eventually solidified by cooling to sufficiently low temperatures (the solidification of helium requires a pressure of 26 bar). This is evidence for van der Waals attractive forces between atoms and molecules which do not otherwise undergo covalent, ionic, or metallic bonding, leading to weak bonds of B.E. <20 kJ/mol. This energy is manifested as energy of sublimation of solids, and as heat of evaporation of liquids. Van der Waals forces exist between all atoms, ions, and molecules, and contribute to the energy of covalent bonds and to the lattice energy of crystals (salts, metals). Van der Waals forces may be divided into three components: dispersion effects, which are always present, and, in polar molecules, additional dipole and induction effects. The dipole effect is most easily understood.

5.3.1. The Dipole Effect

An atom, a molecule or a part of a molecule has a dipole moment, μ, when the centers of positive and negative charge do not coincide, but are separated by the distance l. The dipole moment is the $\mu = e \cdot l$ where e = electron charge and μ is a vector, whose unit is the Debye, D, which may be measured directly via microwave spectroscopy or from the orientation polarization in a condenser field. Two charges, $+e$ and $-e$, separated by 1 Å create a dipole moment of 4.8 D. Polar molecules can arrange themselves in certain lower energy orientations:

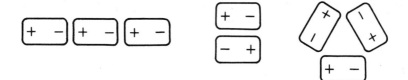

The energy in the first two arrangements is proportional to μ^2/d^3, as long as the distance, d, between the dipoles is large compared with l, and as long as μ^2/d^3 is large compared with the product $k \cdot T$ (rigid orientation; k = Boltzmann's constant; T = temperature in K). At higher temperatures the dipole orientations are perturbed or obliterated by thermal motion and the energy is proportional to μ^4/kTd^6.

Dipolar molecules can also repel, but such arrays are richer in energy, and hence less probable. The effective range of the dipole is about 5 Å, over which the interaction energy, U, decreases to an amount equal to the product kT (2.5 kJ/mol at 25 °C), the energy stored in a vibrational degree of freedom. If kT is larger than U, a vibration leads to dissociation of the bond. The cation-anion interaction has an effective range of *ca.* 500 Å (in vacuo), and for an ion-dipole interaction, *ca.* 14 Å. The dipole-dipole effect is then of little consequence at ≥ 5 Å, owing to the high power of d in the denominator.

5.3.2. The Induction Effect

A molecule with a permanent dipole moment, μ, induces a dipole moment, μ_{ind}, in a neighboring atom or in molecules oriented for attraction. Induction, which is temperature independent, is thus superimposed upon the dipole effect, but is generally much smaller than the dipole or dispersion effect, and is, therefore, generally neglected (cf. Table 18, p. 116).

5.3.3. The Dispersion Effect

The dispersion contribution to the van der Waals force, for molecules with small dipoles ($\mu < 1$D), is greater than the other two effects. In rare gases which crystallize in cubic closest-packed arrays at low temperatures, and in non-polar molecules (CH_4, SF_6, etc.) the dispersion effect alone is responsible for the lattice energy. Rare gas atoms are spherically symmetrical in the time average, but the instantaneous atomic dipole moment will in general be non-zero. This dipole moment in turn induces a dipole moment at the neighboring atoms, resulting in attraction. The binding energy, U, for two like and spherically symmetrical atoms is expressed as:

$$U = -\frac{3}{4} \cdot \frac{\alpha^2}{d^6} \cdot I \qquad \begin{array}{l} \text{I: Ionization energy} \\ \alpha\text{: Polarizability} \\ \text{d: Internuclear distance} \end{array} \qquad (89)$$

as long as d is large compared with atomic diameters. Dispersion is effective to *ca.*

4 Å, and is important in atoms and molecules with high polarizability such as the soft atoms of the heavy non-metals such as Xe, I, Br, Se, etc.; hard atoms with little polarizability are found in the first row of the periodic table.

Attractive forces predominate in the approach of two rare gas atoms to lower the energy. The energy increases again, however, and repulsion forces come into play after a certain distance. The interaction energy must thus contain both attractive and repulsive terms:

$$U = -\frac{a}{d^6} + \frac{b}{d^{12}}\qquad(90)$$

where a and b are atomic constants which are experimentally accessible. Energy curves for rare gases corresponding to the equation above are shown in Figure 55, in which both binding energy and equilibrium internuclear distance increase with atomic weight. Therefore, melting and boiling points and enthalpies of melting and vaporization increase from He to Xe (see Table 17).

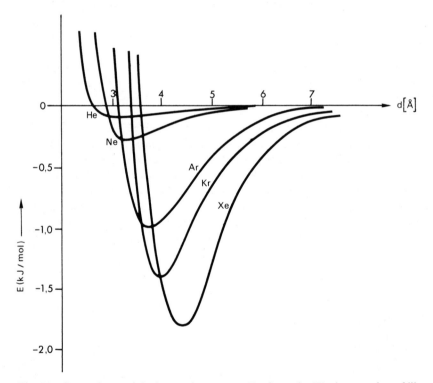

Fig. 55.   Dependence of the interaction energy, E, of van der Waals attraction of like rare gas atoms at distance, d.

Table 17   Melting and Boiling Points and Enthalpies of Melting and Vaporization of Rare Gases.

	Melting Point [K]	Boiling Point [K]	ΔH_s° fusion [kJ/mol]	ΔH° vap. [kJ/mol]
Helium	1.4 (26 bar)	4.2	0.021	0.837
Neon	24	27	0.335	1.805
Argon	83	87.3	1.17	6.53
Krypton	117	120		9.6
Xenon	161	166	2.30	12.6
Radon	202	208		18.0

The three components which contribute to van der Waals forces in six molecules of different dipole moments and polarizabilities are listed in Table 18 with the sum of the three effects shown in column eight. The numerical values for CO, HI, HBr, and HCl correlate with the boiling points. The high boiling points of NH_3 and H_2O cannot be related to van der Waals forces, however, and an additional bond type, through hydrogen bridges must be considered (see Section 6.6.).

5.3.4. Van der Waals Radii

In crystalline xenon the atoms occupy relative positions at the energy minima shown in Figure 55, corresponding to equilibrium between attractive and repulsive forces. Half the smallest internuclear distance in the cubic closest-packed array is the van der Waals radius, which for xenon is 2.20 Å.

Likewise, in molecular crystals, the distance between neighboring, non-bonded atoms gives the van der Waals radii for the respective atoms. For example, a value of 1.80 Å for the sulfur atom is obtained from the structure of rhombic S_8. The values vary between 1.70 and 1.90 Å arising from the complicated crystal structure, the unit cell of which contains 16 S_8 rings (a total of 128 atoms). The accuracy of the data shown in Table 19 thus is not great, since atoms are not rigid spheres with sharp boundaries, but display instead a gradual decrease in electron density with distance. Two atoms can, therefore, have different distances from each other, and yet be in contact.

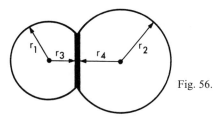

Fig. 56.   Space-filling model of a heteronuclear diatomic molecule (r_1, r_2: van der Waals radii; r_3, r_4: covalent radii).

Table 18   Molecular Properties and van der Waals Forces.

Molecule	Dipole Moment (gas) [D]	Polarizability	Ionization Energy [eV]	Dipole Effect	Induction Effect	Dispersion Effect	Sum of the three Effects	Boiling Point [K]
				$- - - - -$ in 10^{53} J·cm^6 $- - - - -$				
CO	0.11	19.9	14.3	0.0034	0.057	67.5	67.6	82
HI	0.38	54.0	12	0.35	1.68	383	385	238
HBr	0.82	35.8	13.3	7.6	4.45	176	188	206
HCl	1.08	26.3	13.7	22.5	6.0	105	133.5	188
NH_3	1.47	22.1	16	78	9.6	93	181	240
H_2O	1.85	14.8	18	193	9.9	47	250	373

Table 19   Van der Waals Radii of Non-metallic Atoms in Å

H: 1.1–1.3		He: 1.79
N: 1.5	F: 1.35	Ne: 1.59
P: 1.9	Cl: 1.80	Ar: 1.91
As: 2.0	Br: 1.95	Kr: 2.01
Sb: 2.2	I: 2.15	Xe: 2.20
O: 1.40		
S: 1.80		
Se: 2.00		
Te: 2.20		

Van der Waals radii are used in the construction of space-filling molecular models (see Figure 56).

Distances between non-bonded atoms in a molecule or crystal which are substantially smaller than the van der Waals radii sums imply weak covalent bonding, but both atoms must have suitable orbitals to overlap in forming a σ- or π-bond. Geminal atoms are sometimes forced 0.8 Å closer than their van der Waals distances. For example, the chlorine atoms in Cl_2O are only 2.8 Å apart, and in SCl_2 only 3.1 Å (van der Waals distance = 3.6 Å). The fluorine atoms in OF_2 and in SF_6 are only 2.2–2.3 Å apart, although the van der Waals distance is 2.7 Å. Likewise, the oxygen atoms in SO_2, SO_3, NO_2, NO_2^-, ClO_2^-, SO_4^{2-}, and ClO_4^- are 0.2–0.7 Å less than the van der Waals distances apart, and iodine atoms in the I_2O_5 molecule are in a similar situation. Thus non-bonded atoms can interpenetrate to generate repulsive forces balancing the attractive bonding forces.

5.4. Molecular Geometry

5.4.1. Theory of Electron Pair Repulsion

The theory of valence shell electron pair repulsion (VSEPR) developed by R.J. Gillespie and R.S. Nyholm, is based upon Coulombic repulsion of electrons and upon the Pauli principle where two electrons of like spin cannot be at the same location at the same time.

A main group atom is first separated into its rare gas core and the valence electrons which reside on the surface at maximum distance from one another and thus minimum energy. In beryllium, with two valence electrons, a linear arrangement is most probable, in boron the three valence electrons are at the corners of an equilateral triangle and in carbon the four valence electrons are at the corners of a tetrahedron about the core. The electrons are not localized, but share the space on the surface of the core equally. Each electron of a pair occupies a hemisphere, each of three a

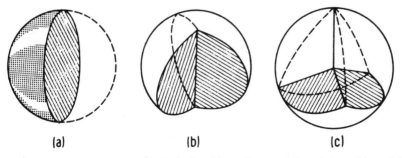

(a)   (b)   (c)

Fig. 57.   Spherical segments as loci of residence for two (a), and three (b), and four (c), electrons or electron pairs on the core surface [R.J. Gillespie, *Angew. Chem.* **79**, 885 (1967)].

spherical segment of 120°, etc., with the electrons treated as if they occupied non-overlapping orbitals.

The four valence electrons of the carbon atom are arranged with like spins tetrahedrally. The electrons of four hydrogen atoms with antiparallel spins also take up a tetrahedral arrangement. Electrostatic repulsion separates the tetrahedra, but electrostatic attraction by the hydrogen nuclei draws the electrons with opposed spins into the same spatial segment, i.e., into the same orbital, and the two tetrahedra become close in orientation, while the four electron pairs separate to maximum distances to form the tetrahedral shape of the CH_4 molecule.

The extension of these concepts requires treating the problem of arranging n points at maximum distance on a spherical surface, the results of which are the polyhedra shown in Figure 58. For n = 5, 7, and 8 several solutions are possible. The important n = 5 case gives a trigonal bipyramid with 2 axial and 3 equatorial positions, as well as a square pyramid with a unique apical position, but most AX_5 molecules are trigonal bipyramidal.

The arrangements in Figure 58 determine the geometry of molecules of type AX_n where A is the center atom and X a substituent. The number of electron pairs is equal to n if only single bonds are present. Examples of such compounds are listed in Table 20.

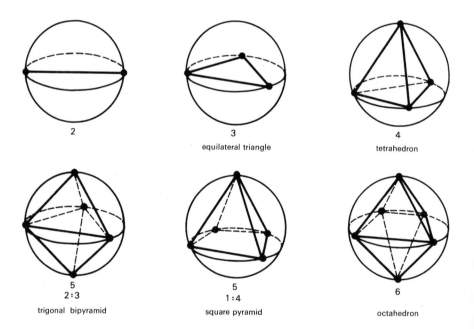

Fig. 58. Arrangements of points (electron pairs) on a spherical core surface maximizing distances between neighboring points. Two possibilities are shown for five points.

Table 20   Shapes of Molecules According to the VSEPR Theory.

Number of Electron Pairs	Arrangement	Type	Shape	Examples
2	linear	AX_2	linear	Hg_2Cl_2, HgX_2, CdX_2, ZnX_2
3	triangular	AX_3	triangular	BX_3, GaI_3, $In(CH_3)_3$
		AX_2E	V-shaped	CF_2, $SiCl_2$
4	tetrahedral	AX_4	tetrahedral	$(BeCl_2)_n$, $(Be(CH_3)_2)_n$, BeX_4^{2-}, BX_4^-, CX_4, NX_4^+, $OBe_4(CH_3CO_2)_6$, BeO, SiX_4, GeX_4, AsX_4^+
		AX_3E	trigonal-pyramidal	NX_3, OH_3^+, PX_3, AsX_3, SbX_3, P_4O_6, As_2O_3, Sb_2O_3, $(AsO_2^-)_n$
		AX_2E_2	V-shaped	OX_2, SX_2, SeX_2, TeX_2
5	trigonal-bipyramidal	AX_5	trigonal-bipyramidal	PCl_5, PF_5, PF_3Cl_2, $SbCl_5$, $Sb(CH_3)_3Cl_2$
		AX_4E	C_{2v}	SF_4, SeF_4, R_2SeCl_2, R_2SeBr_2, R_2TeCl_2, R_2TeBr_2
		AX_3E_2	T-shaped	ClF_3, BrF_3, $C_6H_5ICl_2$
		AX_2E_3	linear	ICl_2^-, I_3^-, XeF_2
6	octahedral	AX_6	octahedral	SF_6, SeF_6, TeF_6, S_2F_{10}, $Te(OH)_6$, PCl_6^-, PF_6^-, $Sb(OH)_6^-$, SbF_6^-, $(SbF_5)_n$, SiF_6^{2-}.
		AX_5E	square pyramidal	ClF_5, BrF_5, IF_5
		AX_4E_2	square	ICl_4^-, I_2Cl_6, BrF_4^-, $ICl_2 \cdot SbCl_6$, XeF_4

Compounds with free electron pairs, E, at the center atom, AX_lE_m, where the number of electron pairs at atom A is equal to $l + m$, behave similarly, i.e., the pairs, E, are active stereochemically, and behave as pseudo-substituents. An example is H_2O (AX_2E_2), in which the four electron pairs at the oxygen atom form a pseudo-tetrahedron with two corners occupied by hydrogen atoms. The molecule NH_3 (AX_3E) also forms a pseudo-tetrahedron with three corners occupied by hydrogens. Examples with two to six valence electron pairs are listed in Table 20, and the structures are shown in Figure 59. Experimental results in all but a few cases agree with the predictions of the VSEPR theory. Both VB and VSEPR theories are equivalent, since VSEPR theory arrives at structures which result from hybridization.

The regular structures shown in Figure 59 are only correct when the substituents are identical. If different substituents or free electron pairs are present, or multiple bonds with substituents exist, then deformation which lowers symmetry will occur. The following rules then apply:

Free Electron Pairs:

The free electron pairs in a molecule, AX_lE_m, unlike the bonding pairs, are influenced by the field of only one nucleus, and their orbitals are larger and take up more space. This has important consequences, for example, in a tetrahedron where the angles between the bonding pairs decrease on substitution by lone pairs:

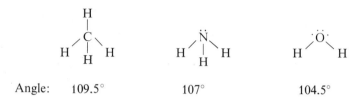

Angle: 109.5°   107°   104.5°

Corresponding decreases are also found with heavier central atoms, e.g., the valence angles in PCl_3 are 100.3°, in $AsCl_3$, 98.4° and in SCl_2, 103°.

Likewise, if a bonding electron pair in a regular octahedron (e.g., in SF_6) is replaced by a free electron pair (e.g., in ClF_5), the valence angles decrease from 90°.

Free electron pairs tend, owing to their greater spacial requirements and stronger repulsion, to occupy those positions with the largest available space. Lone pairs will thus tend to separate in order to minimize interactions with each other. In a molecule, AX_4E_2 (pseudo-octahedron), the lone pairs will occupy *trans*-positions and give a square planar arrangement (cf. XeF_4).

The equatorial and axial positions in a trigonal bipyramid are not equivalent. The angles within the trigonal plane are larger (120°) than between it and the apical positions ($\angle = 90°$). Electron pairs, therefore, tend to occupy equatorial positions:

:SF₄   :ClF₃   :XeF₂

Since free electron pairs require more space than bonding pairs, the angles in AX_4E and AX_3E_3 molecules are smaller than the 90°, 120°, and 180° for an ideal trigonal bipyramid. Thus the angles in SF_4 are 101° ($\angle F_{eq}$—S—F_{eq}) and 173° ($\angle F_{ax}$—S—F_{ax}). In ClF_3 the angles are 87.5° ($\angle F_{ax}$—Cl—F_{eq}) and 175° ($\angle F_{ax}$—Cl—F_{ax}). The molecule XeF_2 is exactly linear. Owing to the difference of

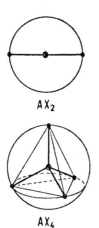 

AX₂   AX₃   AX₂E

AX_2 AX_3 AX_2E

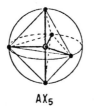

AX_4 AX_3E AX_2E_2

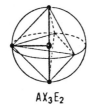 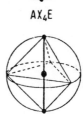

AX_5 AX_4E

AX_3E_2 AX_2E_3

AX_6 AX_5E AX_4E_2

Fig. 59.   Schematic structures of molecules which contain up to six electron pairs in the valence shell of the center atom.
(E: free electron pair,   X: substituent)
R. J. Gillespie, *Angew. Chem.*, **79**, 885 (1967)

equatorial and axial positions, AX_5, AX_4E, and AX_3E_2 molecules also have two different internuclear distances $d(A—X)$ with $d(A—X_{eq}) < d(A—X_{ax})$ by 5–15%. Thus equatorial atoms are bound more tightly.

Substituents of differing Electronegativity:
The electronegativity of an atom or a group is defined as the ability to attract the electrons of a bond. Electronegativity, X_E, increases in the periodic system from the lower left to the upper right (see Section 5.5.5.).
Bonding electron pairs contract under the influence of highly electronegative substituents and their reduced spatial requirements allow other bonding and lone pairs to expand. Valence angles thus decrease with increasing electronegativity of the substituents, as the following examples will illustrate:

H_2O: 104.5° – OF_2: 103.2° because $x_E(F) > x_E(H)$

PI_3: 102° – PBr_3: 101.5° – PCl_3: 100.3° – PF_3: 97.8°   } because $x_E(F) > x_E(Cl)$
AsI_3: 100.2° – $AsBr_3$: 99.7° – $AsCl_3$: 98.7° – AsF_3: 96° } $> x_E(Br) > x_E(I)$

With decreasing electronegativity of the central atom, the free electron pairs on its surface expand; e.g., the angles decrease $H_2O > H_2S > H_2Se > H_2Te$ (cf. Table 14, p. 93).
In molecules with substituents of different electronegativity, the angles between the more electronegative ones are smaller, as illustrated by SiH_3Cl, CHF_3, and CH_2F_2, discussed on p. 102. However, steric crowding can also influence the valence angles.
The most electronegative substituents will occupy the axial positions of a trigonal bipyramid as do the fluorine atoms of PCl_4F and PCl_3F_2. In the molecules CH_3PF_4 and $(CH_3)_2PF_3$, the less electronegative methyl groups are equatorial, as expected.

Multiple Bonds:
The electrons of multiple bonds require more space than those of single bonds. Thus, in AX_3Y molecules with a double bond, $A═Y$, the angles $X—A—X$ are smaller than $X—A—Y$. Molecules with both double bonds and free electron pairs show large deviations from regular shapes, as the following examples show:

$\angle FPF = 101.3°$ $\angle ClPCl = 101.8°$ $\angle OSF = 106.8°$
                                                             $\angle FSF = 92.8°$

In molecules with several double bonds, the angle between these will be the largest in the molecule:

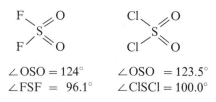

$$\angle OSO = 124°$$
$$\angle FSF = 96.1°$$

$$\angle OSO = 123.5°$$
$$\angle ClSCl = 100.0°$$

Concluding Remarks:
The VSEPR theory furnishes a rational basis for molecular structure which, while derived from available experimental data, has also been successful in predicting unknown structures. Few cases are known in which the molecular structure is predominantly determined by substituent interactions, such as the ions SeX_6^{2-} and TeX_6^{2-} (X = Cl, Br, or I). These AX_6E ions should form distorted octahedral structures, but the free electron pair is apparently located in a spherically-symmetrical orbital (e.g., 4s or 5s), without steric influence, and the ions are regular octahedrons. At a certain size and number of substituents the molecular geometry will also be determined by steric requirements.

5.4.2. Molecular Symmetry and Point Group Symbols

Molecular symmetry is important in interpreting infrared and Raman spectra or in elucidating orbital overlap in polyatomic molecules. The symmetry of a molecule is obtained from its behavior in certain symmetry operations. Consider the two bent molecules H_2O and HOCl:

which may be rotated about the bisector of the valence angle. The H_2O molecule after rotation through 180° is indistinguishable from its starting position, and after rotation through an additional 180° is again at the starting position. Such a rotation is a symmetry operation, and the axis of rotation (symbol: C) is a symmetry element of the molecule. The numbering n of the axis of rotation is written as a subscript (C_n). After n-fold rotation about $2\pi/n$, the molecule is back at the starting position. In H_2O there is the symmetry element, C_2. However, the HOCl molecule possesses only a C_1 symmetry axis, which is present in all molecules and is not counted.
An additional symmetry element of the H_2O molecule is a mirror plane (symbol σ), which bisects the valence angle and is perpendicular to the molecular plane. Reflection of the atoms in this plane exchanges the hydrogen atoms, but the initial

and final states are indistinguishable. This mirror plane is not present in the HOCl molecule.

Both H_2O and HOCl have a mirror plane which coincides with the molecular plane. This symmetry element is common to all planar molecules.

A special symmetry operation is improper rotation, S_n, with an n-fold axis, i.e., rotation about an axis, C_n, followed by reflection in a plane perpendicular to this axis. The sequence followed is optional. Molecules with S_n axes are, e.g., $SiCl_4$ (S_4 bisects $\angle Cl—Si—Cl$) and *trans*-dichloroethylene (S_2, along the carbon-carbon axis).

A fourth symmetry operation is inversion, i.e., reflection of all atoms through a point which is a center of symmetry or inversion center, i. Inversion centers are found in CO_2, XeF_4, SF_6, C_6H_6, etc.

Another symmetry operation possible for all molecules is identity, E. Each molecule is identical with itself.

A molecule is more symmetrical the more symmetry elements it possesses. Thus H_2O with symmetry elements C_2 and 2σ is of higher symmetry than HOCl which only possesses σ.

The four symmetry operations are summarized in Table 21.

Table 21   Symmetry Operations and Symmetry Elements.

Symmetry Operation	Symmetry Element	Symbol	Types
Rotation	Axis of Rotation	C	C_2, C_3, C_4, C_5
Reflection	Mirror Plane	σ	$\sigma_v, \sigma_h, \sigma_d$
Improper Rotation	Improper-Rotation Axis	S	S_2, S_4
Inversion	Inversion Center	i	

Symbols can be derived for any molecule which characterizes its symmetry properties using a scheme as shown in Figure 60. These point group symbols represent certain symmetry classes.

Point group symbols consist of a capital letter and a lower case subscript, which can be a number, a lower case letter or a combination of both. Certain very high symmetries have their own symbols: T_d for tetrahedron, (e.g., CH_4, P_4), O_h for octahedron (e.g., SiF_6^{2-}, PCl_6^-) and I_h for icosahedron (e.g., $B_{12}H_{12}^{2-}$). In these polyhedra (Fig. 61), all corners are considered equivalent. Linear molecules without an inversion center, such as N_2O, ClCN, OCS, SCN^-, etc., are given the symbol $C_{\infty v}$. These molecules have a ∞-fold axis of rotation (C_∞). The axis with the highest rotation is chosen as the vertical axis with the subscript v, as in $C_{\infty v}$. $C_{\infty v}$-molecules have an infinite number of mirror planes which contain the molecular axis and are also called vertical (σ_v), and innumerable improper rotation axes which are perpendicular to the molecular axis. There are no other symmetry elements.

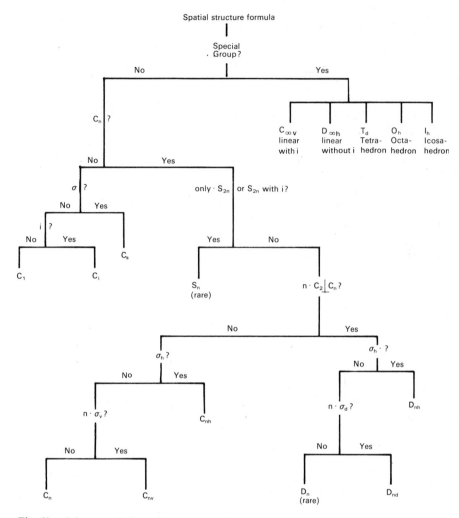

Fig. 60.   Scheme to derive point group symbols.

Linear molecules (CO_2, CS_2, XeF_2, Hg_2Cl_2, C_3O_2, etc.) and ions (NO_2^+, N_3^-, HF_2^-, etc.) with an inversion center are given the symbol $D_{\infty h}$. A $C_{\infty v}$-axis is also present in these molecules. Perpendicular to this axis there are also ∞ C_2-axes, and the molecule possesses dihedral symmetry (symbol: D), a high symmetry found if n twofold axes are perpendicular to the highest-fold C_n axis: $nC_2 \perp C_n$, is present. Other dihedral molecules are, e.g., BF_3, XeF_4, PF_5, and IF_7. The subscript h for horizontal in a point group symbol denotes a mirror plane (σ_h) perpendicular to the highest-fold axis, as in CO_2, giving the symbol $D_{\infty h}$. Symbols for any molecule

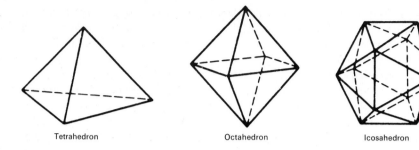

Tetrahedron   Octahedron   Icosahedron

Fig. 61.   Symmetrical polyhedra.
a) Tetrahedron with 24 different symmetry operations.
   ($8\,C_3$, $3\,C_2$, $6\,S_4$, $6\,\sigma_d$ and E)
b) Octahedron with 48 symmetry operations.
   ($8\,C_3$, $3\,C_2$, $6\,C_4$, $6\,C_2'$, i, $8\,S_6$, $3\,\sigma_h$, $6\,S_4$, $6\,\sigma_d$ and E)
c) Icosahedron with 120 symmetry operations.
   (e.g., i, $12\,C_5$, $10\,C_3$, $15\,C_2$ and others)

Table 22   Point Group Symbols of Important Molecular Types.

Molecular Type	Geometry	Examples	Point Group
AB_2	linear	CO_2, XeF_2, C_3O_2	$D_{\infty h}$
	angular	SO_2, NO_2^-	C_{2v}
ABC	linear	COS, N_2O	$C_{\infty v}$
	angular	HOCl, ClNO	C_s
AB_3	trigonal-planar	BCl_3, NO_3^-	D_{3h}
	trigonal-pyramidal	PH_3, XeO_3	C_{3v}
AB_2C	planar	$BClF_2$, $ClNO_2$	C_{2v}
	pyramidal	$ClNH_2$, $OSCl_2$	C_s
AB_4	tetrahedral	ClO_4^-, BF_4^-	T_d
	square-planar	XeF_4	D_{4h}
AB_3C	pseudo-tetrahedral	$OPCl_3$, $S_2O_3^{2-}$	C_{3v}
AB_5	trigonal-bipyramidal	PF_5	D_{3h}
AB_6	octahedral	SF_6	O_h
AB_5C	pseudo-octahedral	$SClF_5$	C_{4v}

others: CFClBrI:C_1 $B_3N_3H_6$: D_{3h}
           H_2O_2: C_2 S_8: D_{4d}
           $S_4N_4F_4$: S_4 C_6H_6: D_{6h}

may be derived from Figure 60. After identifying special groups, a rotation axis, C_n, of at least two-fold order is sought. If C_n axes are absent, then σ or i may be symmetry elements. A molecules with no symmetry is given the symbol C_1 (e.g., CFClBrI). If the molecule has a C_n axis (n > 1) then an evenfold improper rotation axis, S_{2n}, and an inversion center, i, may be the only additional symmetry elements to give Group S_n. In most cases, however, there are additional axes of rotation (e.g., C_2) or mirror planes (σ_v, σ_h, σ_d). The symbol σ_d denotes a diagonal mirror plane, i.e., a vertical mirror plane lying between two C_2 axes in a dihedral molecule. For example, the trigonal-planar molecule BF_3 has the following symmetry elements: C_3, $3C_2 \perp C_3$, σ_h, $3\sigma_d$, to give the symbol D_{3d}. Further examples are shown in Table 22.

5.5. Bond Properties

A bond is part of a molecule, and the nature of the molecules influences the properties of the bond. The electron configuration of a bond depends upon the remaining electrons.

The bond energy (B.E.), dissociation energy (D), internuclear distance (d), and valence force constant (f) are measurable quantities. The overlap integral (S) is a theoretical measure of bond strength. Finally, bond order (b), is fixed by definition, and may be derived from measurable quantities. These parameters and concepts will now be discussed.

5.5.1. Bond Energy and Dissociation Energy

Diatomic Molecules:

The formation of a diatomic molecule, AB, from the gaseous atoms A and B can by symbolized:

$$A(g.) + B(g.) \rightarrow AB(g.) \tag{91}$$

The energy of the system changes as A approaches B according to Figure 26, p. 72. This curve represents the superposition of attractive (large internuclear distances) and repulsive (smaller distances) forces. At the hypothetical equilibrium state, i.e., at the energy minimum, the molecule has internuclear distance, d, and energy corresponding to the bond energy (B.E.) for which in spectroscopy the symbol D_e (e = equilibrium) is employed. When A and B are united to form AB, the total B.E. is not liberated, but a lesser amount called the dissociation energy, D, with a subscript to indicate the absolute temperature (at $0\,K, D_0$). The difference $B.E. - D_0$, the zero point energy, is the vibrational energy remaining in the molecule after all degrees of freedom of translation and rotation are frozen. The zero point energy, $\frac{1}{2}h\nu$, is a direct consequence of the Heisenberg uncertainty principle, where h is

Planck's constant and v the frequency of the ground state vibration of molecule AB which is obtained from vibrational spectra. For the H_2 molecule:

$$\tfrac{1}{2}h\nu = 0.5 \cdot 6.626 \cdot 10^{-34} \cdot 1.25 \cdot 10^{14} \cdot 6.02 \cdot 10^{23} = 25 \text{ kJ/mol} \qquad (92)$$

and:

$$\text{B.E.} = D_0 + \tfrac{1}{2}h\nu = 432 + 25 = 457 \text{ kJ/mol (at 0K)} \qquad (93)$$

The magnitude of $\tfrac{1}{2}h\nu$ is largest for H_2, since the frequency, v, rapidly diminishes with increasing atomic mass (e.g., $H_2 = 1.25 \cdot 10^{14}$; $O_2 = 0.47 \cdot 10^{14}$ s^{-1}), and the difference between B.E. and D_0 is usually neglected. D can be calculated from thermodynamic data, or determined directly by experiment, e.g., from the band spectra of gaseous molecules. B.E. can be calculated from D.

In diatomic molecules, D_0 is identical with the atomic enthalpy of formation. D is temperature-dependent, for example, $D_0(H_2) = 432.3$ kJ/mol, and $D_{300}(H_2) = 436.2$ kJ/mol. This slight difference, often of the order of magnitude of the standard deviation in D is mostly neglected. The reason for temperature dependence is that other oscillations, aside from zero point energy, are frozen even at 300 K. The three degrees of freedom of translation require the energy 3/2 RT, and the two degrees of freedom of rotation require RT (R: universal gas constant $= 8.314$ J $\cdot$ K^{-1} $\cdot$ mol^{-1}). The molecule thus has 5/2 RT thermal energy at its disposal. However, two hydrogen atoms at 300 K have three degrees of freedom of translation each, with a total energy of 6/2 RT. In the dissociation $H_2 \rightarrow 2H$ then, aside from energy, D_0, the thermal energy 1/2 RT ($= 1.25$ kJ/mol at 300 K) has to be provided. In addition the volume expansion demands an energy RT, so that an additional amount of energy of 3/2 RT or 3.75 kJ/mol results. Thus D_{300} for H_2 is that much larger than D_0. At still higher temperatures, the energy stored in the degrees of freedom of vibration of the molecules must also be taken into account.

Polyatomic Molecules:

In polyatomic molecules such as SO_2, H_2O, or BF_3, which contain only like bonds, B.E. is the same for all bonds, but in stepwise dissociation, the energies are unequal:

$$
\begin{array}{lll}
SO_2 & \rightarrow SO + O & D_1 = 549 \text{ kJ/mol} \\
SO & \rightarrow S\ + O & D_2 = 517 \text{ kJ/mol} \\
\hline
SO_2 & \rightarrow S\ + 2O & \Delta H° = 1066 \text{ kJ/mol}
\end{array}
$$

The average bond energy (av. B.E.) which is defined as the arithmetic mean of the dissociation energies is used as a measure of the bond strength in such cases.

$$\text{av.B.E.} = \frac{D_1 + D_2}{2} = \frac{1066}{2} = 533 \text{ kJ/mol}$$

D_1 and D_2 are unequal because the electrons of the second bond reorganize after cleavage of the first, as can be seen in the changes in internuclear distance: the distance $d(SO) = 1.43$ Å in SO_2; however, $d(SO) = 1.48$ Å in SO. The energy needed for the rearrangement is consumed additionally in the first dissociation step. D_1 can in general be either larger or smaller than D_2.

Electron reorganization changes internuclear distances and valence angles. For example, angles change in the dissociation of CCl_4 into CCl_3 and Cl which takes place on heating CCl_4 to ca. $1000°$. The expected pyramidal fragment rearranges simultaneously into the trigonal-planar radical CCl_3, which has been isolated in a matrix and IR and ESR spectra obtained. Corresponding changes take place in the dissociation of CCl_3Br and CCl_3I:

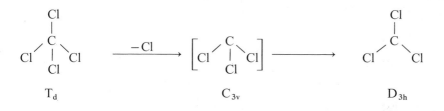

Thus the B.E. and av. B.E. values are complicated by the reorganization energies, whose magnitude is unknown. Averaging the dissociation energies distributes the reorganization energies over all bonds; however, the B.E.'s thus obtained are only approximate, and different B.E.'s may be obtained for bonds of equal strength in different molecules. Thus internuclear distances, d(SO), and valence force constants, f(SO), are nearly equal for SO_2 and SO_3:

	SO_2	SO_3
d(SO):	1.43	1.42 Å
f(SO):	10.1	10.4 mdyn/Å

but the av. B.E.'s are 532 kJ/mol for SO_2 and 469 kJ/mol for SO_3, a difference of $>12\%$. Valence force constants, f, which have a better theoretical foundation than bond energies, are often used to compare bond strength values.

In molecules containing more than two elements, av.B.E. cannot taken as the arithmetic mean of the dissociation energies. The values for analogous binary compounds, as for BCl_2F those of BCl_3 and BF_3, can be used in a first approximation. Bonds of like internuclear distance and valence force constant are chosen, and suitable corrections applied. The sum of the av.B.E.'s equals the negative enthalpy of formation from atoms.

For typical single, double, and triple bonds, the av. B.E.'s lie in the ranges:

E—E: 125–565 kJ/mol

E=E: 420–710 kJ/mol

E≡E: 800–1090 kJ/mol

Table 23   Average Bond Energies (at 25 °C in kJ/mol; zero point energies were neglected).

Single Bond Energies

The bonds N–N, N–O, N–F, O–O, O–F and F–F are partial bonds (cf p. 131). The bonds Si–F, Si–O and P–F, strictly speaking, are multiple bonds (cf p. 122).

	H	C	N	O	F	Si	P	S	Cl	Ge	As	Se	Br	Sb	Te	J
H	435															
C	415	331														
N	389	293	159													
O	465	343	201	138												
F	565	486	272	184	155											
Si	320	281	—	368	540	197										
P	318	264	≈300	352	490	214	214									
S	364	289	247	—	340	226	230	264								
Cl	431	327	201	205	252	360	318	272	243							
Ge	289	243	—	—	465	—	—	—	239	163						
As	247	247	—	—	465	—	—	—	289	—	178					
Se	314	247	—	—	306	289	272	214	251	—	193	193				
Br	368	276	243	—	239	—	—	—	218	276	239	226	193			
Sb	—	—	—	—	—	—	—	—	—	—	—	—	—	126		
Te	268	—	—	—	343	—	—	—	—	—	—	—	—	—	138	
I	297	239	201	201	—	214	214	—	209	214	180	—	180	—	—	151

Multiple Bond Energies

C=C 620	C=N 615	C=O 708	N=N 419
C≡C 812	C≡N 879	C≡O 1072	N≡N 945
O=O 498	S=O 420	S=C 578	
S=S 423	Se=O 425	Se=C 456	

Bonds of lesser energy can be considered partial bonds. Individual av. B.E. values are listed in Table 23.

The energies of single bonds between like atoms within a group of the periodic system vary markedly. Figure 62 shows that $>$N—N$<$, $\diagup$O—O$\diagdown$ and F—F bonds have surprisingly small B.E.'s because of mutual repulsion of their free electron pairs whose orbitals overlap, owing to the small internuclear distances. This effect is absent in $\ggg$C—C$\lll$, and in the higher homologs $>$P—P$<$, $\diagup$S—S$\diagdown$, Cl—Cl, etc., where the larger internuclear distance inhibits such an interaction. The energies of the single $>$N—O—, $>$N—F, $>$O—F, and $\diagup$O—Cl bonds are also comparatively small. Repulsion of nonbonding electrons in these diatomic groups is responsible for the high reactivity of F_2, for the thermodynamic instability of H_2O_2 and N_2H_4, that nitrogen and oxygen do not form stable rings with single bonds like their higher homologues (cf. P_4 and S_8), and that Cl_2O is a very reactive, endothermic compound.

The differing states of elemental oxygen and sulfur can be explained on thermodynamic grounds. In a hypothetical oligomerization of four O_2 molecules to O_8, in analogy to S_8, four O=O double bonds each of 498 kJ/mol would have to be cleaved and eight O—O single bonds with 138 kJ/mol would be formed. The resulting energy balance would be:

$$\Delta H^{\circ}_{300} = 4 \cdot 498 - 8 \cdot 138 = + 887 \text{ kJ/mol } O_8$$

Equilibrium for the reaction

$$4 O_2 \rightleftharpoons O_8$$

will lie on the left if $\Delta G^{\circ} > O$. According to the second law of thermodynamics:

$$\Delta G^{\circ} = \Delta H^{\circ} - T \cdot \Delta S^{\circ}$$

and since $T \cdot \Delta S^{\circ}$ for most reactions is small compared to ΔH° at room temperature, it follows the $\Delta G^{\circ} > 0$ with $\Delta H^{\circ} \gg O$.

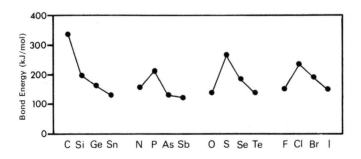

Fig. 62.   Average bond energies of some homonuclear single bonds. (C—C, Si—Si, etc.).

This holds for O_2, whose stability towards polymerization can now be understood. In the reaction:

$$4S_2 \rightleftharpoons S_8$$

the same calculation, owing to the much higher value for B.E. (S—S), leads to a negative value for ΔH°_{300}:

$$\Delta H^\circ_{300} = 4 \cdot 425 - 8 \cdot 264 = -410 \text{ kJ/mol } S_8$$

In this case, $\Delta H^\circ \ll O$ and hence $\Delta G^\circ < O$. The equilibrium at 300 K is completely on the side of S_8, and S_2 molecules are encountered only at higher temperatures in sulfur vapor, and the in equilibrium with S_8 and other S_n species (cf. p. 207). The molecule S_2, like O_2, exists in the triplet state, and both molecules contain double bonds confirming that elements of higher periods can also form stable double bonds. The oligomerization of S_2 to S_8 is not due to the instability of the S_2 bond, but because enthalpy is gained by converting an S=S double bond into two single bonds. Thiothionyl fluoride, S=SF$_2$, which has an S=S double bond, exists at room temperature, even though thermodynamically unstable.

Average B.E. values are well suited for thermodynamic considerations of this kind, but reaction mechanisms, which deal with homolytic cleavages of individual bonds, require dissociation energies. For example, the av. B.E. for S_8 ring is 264 kJ/mol, the arithmetic mean of all eight dissociation energies, but only 138 kJ/mol are necessary for the cleavage of the first S—S bond, i.e., for the ring opening to form the S_8 biradical.

5.5.2. The Internuclear Distance

The internuclear distance is a characteristic property of a covalent bond. Nearly constant values for single, double and triple N—N bonds are obtained, e.g. as shown in Table 24, and a typical nitrogen single, double, and triple bond radius, r, can be defined. For example, the covalent single bond radius of the tetrahedrally coordinated carbon atom is obtained from the C—C internuclear distance in diamond as $r_1(C) = 0.77$ Å, and $r(Cl) = 0.99$ Å from the internuclear distance of the Cl_2 molecule. The covalent radii for the main group elements are listed in Table 25. The covalent radius of nitrogen is not determined using a compound like hydrazine in which weakening of the single bond owing to repulsion of the free electron pairs takes place, but from methylamine, CH_3NH_2, where the value $r_1(N) = 0.70$ Å is obtained by subtracting $r_1(C) = 0.77$ Å from the internuclear distance $d(CN) = = 1.47$ Å. Covalent radii are additive, for example, $r_1(C)$ from diamond and $r(Cl)$ from the Cl_2 molecule, are used to calculate an internuclear distance of $d(CCl) = = 0.77 + 0.99 = 1.76$ Å, in agreement with the experimental value of 1.766 Å for CCl_4.

Table 24   Internuclear Distances, d(NN), in Various Nitrogen Compounds (in Å).

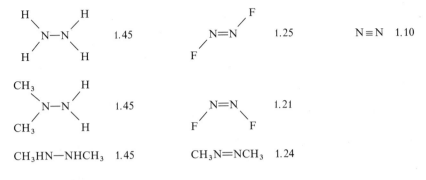

CH$_3$HN—NHCH$_3$ 1.45   CH$_3$N=NCH$_3$ 1.24

However, covalent radii depend on the coordination number of the atoms, and become smaller with decreasing number of ligands. Thus r_1(C) for four-coordination is 0.77 Å, for three-coordination 0.73 Å, and for two-coordination only 0.69 Å.

Table 25   Covalent Radii, r_1, of Non-metallic Atoms in Å.

H	0.28	C	0.77	N	0.70	O	0.66	F	0.64
		Si	1.17	P	1.10	S	1.03	Cl	0.99
		Ge	1.22	As	1.21	Se	1.17	Br	1.14
		Sn	1.40	Sb	1.41	Te	1.37	I	1.33

In addition, single bond radii only apply to bond order 1.0. If the bond is weakened by repulsion of free electron pairs, or enhanced by contribution from multiple bonding from delocalization of electrons, then larger or smaller internuclear distances are to be expected, for example, d(N—N) in N_2H_4 = 1.45 Å, markedly larger than twice r_1(N) (= 1.40 Å), and multiple bonding contribution are noted in BF_3, SiF_4, SiO_2, etc., where the free orbitals of the central atom are partially occupied by the electron pairs of the substituents (coordinate π-bond) (see Section 5.2.4.).
The internuclear distance decreases with increasing bond order. From the internuclear distances of N_2 and C_2H_2 the triple bond radii r_3(N) = 0.55 Å, and r_3(C) = = 0.60 Å are obtained from which d(CN) = 1.15 Å, in agreement with experimentally determined values for nitriles, can be calculated.

The choice of standard compounds containing bonds with orders of 1.0, 2.0, and 3.0 is the central problem in the derivation of covalent radii. There remain exceptions, mainly strongly polar bonds, but good agreement is found when the environment of the respective atoms is similar to that in the standard compounds.

In the AB_5, AB_4E, and AB_3E_2 trigonal-bipyramidal molecules, the axial AB bonds are longer, e.g., in PF_5, d(PF) axial = 1.577 Å, and equatorial = 1.534 Å. Taking the covalent radius of fluorine as 0.64 Å, the axial covalent phosphorous radius is

0.94 Å and the equatorial is 0.89 Å. These relatively small values probably arise from multiple bonding. Likewise, two covalent radii are found for the sulfur atom in SF_4 (of type AB_4E) and for the chlorine atom in ClF_3 (of type AB_3E_2).

5.5.3. The Valence Force Constant

Two covalently bound atoms capable of vibration form a diatomic oscillator. Changes in the internuclear distance are resisted by a counter-force, K, as is seen in the energy curve in Figure 26. In most molecules, K is proportional to the change in internuclear distance Δr according to Hooke's law:

$$K = f \cdot \Delta r \tag{94}$$

The proportionality factor is the valence force constant, and has the dimensions mdyn/Å; f is a parameter reflecting the strength of the covalent bond and lies in the range 1 to 30 mdyn/Å.

Application of Hooke's law implies that the energy curve in the neighborhood of the minimum is approximately parabolic:

$$E = \int K dr = \int f \cdot \Delta r \, d(\Delta r) = \tfrac{1}{2} \cdot f \cdot (\Delta r)^2 \tag{95}$$

but this is only true for small vibrational amplitudes, Δr; however, only hydrogen atoms, owing to their small mass have relatively large amplitudes. Systems describable by Hooke's law are harmonic oscillators.

The frequency of such an oscillator depends only on the atomic masses m_1 and m_2 and on the valence force constant f:

$$v\,[s^{-1}] = \frac{1}{2\pi} \cdot \sqrt{\frac{f(m_1 + m_2)}{m_1 \cdot m_2}} \tag{96}$$

In vibrational spectroscopy, wave number, v, is used:

$$v = \frac{1}{\lambda} = \frac{v\,(\text{frequency})}{c} \tag{97}$$

v wave number in cm^{-1}
λ wave length in cm
c light velocity in cm/s

and lies in the range of 150–4200 cm^{-1} for stretching vibrations. For atomic masses in mass numbers, M_1 and M_2, and f in mdyn/Å:

$$v\,[cm^{-1}] = 1303 \cdot \sqrt{\frac{f(M_1 + M_2)}{M_1 \cdot M_2}} \tag{98}$$

or:

$$f[\text{mdyn/Å}] = 0.589 \cdot 10^{-6} \cdot v^2 \cdot \frac{M_1 \cdot M_2}{M_1 + M_2} \qquad (99)$$

The frequency, v of the stretching vibration measured by infrared or Raman spectroscopy determines f for a diatomic molecule (band analysis of UV spectra also leads to v).

Sometimes pairs of diatomic harmonic oscillators can be assumed for larger molecules. For example, the SO group in SOF_2 may be treated as a diatomic oscillator. If the vibrations are coupled, however, a detailed vibrational analysis of all 3n-6 normal vibrations as well as those of isotopically substituted derivatives leads to the valence force constants of all the bonds.

The valence force constant increases markedly with the bond order, b. For the CC bonds in C_2H_6, C_2H_4, and C_2H_2, e.g., the values listed in Table 26, are compared with the corresponding B.E.'s and d's and b's. Values of f can be determined more easily than internuclear distances, and more exactly than B.E.'s, so that f which is a measure of the curvature of the potential energy curve near the equilibrium value is often used to characterize bond strength. The relative change of f with bond order is much larger than the change of d.

Table 26   Properties of CC-Bonds in Ethane, Ethylene, and Acetylene.

b: bond order   f: valence force constant   B.E.: bond energy
d: internuclear distance

	b	f [mdyn/Å]	B.E. [kJ/mol]	d [Å]
H H \ / H—C—C—H / \ H H	1	4.4	331	1.54
H\ /H C=C H/ \H	2	9.2	607	1.34
H—C≡C—H	3	15.6	837	1.20

5.5.4. The Overlap Integral

The overlap integral, S, defined earlier (see p. 78) is used in VB as well as MO theory as a measure of the strength of a bond. The integration of a product $\psi_A \cdot \psi_B$ over all volume elements presupposes that suitable wave functions for the orbitals of the bonds to be calculated can be found. Wave mechanical treatment of polyelectronic atoms is only possible with the use of considerable simplification; however, some trends can be established with the help of overlap integrals. Overlap integral values in Table 27 show an increase for the CH-bond in the series CH-radical $<C_2H_6 <C_2H_4 <C_2H_2$ arising from increasing the s-contribution to 50% in the hybrid orbitals of the carbon atom which improves overlap. However, with s-content $> 50\%$ overlap again decreases to reach a minimum for the pure s-orbital; the pure s-orbital is less suited for bonding than the s-p hybrids.

Table 27   Properties of the CH Bonds in Simple Hydrocarbons; f = valence force constant, B.E. = bond energy; d = internuclear distance.

Compound	Hybridization at C-Atom	s-Contribution [%]	S	f [mdyn/Å]	B.E. [kJ/mol]	d [Å]
–	s	100	0.59	–	–	–
CH-Radical	p	0	0.49	4.1	352	1.119
C_2H_6	sp^3	25	0.72	4.9	431	1.094
C_2H_4	sp^2	33	0.74	5.1	444	1.079
C_2H_2	sp	50	0.76	5.9	507	1.057

The strength of a bond is thus not only determined by the number of overlapping orbitals (single or multiple bonds) (see Section 5.5.5.) but also by the coordination number.

5.5.5. Definition of Bond Order, b, and its Relation to other Bond Properties

The CC-bonds in C_2H_6, C_2H_4 and C_2H_2 are considered as typical single, double, and triple bonds, of bond orders 1, 2, and 3, respectively. The correlations of b with d, f, and B.E. shown in Figure 63 were constructed from the values listed in Table 26 and may be used to obtain unknown values by interpolation. For example, the internuclear distance d(CC) in benzene is 1.40 Å, corresponding to b = 1.7. From the force constant, f(CC) = 6.7 mdyn/Å, it follows that b = 1.4. The mean value, b = 1.5, reflects the distribution of the three π-bonds over all the six CC-bonds:

Likewise, the CC internuclear distance in the layers of graphite of 1.42 Å (cf. p. 335) gives b = 1.6.

Different bond orders cannot be ascribed to the various CH-bonds in the compounds listed in Table 27, however, and generally it is better to use directly measurable parameters to characterize a given bond. In other cases it is difficult to find model compounds with unequivocal bond orders. Therefore, for a given molecule differing bond orders may be found in the literature.

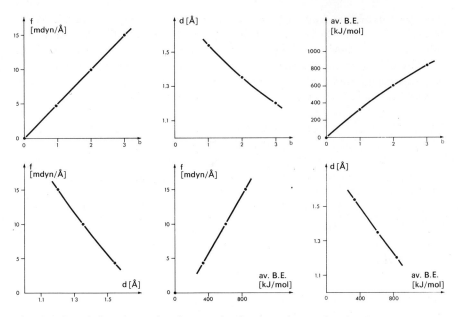

Fig. 63.   Correlations among bond properties, for the carbon-carbon bond.

The bond order can be estimated by the method of H. Siebert, in which the valence force constant, f, of the individual bond is compared with the single bond force constant, f_1, of the respective pair of non-metallic elements, AB. The constant f_1 is calculated from the empirical formula:

$$f_1 = \frac{7.20 \cdot Z_A \cdot Z_B}{n_A^3 \cdot n_B^3} \tag{100}$$

Z = nuclear charge number
n = principal quantum number of the valence electrons

Bond order, b, is found from:

$$b = 0.57 \cdot \frac{f}{f_1} + 0.43 \sqrt{\frac{f}{f_1}} \tag{101}$$

which is derived empirically using model compounds with arbitrarily defined bond orders.

The Siebert bond orders correspond best to the structural formulae used in Part II of this book, and will be used exclusively except when the bond order defined from MO theory, b(MO), (cf. p. 102) is employed. Bond orders should not be written with more than two significant figures.

The use of the Siebert equations will be illustrated by two examples. For the N—N bond, $f_1 = 5.5$ mdyn/Å, however, in the hydrazine molecule f(NN) is only 4.74 mdyn/Å, corresponding to a bond order of 0.9 arising from repulsion of the free electron pairs in the nitrogen atoms (cf. p. 131). The force constant of N_2O_4, which easily dissociates into two molecules of NO_2, is only 1.3 mdyn/Å, corresponding to $b = 0.3$. A pure N—N single bond is found in the hydrazinium cation, $N_2H_6^{2+}$, which possesses no free electrons pairs. Here f(NN) is 5.35 mdyn/Å, and $b = 1.0$.

Low bond orders are also found for H_2O_2 and F_2, but these bonds are usually treated as single bonds (cf. Table 23 and Figure 62).

Almost continuous variation of bond order is observed in the S=O multiple bond in thionyl compounds, for example, dimethylsulfoxide, $(CH_3)_2SO$, b(SO) = 1.5; in the dialkyl esters of sulfurous acid, $(RO)_2SO$, $b = 1.8$, in $SOCl_2$, $b = 2.0$, and in SOF_2, $b = 2.2$. These values are rationalized by a group of resonance structures, in which the relative weights depend on the inductive effects of substituents A and B:

The more electronegative are A and B, the more positively charged the sulfur and oxygen atoms become, producing a rise in b(SO). Similar effects are observed in phosphoryl compounds, R_3PO.

5.5.6. Polarity of Covalent Bonds

A bond is polar if the distribution of electrons is asymmetrical, so that the centers of electronic and nuclear charges do not coincide. All bonds between dissimilar atoms are polar, just as are bonds between like, but non-equivalent atoms, as for the S—S bond in the thiosulfate ion, $S_2O_3^{2-}$. Nonpolar bonds exist only between identical atoms in an identical environment, as in homonuclear diatomic molecules or the N—N bond in N_2H_4, etc.

The concept of bond polarity is of importance in the understanding of properties and reactions of chemical bonds and of compounds, but bond polarities cannot be measured directly, nor derived unambiguously from measurable parameters. Empirical approaches permit understanding of strength and direction of polarity in covalent bonds, and quantum-mechanical calculations have permitted the estimation of charge distributions in small molecules, as well as changes during the formation of molecules from atoms.

The Electronegativity (X_E):

Electronegativity, X_E, was defined by L. Pauling in 1932 as the ability of an atom in a molecule to attract electrons. Despite the impossibility of direct measurement, the concept has had enormous impact upon inorganic chemistry. The atoms in a molecule are in the valence state, and thus values of electronegativity cannot be derived from ground state atomic properties. Pauling's method for the determination of X_E-values uses the B.E.'s of molecules, whereas a method by A.L. Allred and E.G. Rochow is based upon effective nuclear charge and covalent radii. A procedure by R.S. Mulliken employs ionization energies and electron affinities of valence state atoms.

X_E Values according to Pauling:
The bond in a diatomic molecule AB is described by resonance structures:

$$A - B \rightleftharpoons A^- B^+ \rightleftharpoons A^+ B^- \tag{102}$$

The appropriate wave function is equation 73 (cf. p. 77):

$$\psi = N(\psi_{cov} + \lambda \cdot \psi_{ion}) \tag{73}$$

where:

$$\psi_{cov} = \psi_A(1)\psi_B(2) + \psi_A(2)\psi_B(1) \tag{103}$$

and:

$$\psi_{ion} = \psi_A(1)\psi_A(2) + \beta \cdot \psi_B(1)\psi_B(2) \tag{104}$$

where N is the normalizing factor, and β is the polarity parameter. In homonuclear diatomic molecules the weight of the two ionic structures must be equal on grounds of symmetry, i.e., $\beta = 1$. If, however, B is more electronegative than A, the ionic structure $A^+ B^-$ participates more strongly in the ground state than $A^- B^+$, and β must be > 1.

The B.E. of molecules as shown in the VB treatment of H_2 is increased by participation of ionic structures. This increase is small for H_2 because the mixing coefficient, λ, in the wave function of a non-polar molecule is small. But consider, for example, the series of binary fluorides:

$$F_2 \quad OF_2 \quad NF_3 \quad CF_4 \quad BF_3 \quad BeF_2 \quad LiF$$

in which there is an almost continuous transition from the non-polar F_2 through polar bonds, to the ionic LiF. The mixing coefficients, λ, must increase in going from F_2 to LiF, since the LiF molecule is predominantly described by the function, ψ_{ion}. The participation of ionic structures in polar molecules such as HCl results in a considerable gain in B.E., particularly through the resonance between structures I and II.

$$\begin{array}{ccc} H-Cl & \leftrightarrow & H^+ Cl^- \\ I & & II \end{array}$$

neglecting structure $H^- Cl^+$, since Cl is more electronegative than H. Pauling assumed that B.E. for the covalent structure I is the arithmetic mean of the energies of the H_2 and Cl_2 molecules. Using the dissociation energies and neglecting zero point energies:

$$D_{HCl(I)} = \tfrac{1}{2}(D_{H_2} + D_{Cl_2}) = \tfrac{1}{2}(435 + 243) = 339 \text{ kJ/mol}$$

The experimental dissociation energy of HCl is 431 kJ/mol, however. The difference of 92 kJ/mol is ascribed by Pauling as arising from the ionic resonance energy (symbol Δ) from structure II; Δ is obtained from:

$$D_{HCl} = \tfrac{1}{2}(D_{H_2} + D_{Cl_2}) + \Delta$$

or, in general:

$$D_{AB} = \tfrac{1}{2}(D_{A_2} + D_{B_2}) + \Delta \tag{105}$$

Pauling then postulated that Δ was proportional to the square of the difference of the electronegativities of atoms A and B:

$$\Delta = 97|X_E(B) - X_E(A)|^2 = 97|\Delta X_E|^2 \tag{106}$$

where the factor 97 converts the Δ-value from kJ/mol into electron volts. The electronegativity difference ΔX_E results as:

$$\Delta X_E = 0.102 \sqrt{\Delta}$$

If X_E is set for a single element, all other X_E-values are defined correspondingly. Hydrogen is used as a reference with $X_E = 2.2$. The $X_E(Cl)$ value, as an example, is calculated as follows:

$$X_E(Cl) = 0.102 \sqrt{92} + X_E(H) = 0.102 \cdot 9.59 + 2.2 = 3.2$$

Pauling's values are listed in Table 28. Electronegativities in the main groups of the periodic system increase from the lower left to the upper right with fluorine ($X_E = 4.0$) as the most electronegative element. The X_E values of the transition elements fall between 1.2 and 2.5. Rare gases will be discussed in Part II (p. 277).

Electronegativities are useful for estimating bond polarities and inductive effects. Numerical B.E. values are known with only moderate accuracy, and not all increases in bond strength can be ascribed to electronegativity differences. Multiple bonding, present in BF_3, SiF_4, PF_3, etc., can simulate the effect of resonance energy. Therefore, only those molecules should be employed to calculate X_E values in which true single bonds are present, and in which all atoms have a rare gas configuration.

X_E values from the Allred and Rochow treatment:
A.L. Allred and E.G. Rochow proposed a simple electrostatic equation to calculate electronegativities. The attractive force of an atomic core on the bonding electrons can be expressed in a form of Coulomb's law:

Table 28   Electronegativities of the Main Group Elements. Values calculated by (a) Pauling's (heavy type), and (b) Allred and Rochow's method.

H	Li	Be	B	C	N	O	F
2.2	**1.0**	**1.6**	**2.0**	**2.5**	**3.0**	**3.4**	**4.0**
2.2	1.0	1.5	2.0	2.5	3.1	3.5	4.1
	Na	Mg	Al	Si	P	S	Cl
	0.9	**1.3**	**1.6**	**2.1**	**2.2**	**2.6**	**3.2**
	1.0	1.2	1.5	1.7	2.1	2.4	2.8
	K	Ca	Ga	Ge	As	Se	Br
	0.8	**1.0**	**1.8**	**2.3**	**2.2**	**2.6**	**3.0**
	0.9	1.0	1.8	2.0	2.2	2.5	2.7
	Rb	Sr	In	Sn	Sb	Te	I
	0.8	**1.0**	**1.8**	**2.0**	**2.1**		**2.7**
	0.9	1.0	1.5	1.7	1.8	2.0	2.2
	Cs	Ba	Tl	Pb	Bi	Po	At
	0.8	**0.9**	**2 0**	**2.3**	**2.0**		
	0.9	1.0	1 4	1.6	1.7	1.8	2.0
	Fr	Ra					
	0.9	1.0					

a after A. L. Allred, J. Inorg. Nucl. Chem. **17**, 215 (1961) and D. Quane, *ibid.* **33**, 2722 (1971)
b after A. L. Allred and E. G. Rochow, J. Inorg. Nucl. Chem. **5**, 264 (1958)

$$K = \frac{Z^* \cdot e^2}{r^2} \tag{107}$$

Z^* = effective nuclear charge number
r = covalent radius

The bonding electrons are assumed to reside at a covalent radius from the nucleus, and Z^* is calculated from Slater's rules (p. 46).
The ratios Z^*/r^2 for the bonded atoms are compared since the electronegativity of an atom is proportional to Z^*/r^2:

$$X_E \sim \frac{Z^*}{r^2} \tag{108}$$

The calculation of numerical values requires a proportionality factor, which is selected so that an optimal correlation with Pauling's x_E values results:

$$X_E = 0.359 \cdot \frac{Z^*}{r^2} + 0.744 \tag{109}$$

The plot of the Allred-Rochow vs. Pauling electronegativities gives a straight line, implying that both systems are equivalent. The Allred and Rochow values are listed in Table 28 for the main group elements and depicted in Figure 64. The electronegativity increase to the right and to the top in the periodic system is, however, not monotonous, but the discontinuities arise from the interspersed transition elements.

The X_E values of the non-metallic elements are all larger than 1.8 while those of the metals are all smaller than 1.5. The metalloids are found in the range 1.4 to 1.8.

X_E values from the Mulliken treatment:
Electronegativities can be calculated from ionization energies and from the electron affinities of isolated atoms after a proposal by R.S. Mulliken. The ionic structure $A^+ B^-$ is produced from the covalent AB, by expending for the latter the dissociation energy, D_{AB}, and the ionization energy, I_A while the electron affinity of atom $B(A_B)$ and Coulomb attraction energy, E, are gained:

$$A-B \underset{B \xrightarrow{A_B} B^-}{\overset{D_{AB} \quad A \xrightarrow{I_A} A^+}{\Bigl\langle}} \xrightarrow{E} A^+ B^-$$

Taking into account the sign of electron affinity, the total energy change is:

$$D_{AB} + I_A - A_B - E \tag{110}$$

For structure $A^- B^+$, the energy balance likewise is:

$$D_{AB} + I_B - A_A - E \tag{111}$$

Both contributing structures will have the same weight, if they contain the same

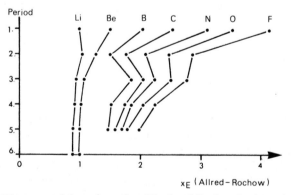

Fig. 64.   Electronegativity values after Allred and Rochow in the main groups of the periodic system.

amount of energy, of differ by an equal amount of energy from the covalent structure. This is the case if:

$$D_{AB} + I_A - A_B - E = D_{AB} + I_B - A_A - E \tag{112}$$

or:

$$I_A + A_A = I_B + A_B \tag{113}$$

If this equation is valid, then the electronegativity difference between A and B is
zero. Thus Mulliken postulated that electronegativity is proportional to $I + A$.

With I and A in electron volts, agreement with Pauling values is obtained with the
following equation:

$$X_E = 0.168(I + A) - 0.204 \tag{114}$$

This simple relation is widely applicable, allowing the calculation of electronegativities directly for the valence states of atoms if I and A in these states can be estimated using the promotion energies. The electronegativities for carbon and sulfur
in Table 29 were calculated for singly-occupied atomic or hybrid orbitals. The
electronegativity depends on the valence state; X_E increases linearly with the s-contribution in the σ-valence orbitals. The electronegativity of σ-orbitals is greater
than that of the π-orbitals, and the polarity of a σ-bond may, therefore, be different
from that of a π-bond.

Table 29 Orbital Electronegativities of Carbon and Sulfur Atoms in Different Valence States;
di: digonally hybridized (sp)   tr: trigonally hybridized (sp^2)   te: tetrahedrally hybridized (sp^3)

| atomic State | carbon | | atomic State | Sulfur | |
	Orbital	X_E		Orbital	X_E
sppp	s	4.84	s^2p^2pp	p	2.28
	p	1.75	te^2te^2tete	te	3.21
didi$\pi\pi$	di	3.3		tr	3.46
	π	1.7	$tr^2tr^2tr\pi$	π	2.40
trtrtrπ	tr	2.8			
	π	1.7			
tetetete	te	2.5			

The electronegativities of unoccupied and of doubly occupied orbitals which are
important in coordinate bonding can also be calculated by Mulliken's method. For
example, the trigonal valence state of boron yields for each of the singly occupied
σ-orbitals, $X_E = 1.93$, and for the empty π-orbital, $X_E = 1.22$. The singly-occupied
orbitals of the tetrahedral valence state of the nitrogen atom yield $X_E = 3.68$, and
for the doubly-occupied orbital, $X_E = 1.32$. The free electron pair is hence more
readily donated to an acceptor than the electron in a singly-occupied orbital.

Atoms in molecules carry partial charges, and it is of interest to investigate the change in electronegativity with charge. This can be done using Mulliken's method, and in agreement with expectations, positive charges increase X_E values, while partial negative charges reduce them.

Electronegativities may also de defined for substituent groups, R. Group electronegativities can be calculated by the Mulliken treatment, or be derived from the chemical shift in NMR spectra, or from vibrational frequencies. While the values depend on how they were determined, the inductive effect of neighboring atoms can be easily recognized, for example, the group electronegativity rises in the series $—CH_3, —CCl_3, —CF_3$ owing to the greater electronegativity of chlorine and fluorine over hydrogen.

The Bond Moment
The permanent dipole moment, μ, of polar bonds can be separated into four parts:

$$\mu = \mu_e + \mu_{at} + \mu_{hom} + \mu_{pol} \tag{115}$$

The first component is the dipole moment which arises from a transfer of the charge δe between atoms because of their electronegativity differences:

$$\overset{\delta\oplus}{A} - \overset{\delta\ominus}{B}$$

If the internuclear distance is d, then $\mu_e = \delta e \cdot d$. Owing to the partial charges, the electronegativity of atom A will increase, and that of atom B decrease. Thus the charge transfer equilibrates the electronegativities of A and B. The second contribution to the moment is the atomic dipole moment (μ_{at}) which results from the asymmetry of charge distribution in the hybrid orbitals. The center of electronic charge in a hybrid orbital (cf. Figure 65) is not at the nucleus as in an AO, but is displaced toward the bond partner, giving rise to a dipole moment in that direction. The homopolar dipole moment, μ_{hom}, arises from the different sizes of the overlapping orbitals (cf. Figure 66). The direction of μ_{hom} is towards the smaller orbital, since the concentration of electron density in the region of overlap is closer to the

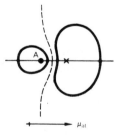

Fig. 65.   The dipole moment of a hybrid orbital. The center of the negative charge is at X; that of the positive charge at nucleus A. The dotted line indicates the nodal plane.

nucleus of the smaller atom. Therefore, μ_{hom} in the HCl molecule is in the opposite direction to the dipole moment, $\mu_e (H \rightarrow Cl)$.

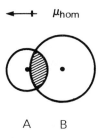

Fig. 66.   The homopolar dipole moment of a bond between two atoms of unequal size, A and B. The center of the positive charge lies in the center between atoms A and B, but negative charge is built up in the cross-hatched region of overlap.

Finally, the polarizing influence of dipole moments μ_e, μ_{at} and μ_{hom} upon the charge distribution in the bond is represented in the component, μ_{pol}. The components of the dipole moment can have different signs and compensate each other. For example, for the CH-bond in methane, μ_e has been calculated as $ca.\ 1\,D$, and μ_{at} as $2\,D$, but with opposite direction. A polarity of $\overset{\delta\ominus}{C}\!-\!\overset{\delta\oplus}{H}$ is expected from the electronegativity differences, but quantum mechanical calculations support a negatively polarized hydrogen. The electronegativity of the σ-orbitals of the sp-hybridized carbon atoms in acetylene is enhanced (cf. p. 143), so that the hydrogen is positively polarized, explaining its hydrogen bonding in acetone and the higher acidity of acetylene vs. methane.

Bond polarity in systems with only small differences in electronegativity is difficult to determine, but where there is great difference in electronegativity, the charge transfer, μ_e, will predominate. Polar bonds of this type are: E—F (where E is H, B, C, N, Si, P, S), E—O (where E = H, B, C, Si, P, S), and E—Cl (where E = B, Si), in which the electronegativity difference is >1.

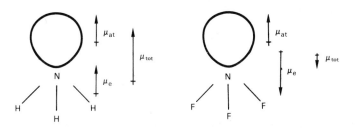

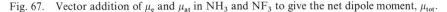

Fig. 67.   Vector addition of μ_e and μ_{at} in NH_3 and NF_3 to give the net dipole moment, μ_{tot}.

The molecular dipole moment is the vector sum of the individual bond and electron moments. Free electron pairs in hybridized orbitals can exert a large influence as in ammonia, in which the three N—H bonds are polarized with the hydrogen atoms positive, and their vector sum is directed with the moment of the free electron pair

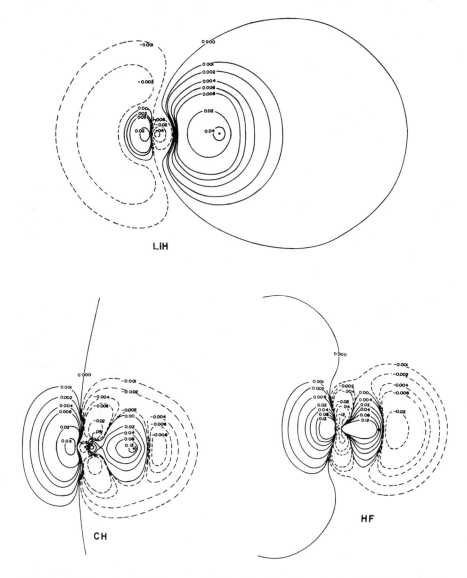

Fig. 68.   Electron density difference charts for the diatomic molecules LiH, CH and HF in the ground state. The hydrogen is on the right.

along its sp³-hybrid orbital. The large dipole moment of 1.468 D for ammonia (cf. Figure 67) and 1.85 D for water arise partly from the moments of their free electron pairs.

In the NF_3-molecule the bond polarity is opposite to that in NH_3, so that the bond moments and μ_{at} compensate, giving a net moment of only 0.235 D, directed toward the three fluorine atoms. In the molecule ONF_3 ($\mu = 0.04$ D) which contains one coordinate NO-bond, compensation of bond moments is nearly complete.

Measured dipole moments are composed of several components combined in a complex fashion, and cannot be interpreted in terms of the polarity of bonds alone.

Calculation of Electron Distribution:
The polarity of a bond is the result of partial charges on the participating atoms. The electron distribution can be calculated for diatomic molecules with the aid of one-electron density functions (cf. p. 70).

Contour diagrams of electron density for H_2^+, H_2, LiH, and N_2 have been shown in Figures 25, 29, 34 and 46 which depict the geometry and the change of electron density along the internuclear axis. Charge reorganization occurring during molecule formation from isolated atoms can be obtained from density difference diagrams derived from subtraction of the atomic densities (p. 83) as in Figure 35 for the N_2 molecule. In the series:

LiH, BeH, BH, CH, NH, OH and FH

the polarity of the bonds decreases and then increases to the right. The density difference contour maps of LiH, CH, and FH are depicted in Figure 68 which shows that in addition to the numbers of electrons in the individual atoms, the spatial changes of the orbitals must be taken into account. Treating the polar covalent bond as a simple dipole with point partial charges on each atom is thus an oversimplification. In the words of R. S. Mulliken,

"The chemical bond is not so simple as some people seem to think."

Part II: Chemistry of the Non-metals

6. Hydrogen

6.1. Elementary Hydrogen

Hydrogen forms compounds with nearly every element. Their number is larger than for any other element. The most common compound, and also the best source of H_2 is water, of which hydrogen constitutes 11.2 wt. %.

Hydrogen is generated as follows:

a) Reduction of H_2O over incandescent coke and catalytic reduction of additional H_2O by the product CO:

$$H_2O + C(s.) \rightarrow H_2 + CO$$
$$H_2O + CO \rightarrow H_2 + CO_2$$

   Carbon dioxide is removed from the gas mixture by water under pressure.

b) Electrolysis of aqueous salt solutions (e.g., in the commercial chlorine-alkali electrolysis for the production of NaOH and Cl_2):

$$NaCl + H_2O \rightarrow NaOH + \tfrac{1}{2} H_2 + \tfrac{1}{2} Cl_2$$

c) Thermal cracking of hydrocarbons in refining crude oil for gasoline production:

$$C_nH_{2n+2} \left\langle \begin{array}{l} \rightarrow C_nH_{2n} + H_2 \\ \rightarrow C_{n-1}H_{2n} + H_2 + C(s.) \end{array} \right.$$

Commercial hydrogen is employed in the manufacture of NH_3, CH_3OH, HCN and HCl, for the hardening of fats, for cutting and welding, as a rocket fuel, a heating fuel, and as reducing agent in the production of metals from their oxides.

Hydrogen is obtained in the laboratory by the electrolysis of 30% KOH using nickel electrodes, or by dissolving metals like zinc or iron in aqueous acid:

$$2 H_3O^+ + Zn \rightarrow H_2 + Zn^{2+} + 2 H_2O$$

Aside from O_2 and N_2, trace impurities of hydrogen sulfide, AsH_3 and even hydrocarbons may be present in electrolytically-produced hydrogen. To obtain pure hydrogen the electrolytically generated product is diffused through heated palladium or nickel tubes which are impermeable to the impurities. Hydrides of uranium or titanium may be formed at higher temperature and pyrolyzed subsequently in vacuo. Hydrogen may also be absorbed by finely divided palladium metal, and then desorbed at 200 °C under reduced pressure to yield 100 ml of hydrogen (25 °C, 1 bar) per gram of palladium.

Natural hydrogen consists of the isotopes ^{1}H (hydronium, H), ^{2}H (deuterium, D) and ^{3}H (tritium, T), in the ratio of $1 : 1.5 \cdot 10^{-4} : 10^{-18}$. Tritium is an electron (β) emitter giving ^{3}He, with a half-life of 12.4 years. Tritium is used as a tracer for radiolabeling hydrogen compounds. Naturally occuring tritium is formed from neutron capture by nitrogen of cosmic radiation:

$$^{14}_{7}\text{N} + ^1_0\text{n} \rightarrow ^3_1\text{H} + ^{12}_{6}\text{C}$$

Artificial production is carried out by neutron bombardment of ^{6}Li:

$$^6_3\text{Li} + ^1_0\text{n} \rightarrow ^3_1\text{H} + ^4_2\text{He}$$

The masses of ^{1}H and ^{2}H differ more from each other than do the isotopes of any other element, as is reflected in deuterium reaction rates and equilibrium constants. Reaction rates of deuterated compounds are slower, and such isotope effects are utilized for the enrichment and isolation of deuterated compounds from natural isotopic mixtures. For example, D_2O is enriched in the electrolysis of ordinary water, since H_2O is more readily reduced at the cathode. Deuterated compounds also have lower vapor pressures and diffuse slower than hydrogen compounds, and so may be separated by fractional distillation or counter-diffusion techniques. Deuterium gas is generated from D_2O by electrolysis, or by reduction with an active metal. Many deuterated compounds can be obtained by isotopic exchange; for example, H_2, CH_4, NH_3 and H_2O exchange with D_2 on the surface of finely divided platinum or nickel metal catalysts. Hydrogen compounds are also dissolved in D_2O and the solutions evaporated to effect deuteration (e.g., ammonium salts, hydrogen salts of polyprotic acids, etc.). Deuterolysis may also be used:

$$\text{LiAlH}_4 + 4D_2O \rightarrow \text{LiOD} + \text{Al(OD)}_3 + 4\text{HD}\uparrow$$
$$\text{SiCl}_4 + 2D_2O \rightarrow \text{SiO}_2 + 4\text{DCl}\uparrow$$
$$\text{Al}_2\text{S}_3 + 6D_2O \rightarrow 2\text{Al(OD)}_3 + 3D_2\text{S}\uparrow$$

Likewise, ND_3 is obtained from Mg_3N_2, D_2SO_4 from SO_3, and D_3PO_4 from P_4O_{10}. Hydrogen is an unreactive, colorless, tasteless and odorless gas. Owing to its high dissociation energy, H_2 decomposes into atoms only at high temperatures. High intensity microwaves produce more extensive dissociation ($> 90\%$) at 0.5 to 5 torr. The atoms recombine after the discharge:

$$\text{H} + \text{H} + \text{M} \rightarrow \text{H}_2 + \text{M}$$

where M is a collision partner, i.e., an atom, molecule, or the vessel wall which can absorb a portion of the recombination energy ($=$ dissociation energy), insuring against renewed dissociation. The half-life of the hydrogen atoms (ca. 0.1 s) is a function of the experimental conditions such as pressure, the dimensions of the vessel, etc. However, the existence of hydrogen atoms several meters from the dissociation site can be demonstrated in high flow velocities in vacuum systems. Hydrogen atoms are very reactive, and can reduce oxides to elements even at room temperature (e.g., SO_2, CuO, PbO, Bi_2O_3, SnO_2) or to lower oxidation states (e.g., NO_2 to NO). Certain elements form volatile hydrides (e.g., oxygen, sulfur, phosphorus, arsenic, antimony, germanium) and halides are reduced with formation of halogen hydrides. Hydrogen atoms play important roles as intermediates in chain reactions; e.g., in the detonation of $H_2 + O_2$ and $H_2 + Cl_2$ mixtures, which have the following reaction mechanisms:

initiation:	$H_2 \rightarrow 2H$	$Cl_2 \rightarrow 2Cl$
chain reaction:	$H + O_2 \rightarrow OH + O$	$Cl + H_2 \rightarrow HCl + H$
	$OH + H_2 \rightarrow H_2O + H$	$H + Cl_2 \rightarrow HCl + Cl$
chain termination:	$OH + H + M \rightarrow H_2O + M$	$Cl + H + M \rightarrow HCl + M$
	$O + H_2 + M \rightarrow H_2O + M$	$2Cl + M \rightarrow Cl_2 + M$
	$2OH + M \rightarrow H_2O_2 + M$	$2H + M \rightarrow H_2 + M$

6.2. Hydrogen Ions, H^+

Hydrogen forms the ions H^+ and H^-. The hydride ion, H^-, is present in hydride salts, LiH. Hydrogen ions, H^+, as free protons are generated in the ionization of hydrogen atoms by high energy electrons present in a glow discharge, or in the ion source of a mass spectrometer.

Free protons, H^+, cannot exist in condensed phases like other ions. Owing to the high electric field strength generated at their small diameters they are immediately solvated:

$$H^+ + H_2O \rightarrow H_3O^+ \qquad\qquad H^+ + H_3PO_4 \rightarrow H_4PO_4^+$$
$$H^+ + NH_3 \rightarrow NH_4^+ \qquad\qquad H^+ + HF \rightarrow H_2F^+$$
$$H^+ + R_2O \rightarrow R_2OH^+ \qquad\qquad H^+ + H_2SO_4 \rightarrow H_3SO_4^+$$

The products of protonization of solvent molecules may be further solvated. The oxonium ion, H_3O^+, produced by protonization of H_2O is further hydrated to a hydronium ion, e.g., $H_9O_4^+$:

$$H_3O^+ + 3H_2O \rightarrow H_9O_4^+$$

The completely hydrated proton is called simply hydrogen ion, and written H^+ (aq). The total hydration enthalpy of the proton:

$$H^+ (g.) + n H_2O(l.) \rightarrow H^+ (aq) \qquad \Delta H^\circ = -1092 \text{ kJ/mol}$$

is much larger than that of other singly charged cations (cf. Table 11, p. 68).
The half-life of an H_3O^+ ion in water is extraordinarily small, ca. 10^{-13} s, because of the exchange equilibrium:

$$H_3O^+ + H_2O \rightleftharpoons H_2O + H_3O^+$$

Hence, the proton is not localized at an individual oxygen atom, but is continually changing its place in a hydrogen bond (see Section 6.6):

$$-O-H\cdots O{<} \rightleftharpoons {>}O\cdots H-O-$$

A similar, rapid proton exchange is also observed in compounds with labile hydrogen atoms, e.g., between NH_3 and NH_4^+.

Pure water dissociates only slightly:

$$2\,H_2O \rightleftharpoons H_3O^+ + OH^-$$

and the equilibrium is strongly temperature dependent. The molar ion product, $K_w = [H^+][OH^-]$ is $1.001 \cdot 10^{-14}$ mol$^2 \cdot$ l^{-2} at $25\,°C$; but at $100\,°C$, is $5.483 \cdot 10^{-13}$ mol$^2 \cdot$ l^{-2}. The concentrations of H^+ and OH^- are thus $1 \cdot 10^{-7}$ mol $\cdot$ l^{-1} at $25\,°C$ and $7.4 \cdot 10^{-7}$ mol $\cdot$ l^{-1} at $100\,°C$. The negative of the common logarithm of the hydrogen ion is the pH:

$$pH = -\log_{10}[H^+]$$

If $[H^+]$ and $[OH^-]$ are equal, the solution is neutral which for $25\,°C$ is at pH $= 7.00$; and at $100°$ at pH $= 6.13$. In alkaline solutions $[H^+] < [OH^-]$; in acid solutions $[H^+] > [OH^-]$.

The aqueous H^+ and OH^- ions have extremely high migration velocities in an electric field; despite the pronounced hydration, both migrate faster than other ions. The reason for this is found in a special mechanism for charge transport. Since water is strongly organized by hydrogen bonds (see Section 6.6), transport of H^+ can take place by an easy positional exchange of protons in the hydrogen bonds:

$$H-\overset{\oplus}{O}-H\cdots O-H\cdots O-H\cdots O-H \rightarrow H-O\cdots H-O\cdots H-O\cdots H-\overset{\oplus}{O}-H$$

Hydroxide ions can likewise migrate:

$$\overset{\ominus}{O}\cdots H-O\cdots H-O\cdots H-O \rightarrow O-H\cdots O-H\cdots O-H\cdots \overset{\ominus}{O}$$

Protonic conductance is also responsible for the extraordinarily high velocity of the reaction:

$$H_3O^+ + OH^- \rightleftharpoons 2\,H_2O$$

Other pure, protonic solvents are also weakly dissociated, e.g., liquid ammonia, liquid hydrogen fluoride, and anhydrous sulfuric acid:

$$2NH_3 \rightleftharpoons NH_4^+ + NH_2^-$$

$$3HF \rightleftharpoons H_2F^+ + HF_2^-$$

$$2H_2SO_4 \rightleftharpoons H_3SO_4^+ + HSO_4^-$$

6.3. Acids

The auto-dissociation of water and like solvents is of the type:

$$HA \rightleftharpoons H^+ + A^-,$$

where ions H^+ and A^- are solvated. Beginning with the theory of electrolytic dissociation by S. Arrhenius (1887), hydrogen ions, i.e., solvated protons were held responsible for the acidity of a solution. A material in water is an acid if after dissolving, $[H^+] > [OH^-]$.

This definition may be generalized for all protonic solvents, HA, with an acid defined as a material which increases the concentration of solvate proton cations (H_3O^+, NH_4^+, $H_3SO_4^+$, etc.). Acids in water, i.e., HCl, HBr, HI, H_2SO_4, HNO_3, H_3PO_4, etc., which apparently dissociate, in fact protonate the solvent molecules:

$$HCl + H_2O \rightarrow H_3O^+ + Cl^-$$

$$H_2SO_4 + H_2O \rightarrow H_3O^+ + HSO_4^-$$

$$H_2SO_4 + 2H_2O \rightarrow 2H_3O^+ + SO_4^{2-}$$

Acids having one acidic hydrogen are monoprotic, H_2SO_4 is diprotic, H_3PO_4 triprotic.

Non-protonic compounds can also increase $[H^+]$ by reaction with the solvent:

$$Cl_2O_7 + 3H_2O \rightarrow 2H_3O^+ + 2ClO_4^-$$

$$SO_3 + 2H_2O \rightarrow H_3O^+ + HSO_4^-$$

These and other acid anhydrides, for example, N_2O_5, P_4O_{10}, SO_2, CO_2, and SeO_2, etc., which behave in water as acids, unlike protic acids, do not contain the functional components of the solvent.

Oxonium Salts:

Proton acids and acid anhydrides not only react with water to form oxonium ions and acid anions, but many also form crystalline hydrates, $HA \cdot nH_2O$ ($n = 1, 2, 3 \ldots$) which have been shown from spectroscopic and structural investigation to be oxonium or hydronium salts, as listed in Table 30.

Table 30   Oxonium and Hydronium Salts

Composition	Structure	Melting Point (°C)
$HX \cdot H_2O$ (X=F, Cl, Br, I)	$[H_3O]X$	
$HCl \cdot 2H_2O$	$[H_5O_2]Cl$	−18
$HCl \cdot 3H_2O$	$[H_5O_2]Cl \cdot H_2O$	−25
$HBr \cdot 4H_2O$	$[H_7O_3][H_9O_4]Br_2 \cdot H_2O$	−56
$HClO_4 \cdot H_2O$	$[H_3O]ClO_4$	+50
$HClO_4 \cdot 2H_2O$	$[H_5O_2]ClO_4$	−18
$HClO_4 \cdot 3H_2O$	$[H_7O_3]ClO_4$	
$H_2SO_4 \cdot H_2O$	$[H_3O]HSO_4$	+8.5
$H_2SO_4 \cdot 2H_2O$	$[H_3O]_2SO_4$	−38

However, the dihydrate of the weak oxalic acid, $(COOH)_2 \cdot 2H_2O$, is not an oxonium salt since its two acidic protons are attached to the carboxyl groups and only bound to water molecules via hydrogen bonds, constituting a true hydrate with isolated H_2O molecules. The distinction between hydrate and the onium salts is established, e.g., by infrared spectroscopy, since H_2O, H_3O^+ and $H_5O_2^+$ have characteristic absorption bands.

The oxonium ion H_3O^+ is isoelectronic with the NH_3 molecule, and like it has C_{3v} symmetry (trigonal pyramidal) as found from the infrared spectrum and the structure of the oxonium salts in which $\angle H—O—H = 112°$ to $117°$. The structure of oxonium perchlorate from x-ray analysis is shown in Figure 69. The monoclinic modification, stable below $-30°C$, has layers of ClO_4^- and H_3O^+ ions which are linked to each other by hydrogen bonds. Each oxonium ion is bound to three different ClO_4 tetrahedra. In the orthorhombic modification, stable above $-30°C$, the oxonium ions rotate freely in their lattice positions.

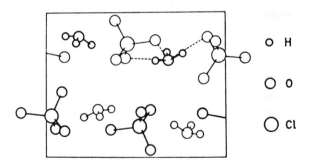

O H

O O

O Cl

Fig. 69.   Crystal structure of the monoclinic modification of oxonium perchlorate, stable below $-30°C$.

Discrete $H_5O_2^+$ ions connected by hydrogen bonds with anions are shown in the structure of the salt $(H_5O_2)^+ ClO_4^-$ in Figure 70. The two water molecules of the

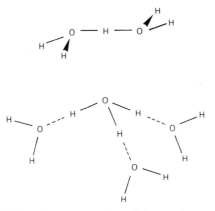

Fig. 70.   Schematic representation of the structures of hydronium ions $H_5O_2^+$ and $H_9O_4^+$.

$H_5O_2^+$ ion are bound by a single proton, located centrally or occupying two closely neighboring positions separated by a small energy barrier. The internuclear distance, $d(OO) = 2.42$ Å, for this bridge is small and argues for a symmetrical hydrogen position as has been found in the cation of the salt $(H_5O_2)^+$ Cl^- · H_2O. The four outer atoms of the $H_5O_2^+$ ion which form the hydrogen bridges with the anions are, as shown in Figure 70, above and below the plane. The structures of $H_7O_3^+$ and $H_9O_4^+$ can be visualized similarly with pyramidal oxygen atoms. The oxonium salts of strong mineral acids are similar in their crystal structures to the corresponding ammonium salts; in fact, $H_3O^+ClO_4^-$ and $NH_4^+ClO_4^-$ are isomorphic. Oxonium salts have markedly lower melting points, however, accounting for their relatively late discovery, and the melting points of the hydronium salts are lower still (cf. Table 30). Oxonium and hydronium salts are acids in water, and like all salts, completely ionized. Alkyl- and alkoxy-derivatives of oxonium salts are formed from alcohols, ROH, $(ROH_2)^+X^-$, and ethers, $(R_2OH)^+X^-$, with strong acids.

Liquid Ammonia:

Liquid ammonia, [see Section 11.4 (p. 291)], is a water-like solvent and behaves like the aquo system. Acids then are materials which raise $[NH_4^+]$, and because of the high proton affinity of NH_3, they can be protic acids, e.g., HCl:

$$HCl + NH_3 \rightarrow NH_4^+ + Cl^-$$

Ammonium salts are acids in the ammono system and the resulting solutions behave like aqueous acids, i.e., they change indicator colors, react with active metals to evolve hydrogen, and can be neutralized with bases. The auto-dissociation of liquid ammonia is weaker than that of H_2O; the ion product, $K_a = [NH_4^+]$ $[NH_2^-] \cong 10^{-29}$ mol^2 · l^{-2}, and the neutrality point is at pH = 14.5.

Anhydrous Sulfuric Acid:

Anhydrous sulfuric acid is a solvent of small proton affinity. In most solvents, H_2SO_4 acts as a protonating agent. The strongest are fluorosulfonic acid, HSO_3F, trifluoromethane-sulfonic acid, HSO_3CF_3, disulfuric acid, $H_2S_2O_7$, as well as mixtures of HSO_3F with SbF_5 and SO_3. These "super acids" can be used to protonate compounds which like H_2SO_4 normally function as acids:

$$HSO_3F + H_2SO_4 \rightarrow H_3SO_4^+ + SO_3F^-$$

J. N. Brønsted in 1923 defined acids as proton donors, and bases as proton acceptors. In this sense, then, H_2SO_4 is a base vs. HSO_3F, and so is H_3PO_4 which is a base vs. H_2SO_4:

$$H_3PO_4 + H_2SO_4 \rightarrow H_4PO_4^+ + HSO_4^-$$

Inorganic acids can be named systematically reflecting their structure, for example, sulfuric acid becomes sulfur dioxide bishydroxide, $SO_2(OH)_2$; orthophosphoric acid is phosphorus oxide trishydroxide, $PO(OH)_3$. Protonated acid molecules, $H_3SO_4^+$, etc., (acidium ions) are isolatable in the form of acidium salts, e.g., $[P(OH)_4]^+ClO_4^-$ (phosphatoacidium perchlorate) and $[Se(OH)_3]^+ClO_4^-$ [selenato (IV) acidium perchlorate], which are readily soluble in nitromethane and ionize to give conducting solutions.

Thus the acid or base strength of a substance depends upon the solvent in which it is dissolved, and, therefore, no absolute definitions can be made.

Other protonic solvents are important in inorganic chemistry, and the concepts developed above can be applied analogously to alcohols, glacial acetic acid, liquid H_2S, HCN, and HF as well as to HSO_3F.

6.4. Bases

According to Arrhenius, the hydroxyl ion OH^- is the carrier of basic properties in water, and in basic solutions $[OH^-] > [H^+]$. The hydroxides of the alkali and alkaline earth metals are typical, yielding OH^- ions which are one of the functional components of water, but other bases lack OH^- ions, yet produce alkaline solutions in water, such as the oxides of alkali and alkaline earth metals, NH_3, N_2H_4, etc., which react with water to give hydroxyl ions:

$$O^{2-} + H_2O \rightarrow 2OH^-$$

$$NH_3 + H_2O \rightarrow NH_4^+ + OH^-$$

A base is defined by Brønsted as a proton acceptor, which applies to OH^-, O^{2-}, and NH_3. Likewise, the solvent anions NH_2^-, HSO_4^-, and HF_2^- are responsible for basic properties in other solvents, for example the salt-like amides, MNH_2, in liquid ammonia, the hydrogen sulfates, $MHSO_4$, in H_2SO_4, and the hydrogen difluorides, MHF_2, in liquid HF correspond to the salt-like hydroxides in the water system. These compounds raise the concentration of solvent anions.

Phosphoric acid is protonated in anhydrous sulfuric acid, and hence acts as a base, like ammonia in water. Protons are transferred to the species of highest proton affinity (basicity). The equation:

$$H_2SO_4 + H_2O \rightarrow H_3O^+ + HSO_4^-$$

means either that H_2SO_4 in the solvent H_2O acts as acid (proton donor), or that H_2O in the solvent H_2SO_4 acts as a base (proton acceptor).

6.5. Relative Acid and Base Strength

6.5.1. Dilute Solutions

Acid or base strength depends upon environment. In dilute aqueous solutions where the law of mass action applies, the acid strength is given by the equilibrium constant:

$$HA + nH_2O \rightleftharpoons H^+(aq) + A^-(aq) \qquad K = \frac{[H^+][A^-]}{[HA]}$$

The larger the dissociation constant of HA, the more the above equilibrium is displaced to the right, the larger $[H^+]$, and the stronger the acid. The base strength is likewise given by an equilibrium:

$$NH_3 + H_2O \rightleftharpoons NH_4^+ + OH^-$$

$$MOH \rightleftharpoons M^+ + OH^-$$

$$M: \text{Metal}$$

Polyprotic acids such as H_2SO_4 and H_3PO_4 have several dissociation constants:

$$H_2SO_4 + nH_2O \rightleftharpoons H^+(aq) + HSO_4^- \qquad K_1$$

$$HSO_4^- + nH_2O \rightleftharpoons H^+(aq) + SO_4^{2-} \qquad K_2$$

which differ by several powers of 10. pK values are often employed:

$$pK = -\log K$$

The relative strengths of aqueous acids are best compared by dividing the compounds into two groups, the binary covalent hydrides such as HF, HCl, etc., and the oxoacids, such as $SO_2(OH)_2$ or $NO_2(OH)$, in which the acidic hydrogen atoms are part of OH groups.

Binary Covalent Hydrides:
Covalent, volatile hydrides are known for all non-metals except the rare gases.

The compounds can be proton donors or acceptors in water. The best known acids are the hydrogen halides whose acid strengths vary. The factors which determine their acid strength are of interest. The equilibrium constant for the equation:

$$HX(aq) \rightleftharpoons H^+(aq) + X^-(aq)$$

can be derived from the free energy, $\Delta G°$, of the reaction, since:

$$\Delta G° = -RT\ln K$$

R = universal gas constant

T = absolute temperature

$\Delta G°$ can be determined from the second law of thermodynamics (equation 61, p. 68):

$$\Delta G° = \Delta H° - T \cdot \Delta S°$$
$$(T = \text{const.})$$

by use of an energy cycle:

The sum of the $\Delta G°$'s for steps 1 through 6 must equal that of step 7. The values of $\Delta G°$, from calculation and experiment, are listed in Table 31.

Table 31   Free Energies, $\Delta G°$, for the Hypothetical Steps in the Dissociation of the Hydrogen Halides in Water with Values of p_K and K ($\Delta G°$ in kJ/mol; T = 298 K)

Reaction	HF	HCl	HBr	HI
1: $HX(aq) \rightarrow HX(g.)$	23.9	−4.2	−4.2	−4.2
2: $HX(g.) \rightarrow H(g.) + X(g.)$	535.1	404.5	339.1	272.2
3: $H(g.) \rightarrow H^+(g.) + e^-$	1320.2	1320.2	1320.2	1320.2
4: $X(g.) + e^- \rightarrow X^-(g.)$	−347.5	−366.8	−345.4	−315.3
5+6: $H^+(g.) + X^-(g.) \rightarrow H^+(aq) + X^-(aq)$	−1513.6	−1393.4	−1363.7	−1330.2
$\Delta G°$ (step 7)	18	−40	−54	−57
p_K	3.2	−7.0	−9.5	−10.0
K	$6 \cdot 10^{-4}$	$1 \cdot 10^7$	$3 \cdot 10^9$	$1 \cdot 10^{10}$
$\mu(HX)$ in Debye	1.91	1.03	0.79	0.38

The acid strengths of the hydrogen halides increase in water or non-aqueous solvents from the weak acid HF to HI which is numbered among the strongest acids (cf. Table 31). This is surprising, since the dipole moments of the molecules decrease in the same direction (cf. Table 31). Proton donor ability is apparently not controlled by the polarity of the element-hydrogen bond. The free energy of dissociation (step 2 in Table 31) seems primarily responsible for the pK values. The strongly polar H—F bond, owing to its high bond energy, retards the dissociation of HF in water.

The same trend in dissociation constants is found in the chalcogen hydrides, $H_2Te > H_2Se > H_2S > H_2O$.

Oxo Acids:

The simple oxo-acids of the non-metals are covalent hydroxides, $EO_m(OH)_n$, which behave as proton donors in water. The oxygen bonds to hydrogen are similar, yet there are large differences in thermodynamic parameters and acid strengths. The pK values for several oxo acids are listed in Table 32.

The extemely strong acids, $HClO_4$, HNO_3, and H_2SO_4 are virtually completely ionized in the first step in water, and hence pK_1 values are difficult to determine, but H_3PO_4 and HNO_2 are only moderately strong acids, and HOCl, H_3BO_3 and H_3AsO_3 are weak acids.

Table 32   pK Values of Some Oxo-acids of Non-metals in Aqueous Solution at 25 °C.

	pK		pK		pK
$HClO_4$	<0	H_3PO_4	2.12	$H_2PO_4^-$	7.2
H_2SO_4	<0	H_2SeO_3	2.57	HOCl	7.50
H_2SeO_4	<0	H_5IO_6	3.29	HOBr	8.68
HNO_3	<0	HNO_2	3.3	H_3AsO_3	9.22
HSO_4^-	1.92	H_3AsO_4	3.5	H_3BO_3	9.22
$HClO_2$	1.94	$HSeO_3^-$	6 60	HPO_4^{2-}	12
$HSeO_4^-$	2.05				

In polyprotic acids, $pK_1 < pK_2$, i.e., $K_1 \gg K_2$, as expected on electrostatic grounds, since the removal of a proton from an anion requires more energy than from a neutral molecule.

For the unknown acids H_2CO_3 and H_2SO_3, K, or pK, is determined by using the dissolved $[CO_2]$ or $[SO_2]$ in the law of mass action, i.e., CO_2 and SO_2 are treated as anhydrides:

$$CO_2(aq) \rightleftharpoons H^+(aq) + HCO_3^-(aq) \quad K = \frac{[H^+][HCO_3^-]}{[CO_2]}$$

The pK values thus obtained do not correspond to the true acid strength of the hypothetical acids H_2CO_3 and H_2SO_3.

Dissolved $[NH_3]$ is also used in lieu of the unknown NH_4OH:

$$NH_3(aq) \rightleftharpoons NH_4^+(aq) + OH^-(aq)$$

These concepts can be extended to non-aqueous solvents in which dissociation equilibrium is determined by measurement of pH, electrical conductivity, or the melting point depression.

6.5.2. Concentrated and Anhydrous Acids

The acids HCl, HNO_3, and H_2SO_4 are virtually completely dissociated in dilute aqueous solution, but are not the most acidic. As the system $H_2O—H_2SO_4$ is concentrated, the acidity, i.e., proton donor strength, increases, although the degree of dissociation decreases. Pure, anhydrous sulfuric acid is only weakly dissociated. The ion product, $[H_3SO_4^+][HSO_4^-]$, is ca. 10^{-4} $mol^2 \cdot l^{-2}$ at 25 °C, yet 100% sulfuric acid is a stronger proton donor than any aqueous acid.

To determine the proton donor strength of a concentrated aqueous solution, or an anhydrous acid one proceeds differently than for dilute solutions. Using the method of L. P. Hammett, a weak base, B, e.g., nitroaniline, is added which is then strongly protonated:

$$H^+ + B \rightleftharpoons HB^+ \qquad K(HB^+) = \frac{[H^+][B]}{[HB^+]}$$

The Hammett acidity function, H_0, defined by analogy with pH as:

$$H_0 = -\log[H^+]$$

is substituted for $[H^+]$ into the expression for the equilibrium constant ($=$ the K of the acid HB^+).

From the equations for H_0 and K:

$$H_0 = p_K(HB^+) - \log\frac{[HB^+]}{[B]}$$

The ratio $[HB^+]/[B]$ can be determined photometrically if B has different colors in the free and protonated forms. The pK value of HB^+ can be determined in dilute solution, where H_0 is identical with the known pH value.

For 100% sulfuric acid, $H_0 = -11.9$, whereas in the 0.1 N aqueous acid pH $= 1.0$, thus the acidity increases by >12 powers of 10.

The H_0 values for a series of common acids are plotted in Figure 71 as a function of water content, expressed in mole fraction $X = n(HA)/n(HA) + n(H_2O)$ ($n =$ = number of moles of component).

Especially high negative values of H_0 (to -19) are obtained for mixtures of HSO_3F with SbF_5 or SO_3. These mixtures and HSO_3F, $H_2S_2O_7$, and H_2SO_4 with $H[B(HSO_4)_4]$ are termed "super acids", because they can protonate substances with little proton affinity which are not bases in other media. For example, HNO_3 is protonated in anhydrous sulfuric acid to $H_2NO_3^+$, which then decomposes into NO_2^+ and H_2O:

$$HNO_3 + 2H_2SO_4 \rightarrow NO_2^+ + H_3O^+ + 2HSO_4^-$$

to give "nitrating acid", used to produce nitro compounds from aromatic hydrocarbons:

$$R—H + NO_2^+ \rightarrow R—NO_2 + H^+$$

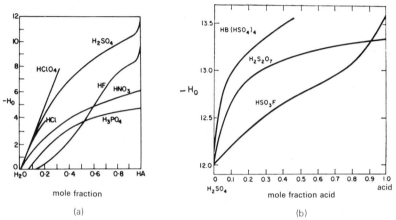

Fig. 71.   Hammett acidity functions, H_0, for solutions of several acids
a) in water   b) in anhydrous sulfuric acid.

The nitronium ion NO_2^+ is too electrophilic to exist in water or other less acidic media:

$$NO_2^+ + 3H_2O \rightarrow 2H_3O^+ + NO_3^-$$

New compounds synthesized in these very acidic liquids will be discussed with their respective elements below.

6.6. Hydrogen Bonds

6.6.1. General

The physical properties of covalent hydrides such as H_2O, NH_3, etc., reflect an intermolecular interaction stronger than the van der Waals force, with typical energies of 4 to 40 kJ/mol, but occasionally ≤ 110 kJ/mol. Hydrogen bonds constitute a special bond type, and have energies between van der Waals and covalent bonds.

Hydrogen bonds are formed between molecules with partially positive hydrogen and electronegative atoms possessing lone electron pairs, such as F, O, and N, and to a lesser degree Cl, S, and P. Hydrogen bonds can also form with bonding electron pairs, such as π-electrons in multiple bonds or aromatic systems. Intramolecular hydrogen bonds are also known.

A simple model illustrates how hydrogen bonds are generated:

$$\overset{\delta(-)}{X} - \overset{\delta(+)}{H} \cdots :Y -$$

by electrostatic attraction between the positive hydrogen atom in X—H and the lone electron pair in an atomic orbital of Y. This attraction depends directly upon the electronegativity difference between atoms X and Y and depends inversely upon the size difference between atoms A and H, since these conditions will maximize the polarization of the X—H bond. Polarized bonds with positively charged hydrogen are, e.g., FH, OH, and NH, but not CH or BH. Furthermore, Y should be smallest and most electronegative to give the smallest lone pair orbital, producing the highest negative charge density and largest atomic dipole moment.

In the liquid state, H_2O, NH_3 and HF form strong hydrogen bonds, while only marginal effects are seen in H_2S, PH_3 or HCl. Higher homologs of these hydrides are bonded only by van der Waals forces.

For a more detailed description of the hydrogen bond, see p. 172.

6.6.2. General Properties of Hydrogen Bonds

a) Most hydrogen bonds are asymmetrical, i.e., the hydrogen atom is not located directly between Y and X, but much closer to X than to Y. Only the strongest hydrogen bonds are symmetrical.

b) Hydrogen bonds are linear or only slightly bent, maximizing attraction between H and Y, and minimizing repulsion of the larger X and Y and thus maximizing bond energy.

c) The valence angle, α, formed between the hydrogen bond and the Y—R bond, usually varies between 100 and 140°:

$$X—H\cdots Y$$
$$\overset{\displaystyle \curvearrowright}{\underset{\alpha}{}} \diagdown$$
$$R$$

d) Normally the hydrogen atom is two coordinated but in nitramide, $H_2N—NO_2$, three-coordination is attained:

$$X—H: \overset{.Y}{\underset{.Y}{}} \qquad X: N$$
$$\qquad\qquad Y: O$$

e) In most cases only one hydrogen bond is directed toward each lone pair of Y, but in crystalline ammonia, three hydrogen bonds extend from each nitrogen atom:

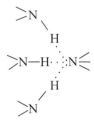

6.6.3. Experimental Evidence for Hydrogen Bonds

a) Physical constants:
Intermolecular interactions in the liquid and the solid states strongly influence physical properties such as melting point and boiling point, the enthalpies of fusion (ΔH°_{fus}), vaporization (ΔH°_{vap}) and sublimation (ΔH°_{sub}), the dipole moment (μ), the dielectric constant (ε) and the viscosity (η). Hydrogen bonds are strong inter-actions, and their presence can be detected by considering the increases in these parameters in comparison with non-associated species. Proof for the existence of hydrogen bonds is provided by spectroscopic or structural investigations.

Figure 72 plots the boiling points in five homologous series. Normally, boiling points increase with molecular weight as with the rare gases and the hydrides of the fourth main group elements. However, the lightest hydrides of the fifth, sixth, and seventh main groups show anomalously high boiling points. Van der Waals

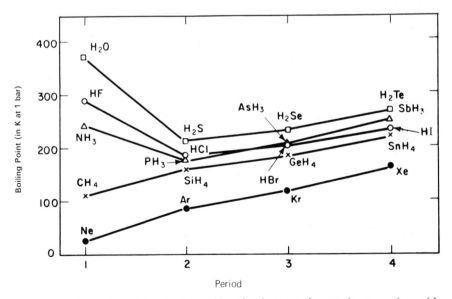

Fig. 72.   Boiling points of the simple hydrides of main group elements in comparison with the rare gases.

forces (see Section 5.3) alone cannot account for this, and strong association through hydrogen bridges must be postulated in the liquid phase. In contrast, CH_4 does not form hydrogen bonds. Hence, in the series CH_4—NH_3—H_2O—HF—Ne, CH_4 and Ne have the lowest boiling points.

Corresponding trends are also found for the enthalpies of vaporization, ΔH°_{vap}, shown in Figure 73. Vaporization is connected with the breaking of hydrogen bonds, and the energy required is in addition to that for the breaking of van der Waals bonds and the volume expansion work.

The large dipole moments and dielectric constants of water and water-like solvents are important for the solubility of salts, since the larger μ and ε, the more soluble are ionic compounds. Hydrogen-bonded liquids such as H_2O, HCN, HSO_3F, etc., have large ε-values.

Water has a higher viscosity than its ether, R_2O, derivatives, which cannot form hydrogen bonds. Particularly high viscosities are found in compounds with several OH-groups such as glycerin or concentrated sulfuric or phosphoric acids, because of the three-dimensional hydrogen bonded network in these liquids.

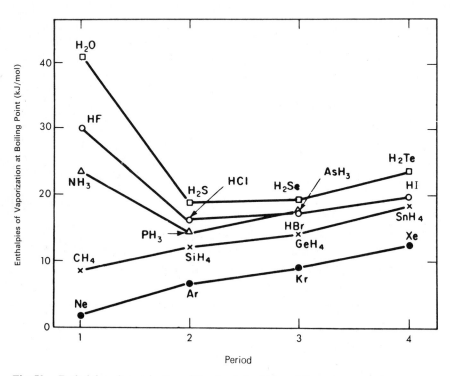

Fig. 73.   Enthalpies of vaporization of the simple hydrides of the main group elements and of the rare gases.

b) Structural analysis

X-ray, electron, and neutron diffraction are used to determine atomic positions and give direct evidence of hydrogen bonding. If the internuclear distance, d(XY), of a group X—H $\cdots$ Y is smaller than the sum of the van der Waals radii of atoms X and Y, then hydrogen bonding is presumed. If the hydrogen atom position can be determined, hydrogen bonding is postulated if the internuclear distance, d(HY) is smaller than the sum of the van der Waals radii of atoms H and Y.

c) Molecular Spectroscopy

The most sensitive detection of hydrogen bonding is afforded by infrared (IR) spectroscopy. The X—H bond in a bridge X—H $\cdots$ Y is polar and gives a strong absorption band in the IR spectrum. The frequency of the non-associated X—H group can be measured in the vapor phase, or in an inert, non-polar solvent at high dilution. The X—H band shifts to lower frequency on association and is markedly broader and more intense because the X—H bond is weakened and is even more polarized by bridge formation.

Raman or NMR spectroscopy, ultrasonic absorption, non-elastic neutron scattering and other techniques can yield information on hydrogen bonding, but in weak interactions no exact demarcation from van der Waals forces is possible, and structural analysis is required to prove the bridge function of hydrogen.

6.6.4. Examples of Special Hydrogen Bonds

Hydrogen fluoride:

Crystalline hydrogen fluoride (m.p. $-83\,^{\circ}$C), consists of zig-zag chains of HF molecules which are associated at $-125\,^{\circ}$ via linear, asymmetrical hydrogen bonds:

d(FF) = 2.49 Å
∠HFH: 120.1°

and the liquid may be similar.

Gaseous HF $<20\,^{\circ}$C (b.p. $19.5\,^{\circ}$C/741 torr) consists mainly of HF and $(HF)_6$ with traces of $(HF)_2$ which co-exist in a temperature- and pressure-dependent equilibrium. The hexamers, according to electron diffraction, form a puckered ring:

d(FF) = 2.53 Å
∠HFH: 104°

Hydrogen bonds are also present in salts of the acids $(HF)_n$. The hydrogen di-fluoride ion, HF_2^-, contains one of the strongest hydrogen bonds. Such salts are synthesized from the fluorides, MF, and HF. The HF_2^- ion in KHF_2 is linear and symmetrical (symmetry $D_{\infty h}$):

$$[F-H-F]^-$$

$$d(HF) = 1.13\,Å, \quad d(FF) = 2.26\,Å$$

Symmetrical HF_2^- ions are connected to NH_4^+ cations via hydrogen bonds in NH_4HF_2. These outer hydrogen bonds increase the internuclear distance, $d(FF)$, in the anion to 2.32 Å, and the hydrogen bonding energy adds to the lattice energy of the salt.

Hydrogen fluoride, owing to the strong mutual interaction of HF molecules and fluoride ions, forms several acid salts, $MX \cdot nHF$ (n = 1, 2, 3 ...). In KH_2F_3, angular anions, $H_2F_3^-$, of symmetry C_{2v} are present.

Ice and Water:

Seven different crystalline phases of H_2O are known to be stable at certain temperatures and pressures (see Figure 74). At 0 °C and 1 bar, water solidifies to ice I with a hexagonal packing of oxygen atoms similar to ZnS (Wurtzite): each oxygen atom is surrounded tetrahedrally by four others. The smallest internuclear distance, $d(OO)$, is 2.74 Å at 0 °C, and decreases with decreasing temperature. The density

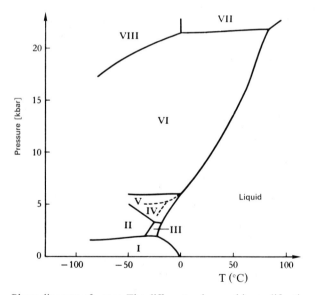

Fig. 74.   Phase diagram of water. The different polymorphic modifications are identified by Roman numerals. The meta-stable phase IV is only encountered in D_2O. 1 Kbar = 987 atm.

at $0\,°C$ is 0.92 g/cm^3. The hydrogen atoms in ice I lie on the oxygen-oxygen axes to give linear hydrogen bridges. The internuclear distance, $d(OH) = 1.01$ Å is larger than in the vapor phase (0.96 Å), and the bond energy is estimated as 20 to 30 kJ/mol.

The other crystalline modifications of water, ice II–VIII, are more dense, but stable only at higher pressures, or meta-stable at liquid nitrogen temperature ($-196\,°C$) at normal pressure. Vitreous ice, from the condensation of water vapor at low temperatures, changes upon warming into ice I via cubic ice Ic. Ice IV is found in D_2O only.

Liquid water has a complex and as yet undetermined structure near its melting point for which the "mixing model" agrees best with experimental results. It is assumed that near $0\,°C$ clusters of ca. 100 H_2O molecules are bonded as in ice I with four hydrogen bonds emanating from each central water molecule. However, molecules on the surface of this ice-like network make only three, two, or a single bridge, (cf. Figure 76). The postulate of ordered domains in liquid water is supported by the similarity of dielectric constants, which at $0\,°C$ for ice = 92, and for liquid water = 88.5, at $100\,°C$ = 55.5, and at $200\,°C$ = 34.5.

Liquid water also contains smaller domains, $(H_2O)_n$, between which polymeri-

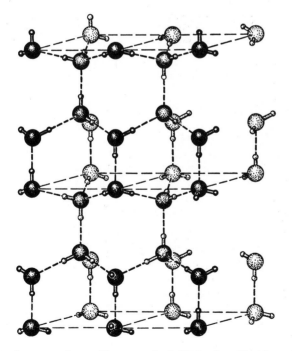

Fig. 75.   Structure of crystalline water, Ice I: [L. Pauling, The Nature of the Chemical Bond, Cornell University Press, Ithaca, New York, 3rd Ed., (1960)].

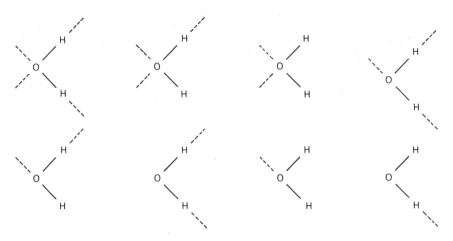

Fig. 76.   Probable bond states of H_2O in liquid water. The hydrogen bonds, shown by dotted lines, are asymmetrical but not necessarily linear.

zation and depolymerization continually take place. The concentration of $(H_2O)_1$ is, however, negligible. The half-life of a hydrogen bond in liquid water is ca. 10^{-11} s, which is a consequence of a bond energy (4–12 kJ/mol) close to the thermal vibrational energy. The median $d(OO)$ at $25\,°C = 2.82\,\text{Å}$, is somewhat larger than in ice I. The average number of hydrogen bonds per oxygen atom in water is estimated as 1.8 at $0\,°C$, and decreases with increasing temperature, as does the cluster size. At $100\,°C$ the clusters are absent.

The proportion of free, non-hydrogen bonded OH groups in liquid water is, from IR spectroscopic measurements, ca. 10% at $0\,°C$; 20% at $100°$, 50% at $260°$, and 100% at the critical point ($374\,°C/216$ bar).

Alternative models for liquid water assume a quasi-continuous, irregular three-dimensional network of H_2O molecules associated by bent hydrogen bonds. Occupation of cage-like cavities in the ice domains by monomeric H_2O molecules explains the higher density of liquid water over hexagonal ice I.

Salts dissolve in water with hydration of the ions. The hydrogen bonds of the H_2O molecules are cleaved by the electrical field of the ions and a first hydration shell with ordered H_2O molecules is created. This structuring through interaction between an ion and polar molecule is accentuated with small, high field-strength ions (e.g., H^+, Li^+, Mg^{+2}, F^-, etc.). The ionic field disturbs the mutual orientation of water molecules at a larger distance where the network structure is broken up, and the H_2O molecules are made more mobile. This disruption takes place in the second hydration sphere of the smaller ions, and in the first sphere of larger, more polarizable ions (e.g., Rb^+, Cs^+, Br^-, I^-, ClO_4^-).

Water vapor at room temperature and pressures $\leq 90\%$ of the saturation vapor pressure is monomeric and behaves as an ideal gas. At higher temperatures near the saturation vapor pressure aggregates are present. These oligomers can be trapped in a matrix at 20K and studied by IR spectroscopy.

Water is sparingly soluble in non-polar solvents, predominantly as the monomer, while in weakly polar solvents, such as partially-chlorinated hydrocarbons, water dissolves as a monomer and oligomers. In polar solvents such as alcohols, ethers, ketones, amines, nitriles, carboxylic acids, sulfoxides, etc., capable of participating in hydrogen bonding, solvent-water complexes are formed, whose composition is temperature- and concentration-dependent.

Ammonia:

Crystalline ammonia has a layered structure in which all atoms participate in hydrogen bonding with three bonds originating from each nitrogen atom (cf. p. 165).

In addition to

$$F—H\cdots F \qquad O—H\cdots O \qquad N—H\cdots N$$

so far discussed, other hydrogen bonds, in which different atoms participate, are listed in Table 33. The structures of many hydrogen compounds are a result of hydrogen bonding, but many compounds lack such bonds which could form them; for example, the solid hydroxides $NaOH$, $Ca(OH)_2$, $Mg(OH)_2$ and $Fe(OH)_2$.

Table 33   Hydrogen Bond Types in Inorganic Compounds

$O—H\cdots O$	$F—H\cdots F$	$O—H\cdots S$	$N—H\cdots Cl$
$(H_2O)_n$	$(HF)_n$	$BaS_2O_3 \cdot H_2O$	NH_4Cl
H_2SO_4	KHF_2	$N—H\cdots F$	$N_2H_6Cl_2$
$B(OH)_3$	KH_2F_3	NH_4HF_2	$Cl—H\cdots O$
K_2HPO_4	$O—H\cdots N$	$(NH_4)_2SiF_6$	HCl in $(C_2H_5)_2O$
$NaHCO_3$	NH_2OH	NH_4BF_4	$C—H\cdots O$
$AlO(OH)$	$2NH_3 \cdot H_2O$	$N—H\cdots O$	$(CH_3)_2SO$ in $CHCl_3$
$CuSO_4 \cdot 5H_2O$	$O—H\cdots Cl$	H_2NNO_2	$C—H\cdots N$
$CaSO_4 \cdot 2H_2O$	$MnCl_2 \cdot 2H_2O$	NH_2OH	HCN
$K_2XeO_4 \cdot 8H_2O$	$H_5O_2Cl \cdot H_2O$	$N—H\cdots N$	$Cl—H\cdots Cl$
H_3OClO_4		NH_3	$[NR_4]HCl_2$
H_2O_2		NH_4N_3	
		N_2H_4	

6.6.5. Theory of Hydrogen Bond Formation

Hydrogen bonds are formed between small, electronegative atoms e.g., fluorine, oxygen or chlorine for which symmetrical bridges are occasionally found. Three types of potential energy functions for hydrogen bonds are shown in Figure 77. The symmetrical X—H—X bridges, the simplest to treat theoretically, are only observed at small internuclear distances, $d(XX)$, where considerable orbital overlap occurs among neighboring atoms to create a partial covalent bond. In the linear ion HF_2^-, for example, the ls orbital of hydrogen overlaps the $2p_z$ of both fluorines to give three molecular orbitals, described by the linear combinations:

$$\psi_3 \text{ (antibonding)} = N_3(\psi_{pA} - c \cdot \psi_s + \psi_{pB})$$
$$\psi_2 \text{ (non-bonding)} = N_2(\psi_{pA} - \psi_{pB})$$
$$\psi_1 \text{ (bonding)} = N_1(\psi_{pA} + c \cdot \psi_s + \psi_{pB})$$

where ψ_1 is a bonding MO since the positive regions of all three orbitals overlap, ψ_2 is a non-bonding orbital, and ψ_3, owing to strongly negative overlap, is antibonding (see Figunre 78). The three orbitals contain four electrons which occupy the bonding and non-bonding MO's, producing a three-center-four-electron bond with each component of order 0.5, in agreement with the HF_2^--ion bonds being weaker than in the HF molecule.

	HF	HF_2^-
d(HF):	0.93	1.13 Å
f(HF):	8.9	2.3 mdyn/Å

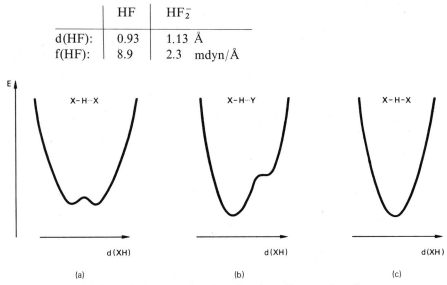

Fig. 77.   Potential energy in hydrogen bonds as function of internuclear distance.
a) Two equivalent positions between identical atoms. Examples: ice I, KH_2PO_4, etc.
b) Non-equivalent positions between atoms of differing size or electronegativity. Example: O—H $\cdots$ N in aqueous ammonia.
c) Single minimum in a symmetrical bond. Example: KHF_2.

The bonds in symmetrical bridges are partially covalent but the situation in weaker, asymmetrical bridges is less equivocal, and the potential energy curve for the O—H···O system shown in Figure 77c only holds for very small d(OO) (2.4–2.6 Å).

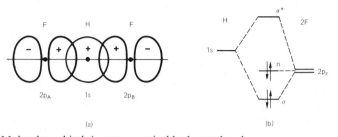

(a)   (b)

Fig. 78.   Molecular orbitals in a symmetrical hydrogen bond
a) Overlap of the atomic orbitals in the bonding state.
b) Energy level diagram.

The curve approximates 77a with increasing internuclear distance, yielding at first a flat minimum and then two energy minima, giving two energetically equivalent positions available for the hydrogen atom which can shift between them (e.g., ice I, cf. Figure 79).

Asymmetrical, X—H···Y, hydrogen bonds typically observed, can be approximated by the electrostatic model, (see p. 164) in which the hydrogen atom is located predominantly in the minimum closer to atom X, shown in Figure 77b, but resonance structures with covalent character may be participating in the ground state:

$$\overset{\ominus}{X}—H\cdots:Y \leftrightarrow \overset{\ominus}{X:} \ \ \overset{\oplus}{H}\cdots:Y \leftrightarrow \overset{\ominus}{X:}\cdots\overset{\oplus}{H}—Y$$
$$\text{(I)} \qquad\qquad \text{(II)} \qquad\qquad \text{(III)}$$

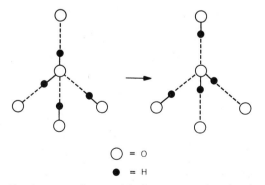

○ = O
● = H

Fig. 79.   Simultaneous change of hydrogen atom location in the asymmetrical hydrogen bond of ice

The proton in structure III occupies a shallower energy minimum at atom Y in Figure 77b. The total formation energy is the sum of electrostatic attraction, van der Waals forces, and participation by structures II and III, less repulsion of the electron shells of Y and HX.

Most of the properties of asymmetrical hydrogen bonds are explained by this model. Only a hydrogen atom lacks inner electron shells which would be repelled by the free electron pair of the Y-atom at sufficiently close proximity.

The short lifetimes of the weaker hydrogen bridges distinguishes them from stronger, partially covalent bonds. These bonds are continually broken and re-established in the liquid and vapor phases, often between different bond partners because of fluctuations in the vibrational energy of the molecules, which is equipartitioned only on the average. Instantaneously, it can surpass the B.E., especially so with smaller B.E.'s and higher temperatures. The mean vibrational energy of 2.5 kJ per mol and per degree of freedom determines at 25 °C a life-time of fractional seconds in hydrogen bonds having a 4 to 40 kJ/mole bond energy.

6.7. Hydrogen Compounds (Hydrides)

Hydrogen compounds are classified into different groups based on their bonding and physical and chemical properties:

a) Covalent Hydrides
   Bonding predominantly covalent, non-polar to strongly polar. Examples: SiH_4, H_2O, B_2H_6, $(BeH_2)_n$, NH_4^+, ReH_9^{2-}, etc.
b) Salt-like Hydrides
   Bonding: predominantly ionic
   Examples: NaH, CaH_2, etc.
c) Metallic Hydrides
   Bonding: metallic, covalent, and ionic
   Examples: PdH_n, UH_3, etc.

There are no sharp boundaries between the three groups, and certain compounds are difficult to classify.

6.7.1. Covalent Hydrides

To this group belong all non-metal hydrides and most organic compounds. Non-metals, except the rare gases, form volatile hydrides, as do the main group metals (e.g., Sn, Sb, Bi, Po). The higher oligomers are liquids of low volatility or are non-volatile (polyboranes, polysilanes, polysulfanes etc.).

Salts of the anions BH_4^-, NH_2^-, and OH^- or of the cations NH_4^+, PH_4^+, H_3O^+ include ionic structures for the non-metallic hydrides, while some metal hydrides such as Be, Mg, Al, and Ga which are polymeric and non-volatile have predominantly covalent bonds to hydrogen. Complex hydrides of the transition metals which will not be considered here are also known, which are volatile, e.g., $MnH(CO)_5$, salt-like e.g., K_2ReH_9, or considered as coordination compounds with the ligand H^-.

6.7.2. Salt-like Hydrides

The alkali metals and the heavier alkaline earths react exothermally with hydrogen gas at high temperatures:

$$2M_{(l.)} + H_{2(g.)} \rightleftharpoons 2MH_{(s.)}$$

to form colorless, salt-like hydrides, MH or MH_2. The equilibrium shifts to the left with increasing temperature, and the preparation of pure compounds, with exception of LiH, is difficult. The dissociation pressure obeys the Arrhenius equation, $\log p = -A/T + B$, so that pure hydrides are obtained by equilibrating at 725 °C for Li, and >500 °C for Ca, and then cooling the system under hydrogen, to maximize the yield of the hydride. Owing to the unfavorable position of the equilibrium at high temperature, a consequence of the small enthalpies of formation of the salt-like hydrides, only LiH (m.p. 691 °C) is fusible without decomposition. The hydrides of the alkali metals crystallize in the cubic system with rock salt lattices; the alkaline earth hydrides are rhombic. The crystals consist of metal cations and hydride ions, H^-, isoelectronic with helium. The electrolysis of a saline hydride eutectic to which the corresponding metal chloride has been added to lower the melting point, evolves hydrogen at the anode:

$$2H^- \rightarrow H_2 + 2e^-$$

analogous to the generation of chlorine in the electrolysis of molten NaCl. Hydride ions react with proton donors:

$$H^+ + H^- \rightarrow H_2$$

and the saline hydrides are vigorously decomposed by water and acids. Calcium hydride is easily handled as it is the least reactive, and is used as a drying agent for inert organic solvents.

Ternary compounds are formed when CaH_2, SrH_2, or BaH_2 are fused with their respective metal halides in a hydrogen atmosphere:

$$CaH_2 + CaCl_2 \rightarrow 2CaHCl$$

These halohydrides crystallize in the tetragonal system with deeper colors with increasing polarizability of cation and of halide anion (CaHCl is colorless, but BaHI is black).

The formation of salt-like hydrides with the anion H^- suggests that hydrogen should be considered as a halogen homologue and listed atop the VIIth main group of the periodic table. Thermodynamic considerations, however, indicate the purely formal character of these analogies. The enthalpy of the reaction:

$$\tfrac{1}{2}X_2(g.) \rightarrow X(g.) \xrightarrow{+e^-} X^-(g.)$$

to convert hydrogen or a halogen to the corresponding anion is:

$$\Delta H^\circ = \tfrac{1}{2}D(X_2) - A_x$$

From the dissociation energies, D, and electron affinities, A, the enthalpy changes, ΔH°, are obtained:

X_2	H_2	F_2	Cl_2	Br_2	I_2	
ΔH°	+151	−193	−126	−155	−167	kJ/mol

Hydride ion formation, by contrast to halide, is endothermic owing to the large dissociation energy of the H_2 molecule and the small electron affinitiy of the hydrogen atom. Thus only the most electropositive metals, with smallest ionization energies are capable of forming salt-like hydrides, otherwise the lattice energies would not compensate the energies of the endothermic evaporation and ionization of the metal (cf. p. 65).

Salt-like hydrides are used as hydrogenation and reducing agents to prepare other hydrides. Of greatest importance is lithium aluminum hydride, $LiAlH_4$ (also abbreviated by organic chemists as LAH), obtained by hydrogenation of $AlCl_3$ or $AlBr_3$ with LiH:

$$4\,LiH + AlX_3 \rightarrow LiAlH_4 + 3\,LiX$$

X: Cl, Br

The product (m.p. 150°) is a colorless, hygroscopic substance whose solutions in ether contain the dietherate, $LiAlH_4 \cdot 2\,R_2O$, and is used to convert halides, such as $BeCl_2$, BCl_3, $SiCl_4$, Si_2Cl_6, $AsCl_3$, etc., into the corresponding hydrides, e.g.,

$$2\,Si_2Cl_6 + 3\,LiAlH_4 \rightarrow 2\,Si_2H_6 + 3\,LiCl + 3\,AlCl_3$$

Lithium aluminum hydride contains the tetrahedral AlH_4^- anion which is isoelectronic with silane, SiH_4, and the phosphonium ion, PH_4^+.

6.7.3. Metal- and Alloy-like Hydrides

Many transition metals form hydrides of non-stoichiometric composition, depending on the pressure and temperature of the hydrogen in exothermic reactions:

$$M + \frac{x}{2}H_2 \rightleftharpoons MH_x$$

Hydrogen content is larger with lower temperature and higher pressure. Suitable metals are, e.g., titanium, uranium, praseodymium, palladium, platinum, etc.

The products are dark-colored powders which retain metallic properties such as electrical conductivity and paramagnetism. The hydrogen content is continuously variable, but stoichiometric compositions are sometimes obtained, for example, TiH_2, UH_3, and PrH_2.

The uptake of hydrogen by metallic palladium, which is of particular interest in hydrogenation catalysis, is shown in Figure 80. Palladium, at 25 °C, and a hydrogen pressure of 1 bar absorbs 0.6 mol H/mol Pd ($PdH_{0.6}$). Maximum hydrogen content is reached at -78 °C with the composition $PdH_{0.83}$. This hydride crystallizes in the cubic system with a defect NaCl lattice with enthalpy of formation of -40 kJ/mol.

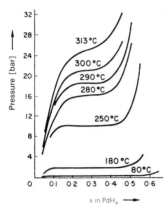

Fig. 80.   Isotherms of the dissociation pressure of the palladium-hydrogen system.

The distance between the layers of palladium atoms expands during hydrogen uptake from 3.890 Å for $PdH_{0.03}$ (α-phase) to 4.018 Å for $PdH_{0.6}$ (β-phase) while the magnetic susceptiblity, χ, decreases linearly with increasing hydrogen content, and disappears for the composition $PdH_{0.65}$ which is diamagnetic. The lattice expansion is ascribed to hydrogen absorption. The paramagnetism of palladium indicates that a portion of its ten valence electrons are in the conduction band, leaving holes in the valence band. The gap between the bands is apparently small, but increases with the distance between the palladium atoms since orbital overlap is diminished, and more electrons are found in the lower energy valence band until diamagnetism occurs once all the valence electrons are transferred. Removal of hydrogen from $PdH_{0.65}$ *in vacuo* yields diamagnetic β-Pd in which the expanded lattice structure is preserved. At high temperatures hydrides of palladium only form under high hydrogen pressures. At red heat, hydrogen diffuses unimpeded through palladium foil, which is utilized for hydrogen purification (cf.p. 151).

Bonding in the Alloy-like Hydrides:
Atomic hydrogen could occupy inter-lattice positions of an expanded metal lattice, forming a quasi-alloy. Two conflicting theories are available, each of which accounts for the variable composition, high hydrogen diffusion velocities, and the electrical and metallic properties of the transition metal hydrides.

In one theory H_2-molecules dissociate into atoms which occupy lattice positions and delocalize their valence electrons in the conductance band of the metal:

$$MH \leftrightarrow M^- H^+$$

The changes in lattice structure and electron distribution are reflected in the properties of the hydrides. Owing to their small size, hydrogen atoms are quite mobile in the lattice, particularly at high temperatures and in the presence of many holes, as in $PdH_{0.6}$.

Atomic hydrogen is also the basis of in the second theory, but as an acceptor of electrons from the conductance band to form H^-:

$$MH \leftrightarrow M^+ H^-$$

The electrical and magnetic properties of the hydride are then a function of the electrons remaining in the conductance band.

These models postulate metallic as well as ionic bonding, but orbital overlap also gives covalent contributions, so that all types of chemical bonds may participate.

7. Oxygen

Natural oxygen, which occurs as O_2, H_2O, oxides, and oxo-salts (silicates, carbonates, etc.) is a mixture of three isotopes: ^{16}O (99.76%), ^{17}O (0.04%) and ^{18}O (0.20%). The ^{18}O isotope, employed as a tracer, is obtained from commercial $H_2{}^{18}O$, which is produced by fractional distillation of water.

7.1. Elemental Oxygen

7.1.1. Molecular Oxygen

Molecular oxygen is manufactured by fractional distillation of liquid air. The components of dry air, 20.95 vol % (23.16 wt %) O_2, 78.09 vol % N_2, 0.93% rare gases (mainly argon) and 0.03% CO_2, are separated by their differences in boiling points.

Purest oxygen is prepared by electrolysis of 30% potassium hydroxide with nickel electrodes, or by catalytic decomposition of 30% aqueous hydrogen peroxide on a platinized nickel foil, which is immersed in the liquid, but commercial oxygen in steel cylinders suffices for most purposes.

Oxygen gas is color- and odorless. In the liquid and solid phases it has a pale blue color (b.p. $-182.97\,°C$, m.p. $-229.39\,°C$). The bonding in the O_2 molecule which is a ground state triplet with two unpaired electrons is discussed on p. 82. The paramagnetism of O_2 is used in its quantitative determination in gas mixtures.

The O_2 molecule as a complex ligand:
Some transition metal complexes react at room temperature to bind O_2 as a ligand by addition or displacement of other ligands:

$$[RuCl_2(AsPh_3)_3] + O_2 \xrightarrow{\text{Benzene}} [RuCl_2(O_2)\,(AsPh_3)_3]\downarrow$$
$$[Pt(PPh_3)_4] + O_2 \xrightarrow{\text{Benzene}} [Pt(O_2)\,(PPh_3)_2]\downarrow + 2\,PPh_3$$

Ph: C_6H_5

and some liberate it reversibly on heating:

$$[Rh(O_2)(diphos)_2]PF_6 \underset{}{\overset{CH_3OH}{\rightleftharpoons}} [Rh(diphos)_2]PF_6 + O_2$$

diphos: $Ph_2P—CH_2—CH_2—PPh_2$

or remove it by chemical means, e.g., reversible conversion to a hydride complex by flushing with hydrogen.

The dioxygen ligand forms a triangle of symmetry C_{2v} with the central atom. Either the O_2 is bound, as in covalent peroxides, via two metal-oxygen-σ-bonds (a), or it functions as π-donor, similar to olefins in corresponding metal complexes (e.g., $[C_2H_4PtCl_3]^-$) (b):

The experimental results agree best with (b). The internuclear distance, d(OO), is larger, the tighter the ligand is bound, and values between 1.30 and 1.63 Å are found. The complexes which liberate O_2 reversibly upon warming have only slightly weakened O—O bonds vs. O_2, (d = 1.21 Å), while a strong interaction, as is present in the complex Ir(O_2) (diphos)$_2$PF$_6$, leads to d(OO) = 1.625 Å. Likewise, the shift of ν(OO), found at 1556 cm^{-1} in O_2 and at 800–900 cm^{-1} in the complexes, indicates a weakened dioxygen bond. The ligand O_2 is bound by a π- and a σ-bond to the metal, M, the σ-bond from the overlap of a doubly-occupied π-MO of O_2 with a vacant metal orbital which may be a d-s-p-hybrid (see Figure 81a), and the π-(back) bond (see Figure 81b) from the overlap of a doubly-occupied d-orbital of M with one of the two π^*-MO's of O_2, from which the unpaired electron migrates to the second, formerly degenerate π^*-orbital of O_2 to form a pair.

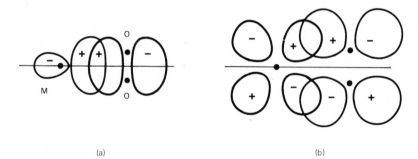

Fig. 81. Overlap of orbitals of the central atom, M with those of ligand O_2 in Dioxo-Complexes.
a) σ-bond   b) π-back bond.

The electron migration away from O_2 and the occupancy of its antibonding π^*-orbitals brings about a weakening of the dioxygen bond which is more pronounced the better the overlap of the metal orbitals with the ligand orbitals, and the stronger the metal-ligand-bond whose strength is a function of the electronegativity, x_E, of the ligands. The larger x_E, the more the ligands draw electrons away from the central atom leaving less for the back bond to O_2. The O—O internuclear distance diminishes with increasing electronegativity of ligands:

$$Ir(O_2)Cl(CO)(PPh_3)_3 \qquad d(OO) = 1.30\,\text{Å}$$

$$Ir(O_2)I(CO)(PPh_3)_3 \qquad d(OO) = 1.51\,\text{Å}$$

The dioxo complexes are models for oxygen transfer in living organisms in which metal complexes bound as prosthetic groups to proteins are active oxygen carriers and transfer agents in respiration. The dioxygen complexes are also catalysts for oxidation by O_2, since metal-bound oxygen is more reactive. For example, $RuCl_2(O_2)(AsPh_3)_3$ reacts with triphenyl-phosphine at room temperature to form PPh_3O and with SO_2 to form a sulfate complex. The analogous S_2-complexes have also been synthesized.

7.1.2. Atomic Oxygen

Atomic oxygen like atomic hydrogen, is generated from O_2 by electric discharge, with the degree of dissociation a function of the strength of the discharge, the gas pressure and the nature of the reaction vessel walls. Certain impurities suppress the recomnation of oxygen atoms, poisoning the catalytic vessel surfaces. Recombination takes place between adsorbed atoms, and the energy liberated is absorbed by the wall:

$$O + O \xrightarrow{\text{wall}} O_2 \qquad \Delta H° = -498 \text{ kJ/mol}$$

Vapor phase oxygen atoms react at room temperature and at pressures of several torr in a three-body collision:

$$O + O_2 + M \rightarrow O_3 + M \qquad \Delta H°_{300} = -109 \text{ kJ/mol}$$

The wall, M, accepts the liberated energy. Ozone formation in $O_2 - O$ mixtures results in only tiny steady state O_3 concentrations, since degradation rapidly occurs:

$$O + O_3 \rightarrow 2O_2 \qquad \Delta H°_{300} = -394 \text{ kJ/mol}$$

because two-body collisions at lower pressures are much more frequent than three-body collisions. Generating ozone from atomic oxygen requires the suppression of the subsequent reaction by removal of the O_3 formed (trapping it on a cold surface).

Oxygen atoms are strong oxidizing agents, reactive at low temperatures, and can be titrated with NO_2:

$$NO_2 + O \rightarrow NO + O_2$$

The NO formed then reacts further:

$$NO + O \rightarrow NO_2^* \rightarrow NO_2 + h \cdot v$$

The NO_2 is produced in an excited state which decays with the emission of yellow-green light. The luminescent $NO-NO_2^*$ reaction can be used as an indicator in the titration of oxygen atoms.

7.1.3. Ozone, O_3

Ozone is an allotropic form of oxygen which is richer in energy than O_2. Its formation is endothermic:

$$\tfrac{3}{2}O_2 \rightarrow O_3 \qquad \Delta H^\circ = 142 \text{ kJ/mol}$$

and pure liquid or solid ozone is explosive. The catalytic decomposition to O_2 in the vapor phase, which takes place with volume expansion, is used to determine O_3 in O_2-O_3 mixtures.

Ozone is prepared by passing O_2 at 25 °C and 1 bar between two concentric metallized glass tubes to which low frequency power of several thousand volts is applied to produce silent electrical discharges. The ozonizer tube, which is cooled to room temperature (it warms by dielectric loss), produces up to 10% of O_3 at moderate flow rates through the formation of oxygen atoms (cf. preceeding page), exited O_2 molecules, and through dissociative ion recombination:

$$O_2^* + O_2 \rightarrow O + O_3$$
$$O_2^+ + O_2^- \rightarrow O + O_3$$

The deep blue-colored, diamagnetic ozone obtained by fractional condensation, m.p. -193 °C, b.p. -112 °C, is poisonous and a strong oxidizing agent which can be determined by iodometric titration, after bubbling through a boric acid-buffered KI solution, liberating iodine:

$$O_3 + 2I^- + H_2O \rightarrow O_2 + I_2 + 2OH^-$$

The O_3 molecule, with C_{2v} symmetry, and $\angle OOO = 117°$, has a dipole moment of 0.5–0.6 D. Bond properties are compared with other O—O species in Table 34 where a bond order of 1.5 is assigned for O_3, similar to the peroxide ion, O_2^- (cf. p. 189). A delocalized π-bond must be assumed:

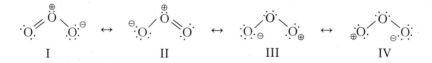

with structures III and IV making little contribution. The terminal oxygen atoms are separated by only 2.18 Å, less than the van der Waals distance of 2.8 Å, not through attraction of the two atoms, but owing to hybridization of the central atom (cf.p. 117).

Table 34   Bond Properties of O—O-Bonded Species.

	O_3	O_2^{2-}	O_2^-	O_2
Bond Order b(MO)	1.5	1.0	1.5	2.0
O—O Internuclear Distance [Å]	1.278	1.49	1.29	1.21
Force Constant [mdyn/Å]	5.7	2.8	6.2	11.4
Average Bond Energy [kJ/mol]	300	138	398	498

The delocalized π-bond extending over three atoms, A, B, and C comes about by the overlap of three p-orbitals perpendicular to the molecular plane which is a node. Three molecular orbitals, bonding, non-bonding, and antibonding, are formed:

$$\psi_3 = N_3 (\psi_{pA} - c \cdot \psi_{pB} + \psi_{pC})$$

$$\psi_2 = N_2 (\psi_{pA} - \psi_{pC})$$

$$\psi_1 = N_1 (\psi_{pA} + c \cdot \psi_{pB} + \psi_{pC})$$

Four electrons reside in the three p-orbitals, occupying the bonding and non-bonding MO with a π-bond order of 0.5, and a total bond order, including the σ-order, of 1.5, consistent with experimental data.

A similar three-center, four-electron bond occurs in isoelectronic NO_2^-.

7.2. Oxygen Bond Types in Covalent and Ionic Compounds

7.2.1. Oxides

Oxides are compounds containing negative oxygen atoms in oxidation state -2, but having no O—O bonds. Ionic (MgO) and covalent oxides are known, which may be polymeric (SiO_2) or monomeric (N_2O).

Ionic Oxides:

Electropositive elements react in exothermic reactions to form oxides which crystallize in ionic lattices (Na_2O, CaO, Al_2O_3) with O^{2-} ions, isoelectronic with Ne:

$$\tfrac{1}{2}O_2(g.) + 2e^- \rightarrow O^{2-}(g.) \qquad \Delta H^\circ = +904 \text{ kJ/mol}$$

The oxide ion is a strong proton acceptor, and ionic oxides react with protonic solvents to form OH^- or H_2O. In water, the equilibrium:

$$O^{2-} + H_2O \rightleftharpoons 2OH^-$$

lies completely to the right.

The ionic hydroxides, M^+OH^-, and their substituted alkoxide derivatives, M^+OR^-, lie between the ionic and covalent oxides, since oxygen employs bonds of both types.

Covalent Oxides:
Non-metallic oxides are covalent, with bond orders of 0.5 to 2.8, corresponding to the electronic arrangements:

$$:\ddot{O}\!\!<\qquad :\ddot{O}=\qquad :O\equiv$$

Oxonium compounds containing the ion R_3O^+ are considered derivatives of co-valent oxides.

The coordination number of oxygen in covalent oxides can be one to three. A coordination number of one is realized in the examples:

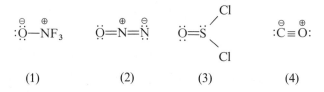

$$(1)\qquad\qquad (2)\qquad\qquad (3)\qquad\qquad (4)$$

The oxygen atom is bonded to a non-metal either by a coordinate σ-bond (1), a σ-plus a p $-$ p-π-bond (2), a σ- plus a p $-$ d-π-bond (3), or a σ-plus two p $-$ p-π-bonds (4). These bond types are encountered in the oxo-anions CO_3^{2-}, SO_4^{2-}, ClO_4^-, and in many molecular oxides, e.g., Cl_2O, NO_2, XeO_3, etc.

The coordination number two is found in water and its derivatives where the oxygen atom is approximately sp^3-hybridized and the valence angles are ca. 105°. Oxygen forms two σ-bonds and has two lone electron pairs as in Cl_2O, $HOCl$, N_2O_5 and higher molecular weight non-metal oxides with bridging oxygen atoms. Oxygen valence angles $>109°$, even approaching 180°, are found in water deriva-tives of type R_2O where additional coordinate π-bonds from oxygen to the unoc-cupied valence orbitals of neighboring atoms are possible, changing the oxygen hybridization from sp^3 to sp^2 to sp and widening the valence angle:

$$E\!-\!\ddot{O}\!-\!E\ \leftrightarrow\ \overset{\ominus}{E}\!=\!\overset{\oplus}{O}\!-\!E\ \leftrightarrow\ \overset{\ominus}{E}\!=\!\overset{2\oplus}{O}\!=\!\overset{\ominus}{E}$$

where E can be silicon, phosphorus, or sulfur, as well as transition metals, all of which have unoccupied orbitals. In disiloxane, $O(SiH_3)_2$, $\angle SiOSi = 144°$, but only 112° in the analogous carbon compound. Similar values are encountered for SiO_2, which crystallizes in several modifications (α-quartz: 142°, α-cristobalite: 150°), and in the metasilicate anions. In the phosphorous oxides P_4O_6 and P_4O_{10} and in several metaphosphates, the $\angle POP = 121$ to 134°. In the disulfate anion, $S_2O_7^{2-}$ $\angle SOS = 124°$ and in the polymeric $(SO_3)_n$ only a little smaller (121°). The $\angle POP$ $= 180°$ in the anion $(O_3P\!-\!O\!-\!PO_3)^{4-}$ of ZrP_2O_7. These examples show that oxygen, like fluorine, is capable of forming intramolecular coordinate π-bonds with an adjacent Lewis acid, surrendering its electrons despite its high electronegativity, since the electronegativity of doubly-occupied orbitals is much lower than that of half-occupied orbitals (see p. 143).

Steric effects may be in part responsible for the widening of valence angles. In angular groups BAB in which a smaller A-atom (N, O) is bonded to two larger B-atoms (Si, P, S, Cl), the internuclear distance, d(BB), can be smaller than the van der Waals distance, giving rise to repulsion of the B atoms, and widening the valence angle. Thus, only large deviations from tetrahedral angles should be ascribed to (p $\rightarrow$ d)-π-bonds, and confirmatory evidence (internuclear distances, Lewis-basicities, spectroscopic data) is required.

The coordination number three is also frequent in oxygen compounds such as H_3O^+, donor-acceptor complexes of ethers, e.g., with BF_3, or in hydroxide-bridged polynuclear complexes which occur as intermediates in the condensation of aquo Fe^{3+} or Cr^{3+} complexes:

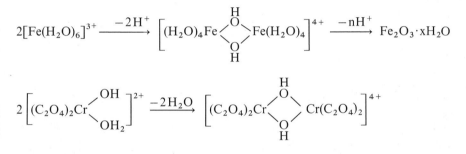

The three coodinated oxygen atom is at the apex of the trigonal pyramidal R_3O^+ group, which is pseudo-tetrahedral with the remaining oxygen electron pair considered the fourth substituent. In rare instances this electron pair can be utilized in coordinate π-bond formation when a planar oxygen uses three sp^2 hybridized orbitals to make three σ-bonds, and the remaining p-π-orbital overlaps with an unoccupied substituent orbital, for example in the following heterosiloxane, where two π-bonds extend over each four Al_2OSi centers:

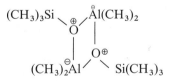

The empty 3d-acceptor orbitals of aluminum and silicon have suitable symmetry for overlap with the oxygen p-π-orbital.

The rarer coordination number four is achieved with the metal atoms in covalent $OM_4(OOCR)_6$ where M = Be or Zn, and R is an organic radical.

7.2.2. Peroxides

Peroxides in which oxygen is in the -1 oxidation state, either in the form of O_2^{2-} or RO_2^- ions, or in covalent derivatives R_2O_2, contain O—O single bonds, which, owing to their low bond energy, are highly reactive (p. 131).

Ionic Peroxides:

Alkali metals and the heavier alkaline earths form salt-like peroxides, e.g., Na_2O_2 and BaO_2.

Sodium peroxide is prepared commercially by heating sodium metal in CO_2-free air at 300° to 400 °C, with the intermediate formation of Na_2O:

$$2\,Na + O_2 \;\rightarrow\; Na_2O_2 \qquad \Delta H° = -507\;kJ/mol$$

The product is a pale yellow, hygroscopic salt which reacts violently with oxidizable substances such as powdered sulfur, carbon, or aluminum. The alkali peroxides generate oxygen with CO_2:

$$Na_2O_2 + CO_2 \;\rightarrow\; Na_2CO_3 + \tfrac{1}{2}O_2$$

Barium peroxide is obtained by heating BaO to 500°–600 °C in air (at a pressure of 2 bar):

$$BaO + \tfrac{1}{2}O_2 \;\rightleftharpoons\; BaO_2 \qquad \Delta H° = -71\;kJ/mol$$

Like Na_2O_2, the product decomposes at higher temperatures into the oxide and oxygen. The hydrolysis of ionic peroxides, which may be considered as the hydrolysis of the salt of a strong base, NaOH or $Ba(OH)_2$, and a weak diprotic acid (H_2O_2), produces H_2O_2, or the HO_2^- anion, since the ion O_2^{2-} is a strong proton acceptor:

$$O_2^{2-} + H_2O \;\rightarrow\; HO_2^- + OH^-$$

Since the decomposition of H_2O_2 to water and oxygen is hydroxyl ion-catalyzed, the mixture is cooled or the peroxide is transferred into dilute acid.

Covalent Peroxides:

Hydrogen peroxide, H_2O_2, (see Section 7.3.2.) is the source of almost all peroxo compounds.

7.2.3. Superoxides

Potassium, rubidium and cesium ignite in oxygen to form compounds MO_2, containing the O_2^- ion, rather than oxides or peroxides. The sodium salt, NaO_2, can be formed from Na_2O_2 and oxygen gas at 300 bar and 500 °C; LiO_2 is formed at lower temperatures from Li atoms and O_2.

The oxygen molecule has a positive electron affinity:

$$O_2(g.) + e^- \rightarrow O_2^-(g.) \qquad \Delta H^\circ = -A_{0_2} \approx -42\,kJ/mol$$

A large cation is a prerequisite for the formation of stable superoxides, and $Ba(O_2)_2$, $Sr(O_2)_2$, and $(CH_3)_4N(O_2)$ are known. The salts KO_2, RbO_2, and CsO_2 crystallize in the calcium carbide lattice, which is a tetragonally distorted rock salt structure. Superoxides are yellow to orange-colored paramagnetic solids which are strong oxidizing agents (the oxidation state of oxygen is $-\frac{1}{2}$). Hydrogen peroxide, oxygen gas and OH^- form on hydrolysis:

$$O_2^- + H_2O \rightarrow HO_2 + OH^-$$
$$2HO_2 \rightarrow H_2O_2 + O_2$$
$$H_2O_2 \xrightarrow{OH^-} H_2O + \tfrac{1}{2}O_2$$

The hydroperoxide radical, HO_2, generated during hydrolysis, can be studied by infrared spectroscopy from the photolytic dissociation of HI at 4 K in an O_2 matrix:

$$HI + h \cdot v \rightarrow H + I$$
$$H + O_2 \rightarrow HO_2$$

HO_2 is also an intermediate in the decomposition of H_2O_2 by gaseous discharge or photolysis. Its bonding is discussed on p. 194.

7.2.4. Ozonides

The tendency to form higher oxides increases with cationic radius, and the more oxygen-rich, paramagnetic, ozonides, MO_3, with the angular O_3^- ion, are known for the larger cations K^+, Rb^+, Cs^+, NH_4^+ and $(CH_3)_4N^+$ only. An MO treatment of O_3^- yields an O—O bond order of 1.25, with the unpaired electron located in an antibonding π-MO (cf. O_3, p. 182). Ozonides are formed in the reaction of ozone with the solid potassium, rubidium or cesium hydroxides:

$$3KOH + 2O_3 \rightarrow 2KO_3 + KOH \cdot H_2O + \tfrac{1}{2}O_2$$

and can be separated by extraction with liquid ammonia from the insoluble hydroxide. The orange-red KO_3 decomposes at room temperature into KO_2 and oxygen gas.

7.2.5. Dioxygenyl Compounds

These compounds which contain oxygen in the $+\frac{1}{2}$ state, mostly in the form of O_2^+ ions, are formed by ionization of the O_2 molecule by removal of one electron from one of the degenerate π^*-orbitals:

$$O_2 \rightarrow O_2^+ + e^- \qquad \Delta H^\circ = 1168\,kJ/mol$$

Owing to the high ionization energy, the process can only be realized chemically with species of high electron affinities, such as platinum hexafluoride, which reacts at room temperature with oxygen gas to form the volatile solid, O_2PtF_6, isomorphous with K_2PtF_6:

$$O_2 + PtF_6 \xrightarrow{25\,°C} O_2PtF_6$$
$$O_2 + Pt + 3F_2 \xrightarrow{450\,°C} O_2PtF_6$$

Pt(VI) oxidizes the oxygen to O_2^+, and is reduced to Pt(V). At $450\,°C$, O_2PtF_6 can be obtained directly from the elements.

Dioxygen difluoride, O_2F_2, which reacts with fluoride ion acceptors to liberate fluorine, is another source of dioxygenyl compounds:

$$O_2F_2 + BF_3 \rightarrow O_2BF_4 + \tfrac{1}{2}F_2$$

$$O_2F_2 + AsF_5 \rightarrow O_2AsF_6 + \tfrac{1}{2}F_2$$

and PF_5 and SbF_3 react analogously. Dioxygenyl fluoroborate, O_2BF_4, is isomorphous with the orthorhombic nitrosyl tetrafluoroborate, $NOBF_4$. The similarities between the O_2^+- and NO^+-salts are based on the similar spatial requirements of the two cations.

The paramagnetic O_2^+ ion contains one unpaired electron in a π^*-orbital. The magnetic moment is a composite of the individual cation and anion moments, if the latter is also paramagnetic (e.g., PtF_6^-).

Covalent dioxygenyl compounds, O_2R, must contain R groups whose electronegativity is greater than O_2, otherwise derivatives of the hydroperoxide, HO_2 (cf. p. 187) would result. Hence, only fluorine and perfluorinated substituents (e.g., $-CF_3, -SF_5$) can qualify. The compound O_2F is formed by the reaction of photochemically generated fluorine atoms at low temperatures:

$$2O_2 + 2F \rightarrow 2O_2F \raisebox{0.5em}{$\nearrow O_4F_2$} \raisebox{-0.5em}{$\searrow O_2 + O_2F_2$}$$

The product has been established in matrix studies by IR and ESR spectroscopies. When diffusion in the matrix occurs, O_2F is in equilibrium with its dimer, tetraoxygenyl difluoride, O_4F_2, which slowly decomposes to oxygen gas and O_2F_2. Oxygen fluorides are discussed on p. 192.

7.2.6. Comparison of Bond Properties of the Ions O_2^+, O_2^-, and O_2^{2-}

Table 35 lists internuclear distances, valence force constants, and dissociation energies for the diatomic ions O_2^+, O_2^-, and O_2^{2-} which indicate that the O—O-bond weakens from O_2^+ to O_2^{2-}. This also follows from the MO treatment in the scheme given on p. 106.

The species differ in the number of electrons which increases stepwise from O_2^+ to O_2^{2-} in the two degenerate, anti-bonding π^*-orbitals, while bond order, b(MO), decreases correspondingly. The occupancy of the inner MO's remains unchanged,

Table 35   Bond Properties of the O—O-Bond in O_2^+, O_2, O_2^-, and O_2^{2-}.

	Valence Electron Number	Bond Order b(MO)	Inter-nuclear Distance [Å]	Force Constant [mdyn/Å]	Dissociation Energy [kJ/mol]
O_2^+	11	2.5	1.123	16.0	626
O_2	12	2.0	1.207	11.4	498
O_2^-	13	1.5	1.28–1.30	6.2	398
O_2^{2-}	14	1.0	1.49	2.8	126

but the splitting of AO's into MO's is a function of the degree of overlap and, therefore, depends on the internuclear distance and ionic charge.

7.3. Hydrides of Oxygen and Peroxo-Compounds

7.3.1. General

Oxygen forms hydrides HO_n and H_2O_n. The only binary hydrides obtainable pure are H_2O and H_2O_2. Hydrides richer in oxygen, H_2O_n, where n = 3 and 4, are formed in the glow discharge decomposition of H_2O_2, and can be studied in a matrix at low temperatures using IR spectroscopy:

$$H_2O_2 \rightarrow H + HO_2; \quad 2HO_2 \rightarrow H_2O_4; \quad H_2O_2 \rightarrow 2OH; \quad OH + HO_2 \rightarrow H_2O_3$$

Hydrogen peroxide, H_2O_2, is a thermodynamically metastable compound which decomposes exothermically to water and oxygen gas.

Hydrides of the series HO_n with n = 1 and 2 are radicals and have been observed as reaction intermediates (studied in a matrix by IR-, ESR- and UV-spectroscopies). They are formed on ignition of hydrogen-oxygen mixtures (p. 153) and on catalytic decomposition of H_2O_2; HO_2 is also an intermediate in the hydrolysis of superoxides (p. 187).

7.3.2. Hydrogen Peroxide, H_2O_2

Preparation:

Hydrogen peroxide is commercially produced by three processes. In the first, sulfate is electrolytically oxidized to peroxodisulfate, which generates H_2O_2 upon hydrolysis:

$$2SO_4^{2-} \rightarrow S_2O_8^{2-} + 2e^-$$

$$\begin{matrix} O—SO_3^{\ominus} \\ | \\ O—SO_3^{\ominus} \end{matrix} + \begin{matrix} HOH \\ HOH \end{matrix} \rightarrow H_2O_2 + 2HSO_4^-$$

In anodic oxidation of sulfate ions in aqueous H_2SO_4 or $(NH_4)_2SO_4/H_2SO_4$ solutions, the formation of O_2 is suppressed by use of high current densities and platinum anodes having a high overvoltage towards oxygen. Hydrogen is liberated at the cathode. Premature hydrolysis of peroxodisulfate ion in the electrolytic cell, which would lead to catalytic decomposition of H_2O_2 at the electrodes, is suppressed by cooling. After electrolysis, the electrolyte is hydrolyzed. The anion of peroxosulfuric acid, $HOOSO_3^-$, is rapidly formed, which hydrolyzes in a rate-determining step to H_2O_2 and HSO_4^-. The hydrogen peroxide produced is distilled *in vacuo*, and concentrated.

A second process for the production of H_2O_2 consists of autooxidation of a substituted 9,10-anthracenediol with atmospheric oxygen:

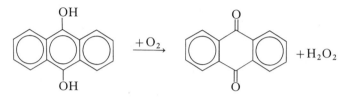

The quinone formed is catalytically reduced by hydrogen gas activated with palladium in a continuous cyclic process occurring in an organic solvent, from which the H_2O_2 produced is removed by water in a counter-current reactor.

In a third process, 2-propanol is partially oxidized to acetone and H_2O_2:

$$
\begin{array}{c}
H_3C \\ \diagdown \\ \qquad C \\ \diagup \quad \diagdown \\ H_3C \qquad OH
\end{array}
\begin{array}{c} H \\ \diagup \end{array}
\; + O_2 \; \rightarrow \;
\begin{array}{c}
H_3C \\ \diagdown \\ \qquad C = O + H_2O_2 \\ \diagup \\ H_3C
\end{array}
$$

without catalyst, either in the liquid phase at $90-140\,°C$ and $15-20$ bar; or in the vapor phase at 350 to $500\,°C$. The hydrogen peroxide produced is separated by distillation.

Properties and Structure of H_2O_2:
Pure hydrogen peroxide, H_2O_2, forms colorless, needle crystals, which melt at $-0.89\,°C$ to a strongly hydrogen-bonded, syrupy liquid, b.p. $150.2\,°C$, which is blue in thick layers. Anhydrous H_2O_2 is stored with cooling in paraffinized or polyethylene bottles, since traces of heavy metal ions, as well as platinum metal, manganese dioxide and bases decompose it to water and oxygen:

$$H_2O_2 \rightarrow H_2O + \tfrac{1}{2}O_2 \qquad \Delta H° = -98 \text{ kJ/mol}$$

$HO\cdot$ and $HO_2\cdot$ radicals have been identified as intermediates.
The H_2O_2 molecule of C_2 symmetry (see Figure 82) is a twisted chain of four atoms with a dihedral angle, φ, of $111.5°$ to minimize the repulsion of the two electron pairs on each oxygen atom and of the two H atoms.

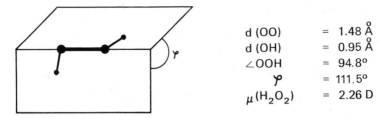

$$
\begin{aligned}
d\,(OO) &= 1.48\ \text{Å} \\
d\,(OH) &= 0.95\ \text{Å} \\
\angle OOH &= 94.8° \\
\varphi &= 111.5° \\
\mu\,(H_2O_2) &= 2.26\ D
\end{aligned}
$$

Fig. 82.   The structure of the H_2O_2 molecule.

This repulsion is, however, still considerable, and the O—O-bond is stretched (bond order <1), accounting for the tendency of H_2O_2 to undergo radical reactions. Rotation about the O—O axis requires ca. 25 kJ/mol.

Reactions of H_2O_2:
The dissociation constant of H_2O_2 in dilute solution is:

$$
H_2O_2 \rightleftharpoons H^+ + HO_2^- \qquad K = \frac{[H^+][HO_2^-]}{[H_2O_2]} = 2 \cdot 10^{-12}\,(20°)
$$

Therefore, H_2O_2 is a very weak acid in water. Alkaline H_2O_2 solutions formed in the hydrolysis of Na_2O_2 or KO_2 (cf. p. 187), contain HO_2^- in addition to OH^- ions. Hydrogen peroxide is an oxidizing agent, both in acid and alkaline solution. The oxygen is reduced to the oxidation state -2:

$$
H_2O_2 + 2e^- \rightarrow 2OH^-
$$

For example, H_2O_2 oxidizes K_2S to sulfur, SO_2 to SO_4^{2-}, NO_2^- to NO_3^-, As_2O_3 to AsO_4^{3-}, Cr(III) salts to chromates, and Fe(II) to Fe(III). Towards stronger oxidizing agents H_2O_2 is a reducing agent:

$$
H_2O_2 \rightarrow O_2 + 2H^+ + 2e^-
$$

This is the case, e.g., with MnO_4^-, Cl_2, Ce(IV) salts and O_3. Permanganate is reduced in acid solution to Mn^{2+}, which is used in the volumetric determination of H_2O_2. Tracer studies of ^{18}O-labelled hydrogen peroxide show that the oxygen generated originates from H_2O_2, so the O—O bond remains intact during oxidation, otherwise isotopic exchange with the water solvent would take place. Similarly, no isotopic exchange occurs in the catalytic decomposition of H_2O_2, which likely proceeds:

$$
\begin{aligned}
H_2O_2 &\rightarrow 2OH \\
OH + H_2O_2 &\rightarrow H_2O + HO_2 \\
HO_2 + OH &\rightarrow H_2O + O_2 \\
HO_2 + HO_2 &\rightarrow H_2O_2 + O_2
\end{aligned}
$$

with oxygen originating from the H_2O_2. The heat liberated accelerates the reaction which may lead to explosive decomposition in concentrated solutions. Phosphate ions or organic acids are added to inhibit the decomposition.

7.4. Fluorides of Oxygen

7.4.1. General

Oxygen forms fluorides of the formula O_nF and O_nF_2, like the hydrides; however, the oxygen atoms are positively polarized and the oxidation state of oxygen is >0, so that the compounds are not fluorine oxides (the mixed compound HOF is known).

In the series O_nF, only $n = 1$ and 2 are known, but both are unstable radicals. The fluorides O_nF_2 with $n = 1,2$ and 4 have been synthesized, and the higher homologs with $n = 5$ and 6 have been postulated, but O_3F_2 proved to be a mixture of O_2F_2 and O_4F_2. With the exception of OF_2, all oxygen fluorides are thermodynamically unstable.

7.4.2. Oxygen difluoride, OF_2

Oxygen difluoride is prepared by bubbling fluorine gas through 2% (0.5 N) NaOH:

$$2F_2 + 2NaOH \rightarrow OF_2 + H_2O + 2\overset{..}{N}aF$$

The maximum yield is 80%, since a portion is lost by hydrolysis:

$$OF_2 + 2OH^- \rightarrow O_2 + 2F^- + H_2O$$

and similar yields are realized in the reaction of fluorine gas with moistened potassium or cesium fluoride. Oxygen difluoride is also prepared by the electrolysis of aqueous KHF_2 solution. All reactions proceed through the intermediate formation of unstable HOF, which has been identified mass spectrometrically in the reaction of elemental fluorine with ice water, and isolated in milligram amounts as colorless crystals (m.p. $-117°C$) containing zero oxidation state oxygen. The compound is hydroxyl fluoride, but the name hypofluoric acid is used in analogy to HOCl. It reacts with fluorine gas to give OF_2 and HF; with water to give H_2O_2 and HF, and with aqueous iodide to give I_3^-, OH^-, and F^-.

Oxygen difluoride is a pale yellow, poisonous gas (b.p. $-145°C$) with $\Delta H° = -17$ kJ/mol which reacts with aqueous hydrogen halides:

$$OF_2 + 4HX \rightarrow H_2O + 2HF + 2X_2 \qquad X: Cl, Br, I$$

Many non-metals are oxidized or fluorinated in OF_2. Steam reacts explosively to form oxygen and HF, as does hydrogen sulfide.

Pure OF_2 is not explosive, but decomposes upon heating to 200–250°C or upon irradiation into oxygen and fluorine. The OF radical is produced by UV photolysis of OF_2 in a matrix, and can be identified by IR spectroscopy:

$$OF_2 \xrightarrow{h \cdot v} OF + F$$

or in the reaction of fluorine atoms with N_2O, so photolysis of OF_2 and N_2O in a matrix is best:

$$F + N_2O \ \rightarrow \ OF + N_2$$

The OF radicals recombine upon warming to form O_2F_2.

7.4.3. Dioxygen difluoride, O_2F_2

Dioxygen difluoride is prepared from oxygen and fluorine gases in a high voltage glow discharge at $77 - 90K$ and a pressure of 10–20 torr:

$$O_2 + F_2 \ \rightarrow \ O_2F_2$$

or by the radiolysis of liquid O_2/F_2 mixtures by x-rays; or by the UV photolysis of O_3/F_2 mixtures at 120–195 K:

$$F_2 \ \xrightarrow{\ h\cdot\nu\ } \ 2F$$
$$F + O_3 \ \rightarrow \ O_2 + OF$$
$$2\,OF \ \rightarrow \ O_2F_2$$
$$OF + F_2 \ \rightarrow \ OF_2 + F$$

Yellow-colored O_2F_2 (m.p. $-154\,°C$) decomposes at $-50\,°C$ exothermically with O_2F as an intermediate. Its structure from microwave studies corresponds to H_2O_2 (cf. Figure 82) with symmetry C_2, $\varphi = 87.5°$ and $\angle OOF = 109.5°$.

Oxygen difluoride is a strong fluorinating and oxidizing agent, converting chlorine to ClF and ClF_3 and hydrogen sulfide to SF_6. With elemental sulfur it explodes even at $-180\,°C$. Fluoride ion acceptors form dioxygenyl salts (cf. p. 188).

7.5. Comparison of Bond Relationships in Hydrides and Fluorides of Oxygen

Table 36 compares internuclear distanses and valence force constants of the oxygen fluorides and hydrides. All OH bonds are normal single bonds, but the O—O bonds are weaker than in O_2.

However, taking the OF bond in the OF radical as a normal single bond, (the internuclear distance is approximately the sum of the covalent radii), the OF bond is markedly weaker in all the other fluorides listed in Table 36, especially in O_2F and O_2F_2, where the OF bond order is ca. 0.3 but the O—O bonds are the same as in O_2, with a bond order of 2.

Table 36   Internuclear Distances, d(in Å), and Valence Force Constants, f (in mdyn/Å), in Oxygen Hydrides and Fluorides.

Com-pound	d(OH)	d(OO)	f(OH)	f(OO)	Com-pound	d(OF)	d(OO)	f(OF)	f(OO)
HO	0.97	—	7.1	—	OF	1.32	—	5.42	—
H_2O	0.96	—	7.7	—	OF_2	1.41	—	3.95	—
HOF	0.96	—	6.9	—	HOF	1.44	—	4.0	—
HO_2	—	—	6.5	5.9	O_2F	—	—	1.43	10.5
H_2O_2	0.95	1.48	7.4	4.6	O_2F_2	1.58	1.22	1.50	10.3
O_2	—	1.21	—	11.4	O_2	—	1.21	—	11.4

Resonance structures from VB theory help rationalize these findings:

and similar structures can be drawn for the angular O_2F molecule which reflect the high electronegativity of the fluorine. Analogous structures play no role in the hydrides. The weak fluorine bond in O_2F_2 is responsible for its decomposition (the dissociation into O_2F and F requires only 71 kJ/mol) and reactivity even at low temperatures.

The MO-treatment of O_2F and O_2F_2 uses multi-center orbitals formed by each fluorine atom from a 2p-orbital, and by each O_2-core structure by one or both π^*-orbitals.

8. Sulfur, Selenium, Tellurium, Polonium

8.1. General

The elements sulfur, selenium, and tellurium are more similar in their chemical properties to each other than to their lighter homologue, oxygen. The covalent radii change more between the first and the second rows of the periodic sytem than between succeeding rows, explaining the abrupt change of electronegativity from oxygen to sulfur:

	O	S	Se	Te	Po
Covalent radius [Å]	0.66	1.03	1.17	1.37	
x_E (Allred-Rochow)	3.5	2.4	2.5	2.0	1.8

Sulfur and selenium are so similar that sulfur chemistry may be considered representative of that of selenium[7].

The paramount difference between the chemistry of oxygen and the remaining chalcogens is the inability of the oxygen valence shell to accept more than 8 electrons. Chalcogen atoms have the valence electron configuration s^2p^4 in the ground state:

and can thus form doubly negative ions, E^{2-}, or make two covalent bonds (type R_2E) in order to achieve the rare gas configuration. For oxygen, additional bonds are formed in the oxonium ion, R_3O^+, which maintains the oxygen rare gas configuration. Sulfur, selenium, and tellurium, however, form many compounds with coordination number 2–6, and the number of valence electrons can be > 8, which involve d-orbitals in the covalent bonds to the substituents (see Table 38). Compounds with coordination number 7 and 8 are known for tellurium.

The central atoms are generally bonded to elements in the first octet, and predominantly with more electronegative elements, in compounds with > 8 valence electrons. The unknown compounds H_6S, SeI_6 and $Te(C_6H_5)_6$ are probably thermodynami-

---

7 This is also true for silicon and germanium; phosphorus and arsenic; and chlorine and bromine.

Table 37 Properties of the Chalcogens.

Element	Valence Electron Configuration	Color	m.p. [°C]	b.p. [°C]	Natural Abundance of Isotopes
S	$3s^2p^4$	yellow	114.5	444.6	^{32}S: 95.0 ^{33}S: 0.76 ^{34}S: 4.22 ^{36}S: 0.014
Se	$4s^2p^4$	gray to black	221	685	^{74}Se: 0.87 ^{76}Se: 9.02 ^{77}Se: 7.58 ^{78}Se: 23.52 ^{80}Se: 49.92 ^{82}Se: 9.19
Te	$5s^2p^4$	metallic lustre brass-colored	452	1087	^{120}Te: 0.01 ^{122}Te: 2.48 ^{123}Te: 0.87 ^{124}Te: 4.61 ^{125}Te: 6.99 ^{126}Te: 18.71 ^{128}Te: 31.79 ^{130}Te: 34.48
Po	$6s^2p^4$		254	962	All isotopes are radioactive

Table 38   Chalcogen Compounds with d-Orbital Participation.

Coordination Number	Valence Electrons at Central Atom	Compound	Approximate Geometry at Center Atom
2	10	SO_2, NSF, $S(NR)_2$	angular
3	10	SOF_2, $SOCl_2$, $SeOF_2$ SO_3^{2-}, $(SeO_2)_n$	trigonal-pyramidal
	12	SO_3, $S(NR)_3$	trigonal-planar
4	10	$(CH_3)_3SO^+$,	tetrahedral
		SF_4, $S(OR)_4$	pseudotrig.-bipyramidal
	12	$(SO_3)_n$, SO_4^{2-}, NSF_3, SeO_2Cl_2	tetrahedral
5	12	SOF_4	trig.-bipyramidal
		SF_5^-	pseudo-octahedral
6	12	SF_6, SeF_6, TeF_6 RSF_5, $Te(OH)_6$	octahedral

cally unstable, as are the SOF_4 and SF_4 analogues $SOBr_4$ and SBr_4, owing to the inability of the d-orbitals to engage in covalent bonds to less electronegative atoms or groups.

8.2. Theory of d-Orbital Participation[8]

Sulfur hexafluoride, which is generated by burning sulfur in fluorine, (see p. 92) has been investigated theoretically for d-orbital participation. The synthesis can be separated into a number of hypothetical steps. The reaction will proceed if the formation of six S—F bonds from sulfur and fluorine atoms in their valence states furnishes more energy than needed to produce these elements from their standard states. Among the endothermic steps, the promotion of the sulfur valence electrons is the most important:

8 D.P.Craig and C.Zauli, J.Chem. Phys., 37, 601, 609 (1962); D.P.Craig and T.Thiruna-machandran, J. Chem. Phys., 45, 3355 (1966); C.A.Coulson and F.A.Gianturco, J. Chem. Soc., (London) (A), 1618 (1968); R.G.A.R.Maclagan, J. Chem. Soc., (London) (A), 2992 (1970), and 222 (1971).

$$S(^3P: s^2p^4) \xrightarrow{+2360\ \text{kJ/mol}} S(^7F: sp^3d^2)$$

The orbital energy is the negative ionization energy of an electron in that orbital (cf. p. 23). Figure 83 shows the relative energies of the levels in sulfur, where $3d > 4p > 4s$. The promotion of an electron to the 4s or 4p level requires 628 and 762 kJ/mol, respectively, vs. 809 kJ/mol for the 3d level. On this basis sulfur would first utilize the 4s- and 4p-orbitals for bond formation.

Further complications arise from the comparison of the sizes of the 3d- and 3s- and 3p-orbitals vs. the S—F internuclear distance. The mean orbital radius, $\bar{r}$, is defined as:

$$\bar{r} = \frac{\int_0^\infty r^3 R^2 dr}{\int_0^\infty r^2 R^2 dr}$$

where R is the radial portion of the wave function describing the orbital. $\bar{r}$ should be approximately the same for all orbitals to be hybridized and should not be larger than the experimental internuclear distance.

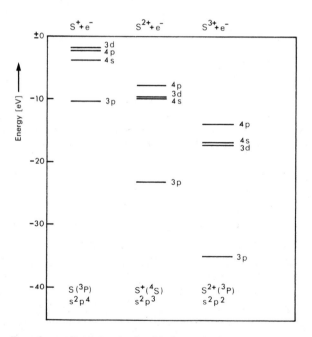

Fig. 83. Experimentally determined orbital energies of the 3p, 3d, 4s, and 4p levels of S, S^+, and S^{2+}. All energies are referred to the ionization limit.

Table 39   Mean Orbital Radii, $\bar{r}$, for the Sulfur Atom in Various States, and Internuclear Distances, d(SF), in the respective Sulfur Fluorides (in Å).

State	Configuration	$\bar{r}(3s)$	$\bar{r}(3p)$	$\bar{r}(3d)$	d(SF)
S(^{3}P)	s^2p^4	0.91	1.09	—	SF_2: 1.59
S(^{5}D)	s^2p^3d	0.88	1.01	4.00	SF_4: 1.55 und 1.65
S(^{7}F)	sp^3d^2	0.86	0.98	1.88	SF_6: 1.59

The results of Hartree-Fock SCF calculations are shown in Table 39 for the sulfur atom in various states, as well as internuclear distances, d(SF), in the fluorides SF_2, SF_4, and SF_6, derived from these states. These values indicate that the 3d-orbitals are larger and more diffuse than the 3s- and 3p-orbitals, or even than the internuclear distances in corresponding compounds. This is especially true if only one d-orbital is occupied. For covalent σ-bonds, the radii should be 0.7–1.8 Å. Similar results are found for phosphorus and chlorine where the d-orbitals are also employed in the σ-bonds of the fluorides PF_5 and ClF_3.

The radii listed in Table 39 are valid for calculations in which the energy of an atom is considered a composite of Coulombic attraction between the nucleus and electrons, kinetic energy of the electrons, repulsion among the electrons, and the exchange interaction of unpaired electrons of like spin, which results in stabilization, i.e., in a contraction of the orbitals. Since the ^{5}D and ^{7}F states of the sulfur atom have a larger number of unpaired electrons of like spin available, the exchange stabilization in these atoms is especially large. This can be verified by comparing the values in Table 39 with calculations in which the exchange interaction is neglected, and larger radii result, for example, for $sp^3 d^2$:

$$\bar{r}(3s) = 0.88 \text{ Å}, \bar{r}(3p) = 1.02 \text{ Å, and } \bar{r}(3d) = 3.00 \text{ Å}.$$

The unpaired electrons are randomly distributed among both spin states in the valence state of an atom, with the number of like spin ca. half in the ^{5}D and ca. half in the ^{7}F configuration, and the exchange interaction becomes weaker. Therefore, the orbital radii listed in Table 39 are lower limiting values for atoms in the valence state, and the d-orbitals are actually somewhat larger.

The Theory of d-Orbital Contraction:
The orbitals of an atom in a molecule are different in energy and size from those in an isolated atom. Calculations for the molecules SF_6, PF_5, PHF_4, PH_2F_3 and ClF_3 lead to the following conclusions:
a) All orbitals contract if the central atom carries a positive charge, the 3d-orbitals most. The energy levels of the ions S^+ and S^{2+} are contrasted with those of the sulfur atom in Figure 83, where the 3d level in S^+ lies below the 4p level, and even below the 4s level in S^{2+}. The sulfur atom carries a high positive charge in SF_6, which stabilizes and contracts the occupied d-orbitals, their mean radii becoming more like those of the 3s- and 3p-orbitals. Thus strongly electronegative atoms like fluorine, oxygen, nitrogen and chlorine favor d-orbital participation.

b) In SF_6, PF_5 or ClF_3 the valence electrons of the central atom are also affected by the field of the attached atoms which changes the geometry and energy of the orbitals even if no bond formation takes place.

If the interaction is purely electrostatic (no bond formation), and all atoms are in the valence state, then the 3d-orbitals are strongly contracted, whereas the 3s- and 3p-orbitals remain nearly unchanged. The contraction is stronger with larger d-orbitals.

These relationships can be illustrated by locating the valence electrons of an $S(^7F)$ atom at various mean distances from the nucleus, as listed in Table 39, exposed to the effective nuclear charges calculated using Slater's rules:

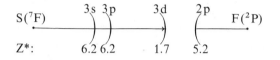

The orbitals are sketched as arcs according to their $\bar{r}$ values. The 2p-electrons of the attached fluorine atoms have an effective nuclear charge of 5.2. At 1.59 Å, the $S(^7F)$ and $F(^2P)$ orbitals interpenetrate, and a mean effective nuclear charge between the two atoms results. For the 3s- and 3p-electrons this charge is similar to that in the isolated sulfur atom. However, the 3d-electrons experience a stronger electric field. The increase of Z^* in the ψ-function for the 3d-electrons is reflected in orbital contraction (cf. p. 45) and the mean radius of the d-orbitals in the fluorides of chlorine, phosphorus and sulfur decreases to half. This contraction is an adaptation of the central atom to the field of the substituents, and favors orbital overlap and the formation of covalent bonds. In PF_5, which is trigonal-bipyramidal (a phosphorus $sp^3d_{z^2}$-hybrid), the d_{z^2} is considerably more contracted than the 4s-orbital, which makes the alternative configuration $3sp^34s$ less favorable, and the d_{z^2} orbital more suited for overlap with the 2p-orbitals of the fluorine atoms.

Orbital contraction stabilizes the central atom, but destabilizes the valence orbitals of fluorine. Since there will be several fluorine atoms over which this destabilization is distributed, the effect on each is small.

c) The contraction of orbitals is dermined by the number and kind of the substituents exerting the field. The contracting power or field strength is fluorine $>$ chlorine $\approx$ carbon $\gg$ hydrogen.

The atoms exerting the contracting field may be bound with other atoms, which do not contribute, for example, in the series PF_5, PHF_4, PH_2F_3, PH_3F_2. The hydride, PH_5, does not exist, but the other fluoro-derivatives are known. Calculations of $\bar{r}(3d)$ in PF_5, PHF_4, and PH_2F_3 suggest that the participating d_{z^2} orbital has the same size, since the fluorine atoms occupy the axial-positions (on the z-axis) of the trigonal bipyramids, and interact strongly with the d_{z^2}-orbital. Likewise, $SClF_5$, $SBrF_5$, and $S(C_6H_5)F_5$ are known, whereas SCl_6, SBr_6, and $S(C_6H_5)_6$ are not.

The contraction depends on the particular d-orbital since on geometric grounds, the interaction with different substituents should be of different strength. Thus, in PF_5 and ClF_3, the d_{z^2} orbital is much contracted, while the other d-orbitals remain unchanged. In octahedral molecules, the $d_{x^2-y^2}$ and the d_{z^2} orbitals are most affected, since the lobes of the d_{xy}, d_{xz}, and d_{yz} lie between the substituents.

d) The energy to promote two electrons to the 3d level remains enermous, even if the d-orbital shapes are strongly modified by substituent fields, or there is a positive charge on the central atom, and increases within a period toward the right, and within a group toward the top. Only substituents which make strong bonds with the central atom can compensate for the promotion energy.

Fluorine can bring about the promotion of two or more electrons to a d-level, but with the heavier central atoms with low promotion energies, oxygen is also suitable (as in SF_6, ClF_5, BrF_5, IF_5, XeF_6, $Te(OH)_6$, etc.).

Fluorine, oxygen, nitrogen, carbon and chlorine allow promotion of single electrons to a d-level (as in the sulfuranes, phosphoranes, PCl_5, $P(C_6H_5)_5$, ICl_3, $XeCl_2$, etc.), but not for chlorine as central atom, since its promotion energy is too large [only fluorine (as in ClF_3) is effective]. Argon forms no compounds for the same reason.

These considerations do not apply if the d-orbitals are employed to form coordinate bonds involving no valence electron promotion (as in $AsCl_4^-$, $SiCl_4 \cdot 2py$, etc.).

In transition metals, the d-orbitals of the penultimate shell are employed in σ- and π-bonding, for example, in iron where stronger nuclear fields than in the excited main group atoms are present, and no further orbital contraction is required. This is easily verified by calculating Z^* values by the Slater method (see p. 46). The d-orbitals of the transition metals can be easily hybridized with the s- and p-orbitals of the outermost occupied shell, as the mean radii are similar.

π-Bonds with d-Orbital Participation:

d-Orbitals are well suited to form π-bonds, since even for large d-orbitals positive overlap with, e.g., a p-π-orbital results (cf. Figure 32). Hence, d-orbital contraction is not required for p $\rightarrow$ d-π-bonding, as it is necessary for σ-bonding. Calculations suggest that substituents of high field strength bring about only negligible contraction of the d-π-orbitals of the central atom, since on geometric grounds, the electrons in the d-π-orbitals are less influenced. Some stabilization does occur, however, owing to exchange interaction.

The atoms of the second row thus employ d-orbitals for π-bonding even for small numbers of substituents (as in SO_2 or NSF) or for weak substituent fields (as in P_4S_{10} or $PSCl_3$).

Two delocalized π-bonds are formed by the sulfur d_{z^2} and $d_{x^2-y^2}$ orbitals in the tetrahedral sulfate ion, which overlap with the p-π-orbitals of the oxygen atoms (Figure 84) to form two 5-center-8-electron-π-bonds, each of wich contributes one to the π-bond order of the S—O bonds, which amounts to 0.5 per bond (total bond order 1.5).

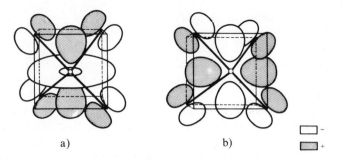

a)   b)

Fig. 84.   Overlap relationships in both 5-center-π-bonds of the SO_4^{2-} ion and in the iso-electronic species ClO_4^-, PO_4^{3-}, SiO_4^{4-}, SiF_4, SO_2F_2 and SO_3F^-.
a) Overlap of four p-π-orbitlas of the substituent atoms with the d_{z^2}-orbital of the central atom.
b) Overlap with the $d_{x^2-y^2}$-orbital.

Evidence for the participation of the outer d-orbitals in σ- and π-bonds comes from molecular geometry, bond strength, and comparisons with calculated molecular dipole moments and ionization energies.

8.3.   Trends in the VIth Main Group

The VIth main group descends from pure non-metals (O, S) via elements with semiconductor properties (Se, Te), to a pure metal (Po), which has a specific resistance of $43 \cdot 10^{-3}$ $\Omega \cdot$ cm, and unlike other chalcogens, a positive temperature coefficient of the electrical resistance, characteristic of metals. The lighter chalcogens are nonconductors, but specific resistances decrease markedly from sulfur to tellurium. Their coordination number increases from one for oxygen in O_2, and two for sulfur in S_8 to six for polonium, and the physical properties of the solids are greatly changed (see p. 204).

The non-metallic chalcogens form homonuclear chains and, with exception of oxygen, rings. The bond energies of the homonuclear single bonds have been discussed above (see p. 131). Owing to the high bond energy of sulfur-sulfur single bonds this element shows the highest tendency of all chalcogens to form chains and rings and is only surpassed by carbon. As a consequence, sulfur has the largest number of allotropes of all elements.

O=C=O	S=C=S	Se=C=Se	Se=C=Te	Te=C=Te
colorless gas	colorless liquid	yellow liquid, polymerizes easily	red-brown, unstable	unknown

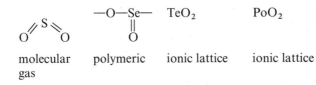

| molecular gas | polymeric | ionic lattice | ionic lattice |

Double bonds are more easily formed the smaller the electronegativity difference of the participating atoms, and the greater the sum of their electronegativities. The stability of chalcogen compounds of higher oxidation state decreases from sulfur to tellurium, and TeO_3 and SeO_3 are stronger oxidizing agents than SO_3, and Se(IV) (for example, $HSeO_3^-$) is reduced by S(IV) (e.g., $HS_2O_5^-$) to Se(0).

8.4. Preparation of the Elements

8.4.1. Sulfur

Deposits of sulfur occur in USA, Mexico, South America, Japan, Poland, USSR and Sicily and are mined by the Frasch process, whereby sulfur is melted (m.p. 114.6 °C) underground with superheated water, extracted from sand, limestone and organic matter, and brought to the surface as 99.99% sulfur with compressed air. It is solidified, or transported as the liquid. About half the world's production of sulfur chemicals originates from native sulfur, the other half from H_2S-containing gases (natural gas, industrial waste gases), and small amounts from pyrite (FeS_2), which is roasted to yield Fe_2O_3 and SO_2.

Recovery of sulfur from H_2S-containing gases is carried out by scrubbing with alkaline solutions, followed by heating the solution to expel H_2S and oxidation on active carbon or Al_2O_3 with air to sulfur in the Claus process:

$$H_2S + \tfrac{3}{2}O_2 \rightarrow SO_2 + H_2O$$
$$SO_2 + 2\,H_2S \rightarrow 3\,S_{(s.)} + 2\,H_2O$$

$$\overline{3\,H_2S + \tfrac{3}{2}O_2 \rightarrow 3\,S_{(s.)} + 3\,H_2O \qquad \Delta H^\circ = -666\ kJ/mol}$$

Pure sulfur is obtained from the catalyst by extraction with CS_2, chlorobenzene, or sulfide solutions or by heating to melt the S_8.

Methods to desulfurize SO_2- or mercaptan-containing industrial waste gases are important in ecology. About 85% of all sulfur production (sulfur, H_2S, and SO_2) goes into H_2SO_4 manufacture.

8.4.2. Selenium and Tellurium

Selenium and tellurium are less widely distributed than sulfur. Traces of selenides and tellurides are present in sulfide ores, which when roasted to oxides, concentrate SeO_2 and TeO_2 in the fly ash and can be isolated. Tellurium is obtained from the anode sludge in electrolytic copper refining, where it is present as Cu_2Te, Ag_2Te, and $(Ag, Au)_2Te$.

The dioxides are easily reduced to the elements in aqueous solution by SO_2:

$$H_2SeO_3 + 2SO_2 + H_2O \rightarrow Se\downarrow + 2H_2SO_4$$

8.4.3. Polonium

All known isotopes of polonium are radioactive with the most stable, ^{210}Po, decaying with a half-life of 138.4 days to ^{206}Pb with α-emission. Polonium-210 is obtained in gram amounts from ^{209}Bi, which is transmuted in a (n, γ)-process to ^{210}Bi, which decays to ^{210}Po with β-emission.

8.5. Modifications

8.5.1. Sulfur

Phase equilibria:
Recovery of native sulfur furnishes S_8-molecules, which are thermodynamically more stable than chains or other ring sizes. The phase diagram of sulfur is shown in Figure 85. Cyclo-octasulfur forms orthorhombic crystals (α-sulfur) with 16 S_8 molecules per unit cell. Figures 86 and 87 show the geometry of the molecules and their arrangement in the crystal. In S_8 the sulfur atoms lie in two parallel planes, giving the ring a crown shape (symmetry D_{4d}). The mean internuclear distance is 2.060 Å (corrected for librations), $<SSS = 108°$, and the dihedral angle 98°. The brittle crystals are light yellow, and dissolve in CS_2, $CHBr_3$ and $1,4-C_6H_4Cl_2$.

At 95.5 °C (0.004 torr vapor pressure, triple point), the thermodynamically stable α-modification changes into monoclinic β-sulfur ($\Delta H° = 32$ kJ/mol S_8) which is likewise made up of S_8 molecules, and can be sublimed *in vacuo* at 100 °C. β-Sulfur melts at 119.3 °C, but this ideal melting point does not correspond to the thermodynamic solid/melt equilibrium which occurs after some time at 114.6 °C, the natural melting point of sulfur.

This melting point depression arises from the slow attainment of equilibrium among smaller and larger rings, formed from S_8 at the melting point. These components, whose size and type (S_n, $n \neq 8$) are unknown (n is probably $\geqslant 6$), produce a melting point depression of about 5°, corresponding to ca. 5.5 mole-% S_n. The average n in molten sulfur is ca. 8, thus there must be molecules with $n > 8$ as well as with $n < 8$. The thermal lability, and their easy interconversion has precluded

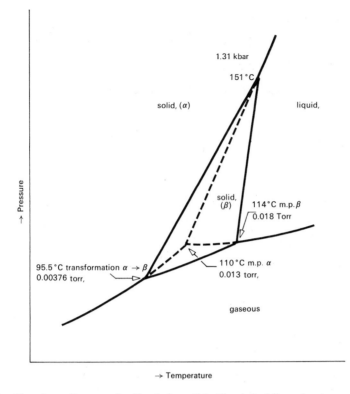

solid, (α)

liquid,

1.31 kbar

151°C

solid,
(β)   114°C m.p. β
0.018 Torr

→ Pressure

95.5°C transformation $\alpha \to \beta$
0.00376 torr,

110°C m.p. α
0.013 torr,

gaseous

→ Temperature

Fig. 85.   The phase diagram of sulfur (schematic). The dotted lines denote equilibria involving metastable solid phases.

separation; however, independent synthesis of ring molecules S_6, S_7, S_9, S_{10}, S_{12}, S_{18} and S_{20} has been accomplished. The equilibrium:

$$x S_8 \rightleftharpoons y S_n \quad (n \neq 8)$$

is strongly temperature-dependent.

Liquid sulfur near the melting point has a light yellow color and low viscosity, but at 159°C the viscosity suddenly increases sharply, and free radicals are present in the reddish-brown melt above 170°C. At higher temperatures, the viscosity slowly decreases again, and at the boiling point (444°C) the melt is of dark red-brown color and of low viscosity.

These phenomena reflect the composition of the melt. The viscosity and other physical properties show a discontinuity at 159° where high-molecular weight ring and chain molecules are spontaneously generated. Their concentration increases with temperature in the region 159 to 200°C, the chain length grows to 10^5

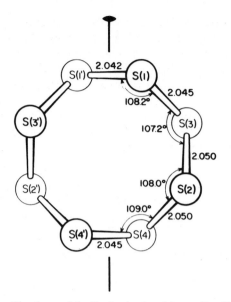

Fig. 86.   The shape of the S_8-ring in rhombic α-sulfur. The twofold axis of rotation changes atom S(n) into S(n') and vice versa (values not corrected for librations).

Fig. 87.   The packing of S_8 rings of rhombic α-sulfur (from B. Meyer, Ed., Elemental Sulfur, Interscience, New York, 1965).

atoms, and the contribution of polymeric sulfur to the total sulfur content at 200 °C is ca. 25%, while smaller ring molecules, S_n, with $n \neq 8$, account for only 15%. Increased temperatures thermally crack the polymers, and molecular size decreases as does viscosity; however, the concentration of biradical short chains, $\cdot S_8 \cdot$, $\cdot S_7 \cdot$, etc., increases, darkening the color. Among the fragments, S_3 (green) and S_4 (red) have been proposed.

The composition of the vapor is as complex as the melt. Mass spectrometric evidence for molecules, S_n, with $n = 2$–10 in pressure- and temperature-dependent equilibria (see Figure 88) has been provided. Sulfur atoms predominate only at >2500 °C, and $<10^{-5}$ torr.

The gaseous molecules, S_n, with $n > 4$ are probably rings, but the red S_4 molecules are probably biradical chains. The green S_3 molecule (thio-ozone) is angular like O_3, SO_2 and S_2O. The blue-violet S_2 has an electron configuration analogous to O_2, and is paramagnetic with a double bond. It is the main component of sulfur vapor at >600 °C and 10^{-3} torr and can be isolated in a matrix.

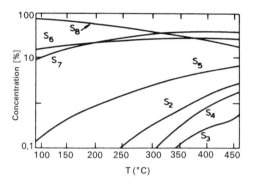

Fig. 88.   Composition of saturated sulfur vapor above the melt.

Monotropic Sulfur Modifications:
The α- and β-forms of S_8 are mutually interconvertible enantiotropes. All other modifications are only accessible via melts, solutions or by chemical reactions.
The stability of the synthetic ring molecules S_n ($n = 6$–7 and 9–12) can be used to assess their potential as components in gaseous and liquid sulfur.
Acidification of aqueous thiosulfate solutions produces, in addition to S_8 and polymeric sulfur, S_6, which canbe extracted from the reaction mixture.

$$Na_2S_2O_3 + 2\,HCl \rightarrow \frac{1}{n} S_n + SO_2 + 2\,NaCl + H_2O$$

The formation of the egg-yellow, light-sensitive, rhombohedral crystals of S_6 goes through the following equilibria among mono- and polysulfonic acids:

$$2\,HS_2O_3^- \rightleftharpoons HS_3O_3^- + HSO_3^-$$

$$HS_3O_3^- + HS_2O_3^- \rightleftharpoons HS_4O_3^- + HSO_3^-$$

etc., and finally:

$$HS_7O_3^- \rightleftharpoons S_6 + HSO_3^-$$

The condensation of polysulfanes with chlorosulfanes in an organic solvent using the dilution principle is of more general utility:

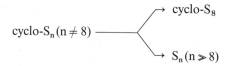

and S_6, S_{12} (m.p. 148 °C), S_{18}, and S_{20} have been synthesized.
Cyclic sulfur molecules can be made from a preformed sulfur ring, at low temperatures, e.g., the red-violet π-complex $(C_5H_5)_2TiS_5$, which contains the sixmembered heterocycle, TiS_5, easily made from $(C_5H_5)_2TiCl_2$ and a polysulfide ($Cp = C_5H_5$, cyclopentadienyl):

$$Cp_2TiS_5 + S_nCl_2 \rightarrow Cp_2TiCl_2 + S_{5+n} \ (n:1, 2, 4)$$

$$2\,Cp_2TiS_5 + 2\,SO_2Cl_2 \rightarrow 2\,Cp_2TiCl_2 + S_{10} + 2\,SO_2$$

The puckered S_6, and S_{12} rings have internuclear distances between neighboring sulfur atoms as in S_8.
Chemical reactivity is directly related to the thermal stability:

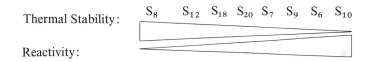

After S_8, S_{12} is most stable, converting to S_8 only at its melting point. All other rings decompose at lower temperatures, often photolytically, yielding S_8 and polymeric, insoluble sulfur which is also formed as a yellow residue by quenching hot sulfur melts and extracting the smaller rings with CS_2:

$$\text{cyclo-}S_n\,(n \neq 8) \longrightarrow \begin{cases} \text{cyclo-}S_8 \\ S_n\,(n \gg 8) \end{cases}$$

The varying stability of the smaller S_n-rings may arise through deviations from ideal valence and dihedral angles; S_n-rings with $n \neq 8$ are less stable than S_8, with the conversion to S_8 arising from the small dissociation energy of the first S—S bond in such rings (cf. p. 132).

Molecules with homocyclic, non-metallic rings are widely known, (as for B, C, Si, Ge, N, P, As, and Te) where $\geqslant 3$, but mostly 4–6, or 8 like atoms are cyclized.

8.5.2. Selenium, Tellurium, Polonium

The thermodynamically stable modification of these elements are polymeric and crystalline. The structures of the gray, hexagonal selenium (m.p. 221 °C) and of metallic tellurium (m.p. 452 °C) both consist of spiral zig-zag chains in which the atoms are in deformed octahedra (Figure 89). There are four atoms in adjacent chains at smaller than van der Waals distances (Table 40). Selenium and tellurium form a continuous series of solid solutions.

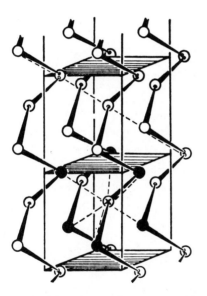

Fig. 89. Lattice of hexagonal selenium or tellurium. The deformed octahedral coordination is indicated for one atom by dotted lines to the six neighboring atoms (after H. Krebs, Grundzüge der anorganischen Kristallchemie, F. Enke Verlag, Stuttgart, 1968).

The coordination in metallic polonium which crystallizes in a NaCl lattice, with all cation and anion positions filled by polonium atoms, is octahedral. Electrical conductivity increases with increasing coordination number, since more orbitals overlap, leading finally to the band structure characteristic of metals. Thus con-

Table 40   Internuclear Distances of Nearest Neighbors (d) and Next-nearest Neighbors (d′) in the Thermodynamically Stable Modifications of Selenium, Tellurium and Polonium.

	d[Å]	d′[Å]	d′/d
Se(hexagonal)	2.374	3.426	1.44
Te(hexagonal)	2.835	3.495	1.23
Po(cubic)	3.359	3.359	1.00

ductivity in selenium and tellurium increases at high pressure, because the ratio d′/d decreases and the mean coordination number increases.

Several high pressure modifications are known for tellurium. Quenching of molten selenium yields a black glassy material. The precipitation of selenium from $HSeO_3^-$ solutions by reduction with SO_2 yields a red, amorphous material soluble in CS_2 from which red crystals of α-Se and darker prisms of β-Se, both made up of Se_8 molecules, result. Heating to 130 °C converts both into gray selenium. Mixed S—Se rings have also been prepared, as in S_7Se and S_6Se_2.

8.6. Positive Chalcogen Ions

These cations, which are closely related chemically and structurally to elemental sulfur, selenium and tellurium:

S_4^{2+}	S_8^{2+}	S_{16}^{2+}	Se_4^{2+}	Se_8^{2+}	Te_4^{2+}
colorless	blue	red	yellow	green	red

are generated during the low temperature oxidation of the respective chalcogens with SO_3, AsF_5, or $S_2O_6F_2$ in a weakly nucleophilic (e.g., HF, SO_2, HSO_3F, H_2SO_4) solvent. The ratio chalcogen/oxidizing agent determines the product:

$$S_8 + 2S_2O_6F_2 \rightarrow 2S_4^{2+} + 4SO_3F^-$$

$$S_8 + S_2O_6F_2 \rightarrow S_8^{2+} + 2SO_3F^-$$

$$2S_8 + S_2O_6F_2 \rightarrow S_{16}^{2+} + 2SO_3F^-$$

The ions Se_8^{2+} and Te_4^{2+} are also formed when the elements are boiled in concentrated sulfuric acid which is utilized in their qualitative identification. With the addition of peroxodisulfate or SO_3, no heating is necessary. The chalcogens dissolve in fuming sulfuric acid with characteristic colors (sulfur: blue, selenium: green, tellurium: red). The positive ions are reduced with $N_2H_6SO_4$:

$$Se(s.) \xrightarrow{S_2O_8^{2-}} Se_8^{2+} \underset{N_2H_4}{\overset{S_2O_8^{2-}}{\rightleftharpoons}} 2Se_4^{2+} \xrightarrow[N_2H_4]{S_2O_8^{2-}} SeO_2$$

gray   green   yellow   colorless

The ions hydrolyze with disproportionation into the chalcogen and the corresponding dioxide.

Several crystalline polychalogen salts have been prepared with large anions, (e.g., AsF_6^-, SO_3F^-, $HS_2O_7^-$, $Sb_2F_{11}^-$ and $AlCl_4^-$). The solids "S_2O_3" (blue), "$SeSO_3$" (yellow) and "$TeSO_3$" (red) prepared from the corresponding chalcogen in liquid SO_3, are likely also such salts with polysulfate anions (e.g., $TeSO_3 = Te_4S_4O_{13}$). Spectroscopic, magnetic, and chemical properties indicate that the polychalcogen cations are cyclic, as in $Te_4(AlCl_4)_2$ and $Se_8(AlCl_4)_2$:

The Te_4^{2+} ion is square-planar (symmetry D_{4h}), but Se_8^{2+} is a puckered ring (symmetry C_s) made bicyclic by a weak bridge bond (d = 2.83 Å), longer than the bonds between adjacent atoms in the five-membered rings (d = 2.32 Å) whose value is that of a single bond; S_8^{2+} has the same structure.

8.7. Chain Initiation and Cleavage Reactions

The formation of chains and rings is characteristic of the chalcogens S, Se, and Te especially sulfur which presents many possibilities for the synthesis of compounds with an S—S backbone:

a) Condensation of sulfur hydride and halides:

$$—S—H + Cl—S— \rightarrow —S—S— + HCl$$

b) Reaction of a metal sulfide and halide:

$$—S^- M^+ + Cl—S— \rightarrow —S—S— + MCl$$

c) Oxidation of hydrides by halide:

$$—S—H + I_2 + H—S— \rightarrow —S—S— + 2HI$$

d) Condensation of hydride and hydroxide (acidic):

$$—S—H + HO—S \rightarrow —S—S— + H_2O$$

e) Sulfurization of covalent or ionic sulfides by S_8:

$$S_8 + R—S—R \rightarrow R—S_n—R \quad R: H, Cl, \text{organic radical, or metal.}$$

Chemical or pyrolytic degradation of compounds with adjacent S—S-bonds eliminates sulfur atoms. Nucleophilic molecules and ions react with S_8 to form sulfur compounds, e.g., triphenylphosphine forms the phosphine sulfide, R_3PS, cyanide, the thiocyanate (SCN^-), ionic sulfides and hydrogen sulfide, the corresponding polysulfides (HS_n^- and S_n^{2-}), and hydrogen sulfite (HSO_3^-), the thiosulfate ($S_2O_3^{2-}$). The reactions occur stepwise by an S_N2 mechanism, for example, nucleophilic attack of CN^- leads to ring opening of the S_8 ring:

$$\dot{S}_8 + CN^- \rightarrow \ ^-S—S—S—S—S—S—S—S—CN$$

The product is rapidly degraded, with S_7CN^-, S_6CN^-, etc. as intermediates until finally:

$$S_8 + 8\,CN^- \rightarrow 8\,SCN^-$$

and all S—S-bonds are cleaved. Sulfur as S_8 can be determined photometrically by reacting the SCN^- ions produced from CN^- with Fe^{3+} to give the red $Fe(SCN)_3$. Organic polysulfides containing adjacent S—S-bonds react analogously:

$$R—S—S—S—R + CN^- \rightarrow R—S—S—R + SCN^-$$

Compounds with sulfur chains are thus sensitive towards nucleophilic agents like alkali, ammonia and the amines.

8.8. Hydrides

Volatile, covalent, binary hydrides, H_2E (E = S, Se, Te, and Po) and, for sulfur, H_2S_n, are known along with organic and inorganic derivatives of type REH and RS_nH.

8.8.1. The Hydrides H_2E

The hydrides H_2S, H_2Se and H_2Te are colorless, poisonous, noxious gases (H_2Po is a liquid) which are obtained:

a) From the elements after establishing equilibrium on heated pumice chips:

$$H_2 + S(g.) \ \underset{}{\overset{600\,°C}{\rightleftharpoons}} \ H_2S$$

$$H_2 + Se(g.) \ \underset{}{\overset{350-400\,°C}{\rightleftharpoons}} \ H_2Se$$

but H_2Te requires the electrolysis of sulfuric acid at $-70\,°C$ between a tellurium cathode and a platinum anode.

b) From metal chalcogenides by protonation of the anions:

$$NaHS + H_3PO_4 \ \overset{H_2O}{\longrightarrow} \ H_2S + NaH_2PO_4$$

$$Al_2S_3 + 6\,H_2O \longrightarrow 3\,H_2S + 2\,Al(OH)_3 \qquad \text{analogously: } D_2S,\ H_2Se$$

$$Al_2Se_3 + 6\,HCl \ \overset{H_2O}{\longrightarrow} \ 3\,H_2Se + 2\,AlCl_3 \qquad \text{analogously: } H_2Te$$

Hydrogen selenide and telluride oxidize in air to water and the respective elements. The chalcogen hydrides, except for water and hydrogen sulfide, are endothermic compounds.

Aqueous solutions of hydrogen sulfide, selenide or telluride are acidic, and the acid strength increases $H_2S < H_2Se < H_2Te$, as exemplified by the dissociation constants (at 25 °C):

		H_2S	H_2Se	H_2Te
$H_2E \rightleftharpoons H^+ + HE^-$	K_1:	$1.3 \cdot 10^{-7}$	$1.9 \cdot 10^{-4}$	$2.3 \cdot 10^{-3}$
$HE^- \rightleftharpoons H^+ + E^{2-}$	K_2:	$\approx 10^{-17}$	$\approx 10^{-11}$	$\approx 1.6 \cdot 10^{-11}$

This unexpected sequence in dissociation constants is the result as with the hydrogen halides (see p. 160) of the decrease of mean H—E bond energy from sulfur to tellurium. The water-soluble ionic chalcogenides, MHE and M_2E, hydrolyze owing to the small values of K_1 and K_2, and the hydrides are prepared by adding water to the barium or aluminum salts. Only insoluble metal sulfides can be isolated from aqueous solutions, but the soluble portions hydrolyze and the solubility of sulfides, selenides and tellurides is strongly pH-dependent. The hydrolysis-sensitive sulfides, selenides, and tellurides of the alkali and alkaline earth metals and aluminum are prepared dry or in a weak proton donor solvent such as alcohol or liquid ammonia.

The hydrides, H_2E, react as weak reducing agents or in condensation reactions with sulfur, phosphorous, and silicon halides, for example, to form the cyclic S_7Se:

$$H_2Se + S_7Cl_2 \rightarrow S_7Se + HCl$$

Further reactions, particularly those of hydrogen sulfide are discussed elsewhere.

8.8.2. Polysulfanes, H_2S_n

Hydrogen sulfide dissolves in liquid sulfur with sulfurization:

$$H_2S + S_n(l.) \rightleftharpoons H_2S_{n+1}$$

The solubility of H_2S at 200°–400 °C is 0.2 g/100 g of sulfur and is less at higher and lower temperatures. The resulting polysulfanes, H_2S_n, can be identified by proton nmr since the chemical shifts of the first 10 members of the series H_2S_n can be distinguished (but not n > 10).

The reaction of H_2S with S_8 at 25 °C is endothermic with $\Delta G° > 0$, and the sulfanes, H_2S_n, are unstable, and catalytically decomposed by traces of NH_3, hydroxides, or powdered quartz to H_2S and S_8. The composition of the polysulfane can be determined from the evolved H_2S.

Well-defined polysulfanes can be prepared by:
a) Adding aqueous sodium polysulfide (Na_2S_n) at −10 °C to dilute HCl, where-

upon the sulfane mixture, H_2S_n, separates as a heavy yellow oil which is cracked and fractionated to give the sulfanes, H_2S_2 and H_2S_3, and the $n \leq 6$ oligomers from vacuum distillation and selective extraction.

b) Condensation of sulfanes ($n = 1,2$) with chlorosulfanes ($n = 1$ to 4) at $-50\,°C$ yields higher sulfanes with $n \leq 8$.

$$2\,H_2S + SCl_2 \;\rightarrow\; H_2S_3 + 2\,HCl$$

$$2\,H_2S + S_2Cl_2 \;\rightarrow\; H_2S_4 + 2\,HCl$$

$$2\,H_2S_2 + SCl_2 \;\rightarrow\; H_2S_5 + 2\,HCl$$

Excess sulfane suppresses formation of longer chains. The reaction mixtures are separated by vacuum distillation.

The polysulfanes are liquids whose yellow color intensifies with molecular weight while viscosity increases and volatility decreases. The higher members ($n > 6$) cannot be separated from each other or S_8.

Polysulfanes dissolve in CS_2, the lower members also in $CHCl_3$ and C_6H_6. Vacuum distillation is possible only to $n \leq 4$, since higher members decompose on warming. The molecular structure of H_2S_2 corresponds to H_2O_2; in higher sulfanes zig-zag chains as in polysulfide anions are postulated.

The polysulfanes participate in condensation reactions (see p. 211), for example, H_2S_7 with thionyl chloride at $-40\,°C$ using the dilution principle, yields cyclo-octasulfur oxide:

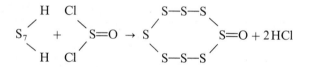

which crystallizes in intensely yellow colored needles which decompose to SO_2 and polymeric sulfur. The ring is a puckered crown as in S_8.

8.9. Metal Chalcogenides

All metals form chalcogenide compounds, which may be binary or ternary (e.g., $CuFeS_2$) or more complex. Sometimes several phases exist in a metal-chalcogen system.

Metal chalcogenides, especially sulfides, occur in nature as minerals and ores (e.g., zinc blende, ZnS; galena, PbS; argentite, Ag_2S). Iron pyrite (FeS_2) is mined for its sulfur content.

The structures and chemical properties of the oxides of a metal are different from its sulfides, selenides or tellurides because of the lower electronegativity and higher polarizability of the heavier chalcogen atoms, and their formation of stable chains and rings, e.g., in $(C_2H_5)_2TiS_5$, and $(NH_4)_2Pt(S_5)_3 \cdot 2H_2O$, which contain hetero-cyclic MS_5 rings, the latter containing three such rings with a platinum atom common to all three. Analogous seven- and five-membered rings are also known.

The ionic chalcogenides of the alkali metals will be considered here, while the chalcogenides of non-metals will appear with these elements.

Chalcogenides of Alkali Metals:

When aqueous sodium or potassium hydroxide is saturated with H_2S, solutions of NaHS and KHS, respectively, are formed from which on addition of equivalent amounts of base and cooling hydrated sulfides crystallize:

$$NaOH + H_2S \rightarrow NaHS + H_2O$$

$$NaHS + NaOH \rightleftharpoons Na_2S + H_2O$$

The colorless salts $Na_2S \cdot 9H_2O$ and $K_2S \cdot 5H_2O$ oxidize slowly in air with yellow discoloration to polysulfides and thiosulfate. Dehydration is accompanied by partial decomposition, and so anhydrous Na_2S, for example, is produced by reducing Na_2SO_4 with carbon at red heat:

$$Na_2SO_4 + 4C \rightarrow Na_2S + 4CO$$

Sodium and potassium sulfides, selenides and tellurides are precipitated by the reduction of the corresponding chalcogen with the alkali metal in liquid NH_3:

$$16\,K + S_8 \xrightarrow{NH_3} 8\,K_2S\downarrow$$

Solvated electrons are the active reducing agents (see p. 294), and the equilibrium is displaced to the right by precipitation of the chalcogenides which crystallize in the antifluorite lattice where cation positions are occupied by chalcogenide ions, and anion positions by the metal cations.

Dry, anerobic heating of sodium or potassium chalcogenides with additional chalcogen at 500–600 °C produces the salt-like polychalcogenides:

$$Na_2S + \tfrac{1}{8}S_8 \xrightarrow{500°} Na_2S_2$$

$$K_2S + \tfrac{3}{8}S_8 \xrightarrow{500°} K_2S_4$$

Na_2S_2 (light yellow), K_2S_{2-6} (yellow to red), Na_2Se_2 (gray), and Na_2Te_2 (gray-black, glossy metallic) can be prepared this way. Heating aqueous sulfide solutions with sulfur produces mixtures of hydrogen polysulfide ions:

$$HS^- + xS_8 \rightleftharpoons HS_n^-$$

Pure polychalcogenides can be prepared in liquid ammonia to which appropriate amounts of the elements are added.

Polysulfides of alkali- and alkaline earth metals contain zig-zag S_n^{2-} ions, iso-electronic with the corresponding chlorosulfanes:

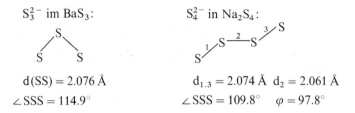

S_3^{2-} im BaS_3:

d(SS) = 2.076 Å

∠SSS = 114.9°

S_4^{2-} in Na_2S_4:

$d_{1.3}$ = 2.074 Å d_2 = 2.061 Å

∠SSS = 109.8°   φ = 97.8°

Anhydrous hydrogen sulfides are precipitated by:

$$C_2H_5ONa + H_2S \rightarrow NaHS + C_2H_5OH$$

$$NH_3 + H_2S \xrightarrow[0\,°C]{(C_2H_5)_2O} NH_4HS\downarrow$$

bubbling H_2S into alcoholic sodium ethanolate and adding ether, or bubbling ammonia and H_2S into ether.

8.10. Oxides

The chalcogen dioxides and trioxides are monomeric or polymeric:

$$SO_2 \quad SeO_2 \quad TeO_2 \qquad SO_3 \quad SeO_3 \quad TeO_3$$

Monoxides of all the elements are known, at high temperatures in equilibrium with their degradation products (dioxides and chalcogens), or in electrical discharges, but cannot be isolated in pure form.

Various other sulfur oxides are known:

S_2O	S_8O	·S—(S—S—S)—S—S·	$(-\overset{\overset{O}{\|}}{\underset{\underset{O}{\|}}{S}}-O-O-\overset{\overset{O}{\|}}{\underset{\underset{O}{\|}}{S}}-O-)_n$
Disulfur monoxide	Octasulfur monoxide	Polysulfuroxide	Polysulfur-peroxide
(p. 220)	(p. 214)	(p. 221)	

8.10.1. Dioxides

Sulfur dioxide is produced by burning native sulfur, or H_2S from natural gas, by roasting sulfidic ores, primarily pyrite (FeS_2), by reduction of $CaSO_4$, by thermal decomposition of $FeSO_4$ or technical waste H_2SO_4. Huge amounts are generated in burning fossil fuels which contain from traces to several per cent sulfur. The concentration of SO_2 in city air can reach 0.1 ppm. Sulfur dioxide is a colorless, poisonous and corrosive gas (m.p. $-75°$, b.p. $-10°C$) which is monomeric in all phases. Its gas phase properties are:

$$d(SO) = 1.43\,\text{Å} \qquad < 119.5°$$
$$f(SO) = 10.1\,\text{mdyn/Å},\ \mu = 1.62\,\text{D}$$

The bonds in SO_2 are similar to those in SO_3 (see p. 129), with nearly identical d and f values. From the $\angle OSO$, sulfur sp^2-hybrids (from 3s, $3p_x$ and $3p_y$) are utilized for the σ-bonds and free electron pair. The four π-electrons, as in SO_3, occupy two of the three pd^2-hybrid orbitals which overlap with the p-π-orbitals of the oxygen atoms.

A partial negative charge of 0.18e at the oxygen atoms is calculated from the dipole moment of SO_2.

In water, SO_2 is a reducing agent, reducing selenites and tellurites to the respective elements, and chlorites to ClO_2, with SO_2 oxidized to H_2SO_4.

With fluorine and chlorine gas, SO_2 forms the corresponding sulfuryl halides, SO_2X_2. Certain chlorides exchange oxygen for chlorine to form thionyl chloride:

$$PCl_5 + SO_2 \ \rightarrow\ POCl_3 + SOCl_2$$

$$COCl_2 + SO_2 \ \rightarrow\ CO_2 + SOCl_2$$

$$UCl_6 + 2\,SO_2 \ \rightarrow\ UO_2Cl_2 + 2\,SOCl_2$$

Towards Lewis acids and bases, SO_2 is amphoteric, forming crystalline complexes with tertiary amines, such as $(CH_3)_3\overset{(+)}{N}\!\!-\!\!\overset{(-)}{S}O_2$, and with transition metal complexes, such as, e.g., $Fe_2(CO)_8SO_2$.

Liquid SO_2 is a solvent resembling water, but undissociated, and a reaction medium for redox and complexation reactions, especially for double decompositions, where O^{2-} transfer plays a role as in the neutralization of sulfites with thionyl chloride:

$$SOCl_2 + M_2SO_3 \ \rightarrow\ 2\,MCl + 2\,SO_2$$

which proceeds via the ions $SOCl^+$ and SO_3^{2-}:

$$SOCl_2 \rightleftharpoons SOCl^+ + Cl^-$$

$$M_2SO_3 \rightleftharpoons 2M^+ + SO_3^{2-}$$

$$SOCl^+ + SO_3^{2-} \rightarrow [SO_2Cl^-] + SO_2$$
$$\downarrow$$
$$SO_2 + Cl^-$$

Selenium and tellurium dioxides are formed by burning the elements or in their oxidation with concentrated nitric acid after which the solution is evaporated and the residue heated to 300° (SeO$_2$) or 400° (TeO$_2$). Selenium dioxide forms colorless crystals, soluble in water, benzene, and glacial acetic acid, and sublimes at 315 °C to give a vapor of molecules of symmetry C$_{2v}$ (d(SeO) = 1.61 Å $\angle$OSeO = 125°). In the crystal, however, SeO$_2$ forms non-planar chains:

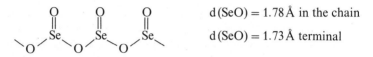

d(SeO) = 1.78 Å in the chain

d(SeO) = 1.73 Å terminal

in which the double bonds resonate among all Se—O linkages. The sums of covalent radii give d(Se—O) = 1.83 Å, and d(Se=O) = 1.6 Å.

Selenium dioxide dissolves in water as selenous acid, H$_2$SeO$_3$ or H$_2$Se(OH)$_6$. Vacuum evaporation of the solution gives crystalline H$_2$SeO$_3$ which delinquesces to SeO$_2$.

Selenium dioxide is an oxidizing agent being reduced to selenium. It forms the addition compound SeO$_2 \cdot 2$HCl, which is dehydrated by concentrated sulfuric acid to SeOCl$_2$.

Tellurium dioxide (m.p. 733 °C) forms colorless crystals with a rutile lattice, d(TeO) = 1.91, and 2.09 Å, sparingly soluble in water, more soluble in SeOCl$_2$, and forming TeCl$_6^{2-}$ ions in HCl. Reduction by carbon, aluminum, or zinc yields the element.

8.10.2. Trioxides

Sulfur trioxide is prepared by catalytic oxidation of SO$_2$ with air in the contact process:

$$SO_2 + \tfrac{1}{2}O_2 \rightleftharpoons SO_3 \qquad \Delta H^\circ_\circ = -95.7 \text{ kJ/mol}$$

The reaction is slow, but owing to a negative enthalpy, the equilibrium lies on the left at higher temperatures, and so a contact catalyst, V$_2$O$_5$ on SiO$_2$ or silicate as carrier, doped with K$_2$SO$_4$ as an activator, is used:

$$SO_2 + 2V^{5+} + O^{2-} \rightarrow SO_3 + 2V^{4+}$$

$$2V^{4+} + \tfrac{1}{2}O_2 \rightarrow 2V^{5+} + O^{2-}$$

Sulfur trioxide is scrubbed from the reaction mixture with H_2SO_4, giving poly-sulfuric acids (oleum, or fuming sulfuric acid), which are hydrolyzed to H_2SO_4.

$$(n-1)SO_3 + H_2SO_4 \rightleftharpoons H_2S_nO_{3n+1} \xrightarrow{+(n-1)H_2O} nH_2SO_4$$

Pure SO_3 is obtained by distillation of oleum.

Modifications of SO_3:
Gaseous sulfur trioxide consists of SO_3 and S_3O_9 molecules in a pressure- and temperature-dependent equilibrium:

$$3SO_3 \rightleftharpoons S_3O_9 \qquad \Delta H_o^\circ = -126\ kJ/mol\ S_3O_9$$

Condensation at the boiling point of 44.5 °C gives a colorless liquid of low viscosity which consists predominantly of S_3O_9. Ice-like crystals of rhombic γ-SO_3 form at <16.9 °C (the m.p.) consisting of S_3O_9 molecules with monomeric SO_3 (see p. 96) in trace amounts. The puckered S_3O_3 heterocycle with six *exo*-, double-bonded, oxygen atoms:

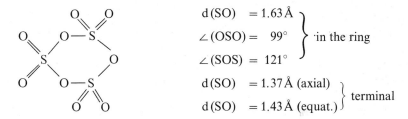

$$d\,(SO) \quad = 1.63\,\text{Å} \left.\begin{array}{c} \\ \\ \end{array}\right\}$$
$$\angle\,(OSO) = \quad 99° \quad \right\} \text{ in the ring}$$
$$\angle\,(SOS) = \quad 121°$$

$$d\,(SO) \quad = 1.37\,\text{Å (axial)} \left.\begin{array}{c} \\ \end{array}\right\} \text{ terminal}$$
$$d\,(SO) \quad = 1.43\,\text{Å (equat.)}$$

may be considered as three distorted SO_4 tetrahedra joined by common corner atoms. Similar structures are encountered in two other crystalline modifications, polymeric, monoclinic β-SO_3, formed by catalytic traces of water at < 30 °C, which is a polysulfuric acid consisting of a mixture of long chain molecules:

$$HO-(S-O)_n-H \qquad d(S-O) = 1.61\,\text{Å}, \quad d(S=O) = 1.41\,\text{Å}$$

Depolymerization to SO_3 and S_3O_9 occurs at the m.p., 32–45 °C. Polymerization in liquid SO_3 is inhibited by trace amounts of stabilizers.
The thermodynamically stable, asbestos-like α-SO_3, is formed by condensation of gaseous SO_3 on cold surfaces, and subsequent warming to 25 °C and irradiation with x-rays. The structure is thought to be similar to that of β-SO_3, but depolymerization occurs at 62 °C, indicating a higher degree of polymerization.

Reactions of Sulfur Trioxide:

Sulfur trioxide behaves as an oxidizing agent, Lewis acid and, more rarely, as a Lewis base, oxidizing sulfur to SO_2, SCl_2 to $SOCl_2$ and SO_2Cl_2, PCl_3 to $POCl_3$, and phosphorus to P_4O_{10}, forming donor-acceptor complexes with Lewis bases, e.g., crystalline 1-pyridinyl sulfur trioxide ($C_5H_5N^+$—SO_3^-) from pyridine, sometimes followed by rearrangement, as with H_2O to give H_2SO_4:

$$H_2O + SO_3 \; \rightarrow \; \begin{bmatrix} & O & \\ & \overset{\oplus}{\underset{|}{H-O}}-\overset{\|}{\underset{\|}{S}}-\overset{\ominus}{O} \\ & H & O \end{bmatrix} \; \rightarrow \; H-O-\overset{\overset{O}{\|}}{\underset{\underset{O}{\|}}{S}}-O-H$$

With HCl, HSO_3Cl, and with H_2S at low temperatures, $H_2S_2O_3$ is obtained. The SO_3 group inserts as (—O—SO_2—) between the existing bonds, even if no hydrogen atoms participate. With inorganic acid SO_3 adds to the anion:

$$KF + nSO_3 \; \rightarrow \; KS_nO_{3n}F$$
$$Na_2SO_4 + (n-1)SO_3 \; \rightarrow \; Na_2S_nO_{3n+1}$$

or reacts with displacement of the anhydride, e.g.:

$$K_2SeO_4 + nSO_3 \; \rightarrow \; K_2S_nO_{3n+1} + SeO_3$$

$$2KClO_4 + 3SO_3 \; \rightarrow \; K_2S_3O_{10} + Cl_2O_7$$

$$K_2CO_3 + 3SO_3 \; \rightarrow \; K_2S_3O_{10} + CO_2$$

Selenium and Tellurium Trioxide:

Selenium trioxide, prepared as above from K_2SeO_4 and liquid SO_3, forms colorless, hygroscopic crystals (m.p. $118\,°C$) consisting of 8-membered S_4O_{12} rings, which in the vapor are in equilibrium with monomeric SeO_3. Decomposition occurs at $>160°$ to Se_2O_5 and O_2; SeO_3 is a stronger oxidizing agent than SO_3.

Dehydration of ortho-telluric acid at $300–360°$ yields the strong oxidizing agent, TeO_3:

$$Te(OH)_6 \; \rightarrow \; TeO_3 + 3H_2O$$

which is insoluble in water, and decomposes $>400\,°C$ into Te_2O_5 and O_2.

8.10.3. Lower Sulfur Oxides

Disulfur monoxide, S_2O, which occurs as an intermediate in the combustion of sulfur, is obtained by passing $SOCl_2$ at low pressure over powdered Ag_2S at $160\,°C$:

$$OSCl_2 + Ag_2S \; \rightarrow \; S_2O + 2AgCl$$

The S_2O molecule, like O_3, SO_2, S_3, is angular and has an electron distribution comparable to SO_2:

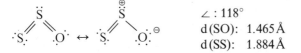

∠ : 118°
d(SO): 1.465 Å
d(SS): 1.884 Å

Gaseous S_2O polymerizes at >1 torr or upon condensation at 77 K and subsequent warming to 25 °C in a radical chain reaction with disproportionation to SO_2 and insoluble polysulfur oxides:

$$S{=}S + S{=}S + S{=}S + S{=}S + \ldots \rightarrow \cdot S{-}S{-}S{-}S{-}S{-}S{-}S{\cdot} + SO_2$$

corresponding to polymeric sulfur (see p. 208), which decompose at 100 °C into SO_2 and S_8. Similar ring- or chain-forming oxides can also be obtained from $SOCl_2$ and H_2S or H_2S_n. For example, the cyclic oxide S_8O (see p. 214) is formed from $SOCl_2$ and H_2S_7.

8.11. Oxo-, Thio-, and Halo-Acids

8.11.1. General

Sulfur in particular forms a large number of oxo-acids. Replacement of oxygen in these acids by sulfur gives the thio-acids:

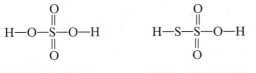

Sulfuric Acid
[Tetraoxosulfur(VI)
Acid]

Thiosulfuric Acid
[Trioxochlorosulfur(VI)
Acid]

and substitution of OH groups by halogen atoms leads to halosulfuric acids:

Chlorosulfuric Acid
[Trioxomonochlorosulfur(VI)
Acid]

Fluorosulfurous Acid
[Dioxofluorosulfur(IV)
Acid]

Some chalcogen acids cannot be isolated, and are only known as their anion salts, organic esters or in solution.

Most chalcogen acids have trivial names (sulfate, sulfite, dihionate, etc.), but a rational nomenclature is one in which the salts of all sulfur-oxygen acids are named as sulfates, provided that the oxidation number of the sulfur atom(s) also appears in brackets.

8.11.2. Sulfurous Acid (H_2SO_3)

Sulfur dioxide dissolves in water (*ca.* 45 vol SO_2/vol H_2O at $15\,^\circ C$) to give a solution of the hypothetical, non-isolable acid H_2SO_3 with reducing properties.
Removal of the solvent regenerates SO_2 and water, and the hydrate $SO_2 \cdot 6H_2O$ cyrstallizes on cooling. The solution, based on Raman spectra, contains SO_2, as well as the ions H_3O^+, HSO_3^-, and at higher concentrations $S_2O_5^{2-}$:

$$SO_2(aq) \rightleftharpoons H^+ + HSO_3^- \rightleftharpoons 2H^+ + SO_3^{2-}$$

$$2HSO_3^- \rightleftharpoons H_2O + S_2O_5^{2-}$$

Bubbling SO_2 into aqueous sodium hydroxide to saturation produces $NaHSO_3$, which can be neutralized with additional NaOH or boiled with Na_2CO_3 solution to obtain Na_2SO_3:

$$NaOH + SO_2 \rightarrow NaHSO_3 \xrightarrow{+NaOH} Na_2SO_3 + H_2O$$

Dry heating of $NaHSO_3$, or concentration of a solution produces sodium disulfite (metabisulfite):

$$2NaHSO_3 \rightarrow Na_2S_2O_5 + H_2O$$

Anions of these colorless salts have the structures:

Sulfite (Symm. C_{3v})	Hydrogensulfite	Disulfite
d(SO) = 1.53 Å	(Symm. C_{3v})	d(SS) = 2.209 Å
∠OSO = 107.4°		

The long, weak S—S bond in bisulfite is responsible for the decomposition of $Na_2S_2O_5$ at $400\,^\circ C$ into Na_2SO_3 and SO_2. Hydrogen sulfite anions are protonated at sulfur. The tautomeric form, $HOSO_2^-$, is unknown, although both organic derivatives of sulfurous acid are known:

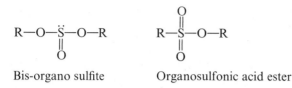

Bis-organo sulfite   Organosulfonic acid ester

Sulfites, hydrogen sulfites and disulfites behave similarly in oxidation to sulfate in aqueous solution, e.g., by air, iodine, and Fe^{3+} ions. Dithionates are formed with MnO_2, and dithionites are obtained by reduction with base metals, NaH and by cathodic reduction.

Boiling of aqueous sulfite with sulfur leads to nucleophilic degradation yielding thiosulfate:

$$8\,Na_2SO_3 + S_8 \rightarrow 8\,Na_2S_2O_3$$

The sulfurous acid derivatives, fluorosulfurous acid, HSO_2F, thiosulfurous acid $H_2S_2O_2$, and dithiosulfurous acid H_2S_3O are unstable and two are known in the form of derivatives only:

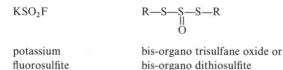

potassium   bis-organo trisulfane oxide or
fluorosulfite   bis-organo dithiosulfite

The former, generated by the action of SO_2 on powdered KF, dissociates upon heating, and serves as a fluorinating agent. Dithiosulfites are made by condensing thiols, RSH, with thionyl chloride, $SOCl_2$.

8.11.3. Selenous and Tellurous Acids (H_2SeO_3, H_2TeO_3)

Selenium dioxide dissolves in water to give the weak acid, H_2SeO_3, which is obtained on evaporation as colorless crystals that decompose into SeO_2 and water on warming. Selenous acid is reduced by N_2H_4, SO_2, H_2S and HI to yield red selenium, e.g.:

$$H_2SeO_3 + N_2H_4 \rightarrow Se\downarrow + N_2\uparrow + 3\,H_2O$$

Neutralization of selenous acid gives hydrogen selenites, $MHSO_3$, and selenites M_2SeO_3. Concentrated hydrogen selenite solutions also contain biselenite ions, $Se_2O_5^{2-}$ which, unlike bisulfite ions, have a Se—O—Se bond.

Strong oxidizing agents like F_2, O_3, MnO_4^- or H_2O_2, convert H_2SeO_3 or selenites into selenic acid, H_2SeO_4, or selenates.

Owing to its large lattice energy, TeO_2 is insoluble in water, and the hydrolysis of $TeCl_4$, or acidification of tellurite solutions, does not yield H_2TeO_3, but materials which decompose to TeO_2 and water. Tellurites are obtained by dissolving TeO_2 in strong alkali. Polytellurites, $M_2Te_nO_{2n+1}$, are also known.

8.11.4. Sulfuric Acid (H_2SO_4)

Sulfuric acid, the most important sulfur compound ($>40 \times 10^6$ tons annually, world-wide), is produced almost exclusively by the contact process (see p. 218) and used in the synthesis of fertilizers, dyestuffs, and other chemicals.

Pure sulfuric acid is a colorless, oily liquid (m.p. 10.4 °C) which boils at 290–317 °C with partial decomposition to water and SO_3. Solid and liquid H_2SO_4 are hydrogen-bonded. The liquid contains various ions and molecules:

$$2H_2SO_4 \rightleftharpoons H_3SO_4^+ + HSO_4^- \qquad \text{Ion product } K = 2.7 \cdot 10^{-4}$$
$$2H_2SO_4 \rightleftharpoons H_3O^+ + HS_2O_7^- \qquad \qquad \text{mol}^2 \cdot \text{l}^{-2}\,(25°)$$
$$2H_2S_2O_7 \rightleftharpoons H_2SO_4 + H_2S_3O_{10}$$

Sulfur trioxide dissolves in H_2SO_4 to give disulfuric acid, $H_2S_2O_7$, and higher polysulfuric acids, $H_2S_nO_{3n+1}$ (n = 3,4):

$$2H_2SO_4 + 2SO_3 \rightleftharpoons 2H_2S_2O_7 \rightleftharpoons H_2SO_4 + H_2S_3O_{10}$$

Mixtures with melting point minima (the others are solids), containing 20% and 65% of free SO_3, are sold as fuming sulfuric acid or oleum. The acidity (see p. 161) of polysulfuric acids increases with increasing chain length. Water reacts with H_2SO_4 to form oxonium salts and hydrates (see p. 156). In water H_2SO_4 is completely dissociated into HSO_4^- with the dissociation constant for the second step *ca.* 10^{-2} mol/l. Sulfates and hydrogen sulfates of most electropositive elements can be isolated. The ions and acids have the structures:

Sulfate
d(SO) = 1.51 Å (4x)

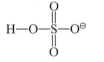

Hydrogen sulfate
d(S—OH) = 1.56 Å (1x)
d(S=O) = 1.47 Å (3x)

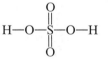

Sulfuric acid
d(S—OH) = 1.535 Å (2x)
d(S=O) = 1.426 Å (2x)

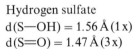

Disulfuric acid

Salts of the polysulfuric acids can be prepared:

$$2\,KHSO_4 \xrightarrow{\Delta T} K_2S_2O_7 + H_2O$$

$$K_2SO_4 + (n-1)SO_3\,(g.) \rightarrow K_2S_nO_{3n+1}$$

Polysulfates are hydrolyzed to SO_4^{2-}, and thermally decompose to sulfate and SO_3.

Concentrated sulfuric acid (98%) is reduced by metals to SO_2, and is strongly dehydrating, carbonizing most carbohydrates. Anhydrous sulfuric acid is a strongly protonating solvent (cf. p. 158). Owing to its high dielectric constant (100 at 25 °C) it dissolves electrolytes, but non-electrolytes can dissolve on protonation.

8.11.5. Selenic and Telluric Acids [H_2SeO_4, $Te(OH)_6$]

Selenic acid, which forms colorless, hygroscopic crystals, m.p. 58 °C, is prepared by oxidation of SeO_2 with 30% H_2O_2, bromine or chlorine (in H_2O) and dehydration of its solutions at 160 °C/2 torr. At > 260 °C, H_2SeO_4 decomposes into SeO_2, water, and oxygen. Selenates are prepared by neutralization of the acid with carbonate, or by oxidation of selenites.

Selenic acid is as strong an acid as H_2SO_4 and a stronger oxidizing agent. Most selenates and hydrogen selenates are isomorphous with the corresponding sulfates and hydrogen sulfates. Nitrosyl, NO^-, and nitryl, NO_2^-, salts are known for H_2SeO_4, like H_2SO_4, as well as fluoro-, chloro-, amido-, and peroxoselenic acids. Diselenates, with the anion $Se_2O_7^{2+}$, also exist.

Telluric acid, with the formula $Te(OH)_6$ (orthotelluric acid), in which tellurium atoms are octahedrally coordinated by OH groups, unlike selenic and sulfuric acids, is formed as colorless crystals by oxidation of TeO_2 with $HClO_3(aq)$, H_2O_2 (30%), or $KMnO_4/HNO_3$. Similarly highly hydrated octahedral compounds are known:

$Sn(OH)_6^{2-}$	$Sb(OH)_6^-$	$Te(OH)_6$	$IO(OH)_5$
Hydroxostannate	Hydroxo-antimonate	Orthotelluric acid	Periodic acid

In water, $Te(OH)_6$ is a weak, diprotic acid, from which $MTeO(OH)_5$ and $M_2TeO_2(OH)_4$ can be isolated. Precipitation with silver ion, however, gives Ag_6TeO_6, and fusion with NaOH yields Na_6TeO_6.

Heating $Te(OH)_6$ successively yields polytelluric acid, $(H_2TeO_4)_n$ at 100 to 200 °C; TeO_3 at > 220°; Te_2O_5 at > 400 °C; and finally TeO_2. Telluric acid is a strong, but slowly acting oxidizing agent.

8.11.6. Peroxosulfuric acids (H_2SO_5, $H_2S_2O_8$)

Both acids:

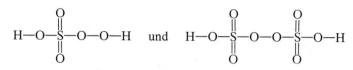

Peroxomonosulfuric acid   Peroxodisulfuric acid

form hygroscopic crystals which can be prepared on cooling from 100% H_2O_2 and HSO_3Cl:

$$HSO_3Cl + HOOH \xrightarrow{-HCl} HSO_3OOH \xrightarrow[-HCl]{+HSO_3Cl} HSO_3OOSO_3H$$

the former (m.p. 45 °C) hydrolyzes to form H_2SO_4 and H_2O_2. $H_2S_2O_8$ (m.p. 65 °C) is, like H_2SO_5, a strong oxidizing agent which hydrolyzes exothermically to give H_2SO_4 and H_2SO_5.

Peroxodisulfates, prepared by the anodic oxidation of sulfates (p. 189), are strong oxidizing agents when catalyzed by silver ion, converting Mn^{2+} ions into MnO_4^-, and Cr^{3+} into CrO_4^-. The acid is prepared from its salts.

8.11.7. Halosulfuric acids ($HS_nO_{3n}X$)

Sulfur trioxide reacts with HF, HCl, and HBr to form halosulfuric acids (HI is oxidized to I_2):

$$SO_3 + HX \rightarrow H-O-\overset{\overset{O}{\|}}{\underset{\underset{O}{\|}}{S}}-X$$

X = F, Cl, Br

The fluoro- and chlorosulfuric acids are colorless, fuming liquids, but HSO_3Br, which is prepared at -35 °C in liquid SO_2 decomposes at its m.p. (8 °C) into bromine, SO_2 and H_2SO_4. Analogous decomposition occurs for HSO_3Cl upon heating. Fluoro- and chlorosulfates are readily prepared from the halides and SO_3:

$$CaF_2 + 2SO_3 \rightarrow Ca(SO_3F)_2$$

Fluorosulfuric acid hydrolyzes vigorously to H_3OSO_3F which hydrolyzes slowly. The anhydrous acid is a fluorinating agent as well as an acidic solvent (m.p. -89 °C, b.p. 163 °C, dielectric constant at 25°: 120). The anion SO_3F^- is an isostere of perchlorate, and about equal in size, and fluorosulfates and perchlorates form solid solutions and have similar solubilities.

Chlorosulfuric acid hydrolyzes explosively, and acts as a sulfonating agent:

$$RH + HSO_3Cl \rightarrow RSO_3H + HCl$$

8.11.8. Thiosulfuric Acid ($H_2S_2O_3$) and Sulfanesulfonic Acids

Thiosulfuric acid is unstable to disproportionation, but can be isolated at low temperatures from H_2S and SO_3 in ether as the $H_2S_2O_3$-etherate, or unsolvated from H_2S and HSO_3Cl with liberation of HCl.

Thiosulfates are prepared by boiling aqueous sulfite with sulfur (see p. 223). Sodium thiosulfate pentahydrate is used in photography as a fixer, since it dissolves unexposed AgBr as $[Ag(S_2O_3)_2]^{3-}$ from the photographic emulsion.

The $S_2O_3^{2-}$ anion in anhydrous $Na_2S_2O_3$ is similar to the SO_4^{2-} ion but of C_{3v} symmetry.

Iodine acts on thiosulfates to form tetrathionates (used for the quantitative determination of either ion):

$$2S_2O_3^{2-} + I_2 \rightarrow S_4O_6^{2-} + 2I^-$$

No salts of di-, tri-, or tetrathiosulfuric acids are known, but acids rich in sulfur are obtained when SO_3 reacts with polysulfanes at low temperatures in ether:

$$H-S_n-H + SO_3 \rightarrow H-S_n-SO_3H \xrightarrow{SO_3} HO_3S-S_n-SO_3H$$

Sulfanesulfonic	Polythionic acids or
Acids	Sulfanedisulfonic Acids

The acids HS_nSO_3H and their salts are unstable, but polythionic acids can be prepared in aqueous solution and their salts are stable compounds:

Trithionate ($n = 1$):   Condensation of SCl_2 with HSO_3^--ions

$$SCl_2 + 2HSO_3^- \xrightarrow{H_2O} S_3O_6^{2-} + 2HCl$$

Tetrathionate ($n = 2$):   Oxidation of $S_2O_3^{2-}$ solution with I_2 (see above) or Condensation of HSO_3^- with S_2Cl_2.

Pentathionate ($n = 3$):   Condensation of SCl_2 with $HS_2O_3^-$-ions

$$SCl_2 + 2HS_2O_3^- \xrightarrow{H_2O} S_5O_6^{2-} + 2HCl$$

Hexathionate ($n = 4$):   Condensation of S_2Cl_2 with $HS_2O_3^-$-ions

$$S_2Cl_2 + 2HS_2O_3^- \xrightarrow{H_2O} S_6O_6^{2-} + 2HCl$$

These anions form colorless cyrstalline alkali or alkaline earth salts, and selenium or tellurium derivatives (e.g., $[Se(SSO_3)_2]^{2-}$, and $[Te(SSO_3)_2]^{2-}$) have been prepared.

8.11.9. Dithionic Acid ($H_2S_2O_6$)

Dithionic acid is only stable in aqueous solution, but its salts can be prepared by oxidation of hydrogen sulfite with MnO_2:

$$2 HSO_3^- \longrightarrow S_2O_6^{2-} + 2 H^+ + 2 e^-$$

$$2 MnO_2 + 3 SO_2 \xrightarrow{H_2O} MnS_2O_6 + MnSO_4$$

the resulting anion, $S_2O_6^{2-}$, has a symmetrical structure:

Treating BaS_2O_6 with sulfuric acid yields $H_2S_2O_6$, a strong, diprotic acid, whose solutions decompose above $50\,^\circ C$:

$$H_2S_2O_6 \rightarrow H_2SO_4 + SO_2$$

Dithionates decompose in an analogous manner above $200\,^\circ C$.

8.11.10. Dithionous Acid ($H_2S_2O_4$)

Dithionites are obtained by reduction of hydrogen sulfites in aqueous solution with sodium amalgam, formic acid, zinc dust or by cathodic reduction:

$$2 HSO_3^- + 2 e^- \rightarrow S_2O_4^{2-} + 2 OH^-$$

The sodium salt crystallizes from aqueous solution as the hydrate, and can be dehydrated *in vacuo*. This salt is a reducing agent. Dithionous acid cannot be isolated since dithionite solutions decompose when acidified:

$$2 S_2O_4^{2-} + H_2O \xrightarrow{H^+} 2 HSO_3^- + S_2O_3^{2-}$$

The anion, $S_2O_4^{2-}$, has a structure:

in which the S—S bond is unusually long (2.39 Å) and easily cleaved, explaining the high reactivity and solution equilibrium:

$$S_2O_4^{2-} \rightleftharpoons 2SO_2^-$$

The aqueous dissociation constant at $25\,^\circ C$ is $0.63 \cdot 10^{-9}$ mol $\cdot$ l^{-1}, and 1 M $S_2O_4^{2-}$ contains $2.5 \cdot 10^{-5}$ mol $\cdot$ l^{-1} of SO_2^- radical anions, isoelectronic with the ClO_2 molecule, and observable in the ESR spectrum.

8.12. Halides and Oxyhalides

8.12.1. General

Table 41 lists the halides of the chalcogens. From the electron configuration s^2p^4, chalcogens form halides of the type EX_2. Tetra- and hexahalides are derived from the excited states s^2p^3d and sp^3d^2. Chalcogens also form halides with E—E bonds (e.g., S_2Cl_2, Se_2Br_2, Te_2F_{10}).

The apparently analogous compounds listed in Table 41 may have different structures. For example, SF_4 is a pseudo-trigonal bipyramidal gaseous monomer; SCl_4, stable only below $-30\,^\circ C$, is a crystalline salt of formula $(SCl_3^+)Cl^-$; and the monoclinic crystals of $TeCl_4$ are made up of Te_4Cl_{16} units. Few iodides have been synthesized. Most halides are prepared from the elements.

Table 41   Chalcogen Halides (mixed halides are not listed). Compounds in parentheses cannot be prepared pure.

Oxidation State	$< +2$	$+2$	$+4$	$+5$	$+6$
Sulfur:	S_2F_2	(SF_2)	SF_4	S_2F_{10}	SF_6
	S_2Cl_2	SSF_2	SCl_4		$SClF_5$
	S_2Br_2	SCl_2			$SBrF_5$
	S_nCl_2				
	S_nBr_2				
Selenium:	Se_2Cl_2	$(SeCl_2)$	SeF_4		SeF_6
	Se_2Br_2	$(SeBr_2)$	$SeCl_4$		
			$SeBr_4$		
Tellurium:	Te_3Cl_2	$TeCl_2$	TeF_4	Te_2F_{10}	TeF_6
	Te_2Br	$TeBr_2$	$TeCl_4$		
	TeI		$TeBr_4$		
	Te_xI		TeI_4		

8.12.2. Sulfur Halides

Fluorides:

Sulfur burns in fluorine to form SF_6:

$$\tfrac{1}{8}S_8 + 3F_2 \rightarrow SF_6 \qquad\qquad \Delta H^\circ = -1210\,\text{kJ/mol}$$

which is also generated from SO_2 and F_2:

$$SO_2 + 3F_2 \rightarrow SF_6 + O_2$$

Sulfur hexafluoride is a colorless, odorless, non-toxic gas (b.p. $-64\,^\circ\text{C}$) whose low dielectric constant and chemical inertness recommend it as an insulator (e.g., SF_6 is resistant to molten KOH at $410\,^\circ\text{C}$, superheated steam at $500\,^\circ\text{C}$, and oxygen in an electric discharge). However, the hydrolysis equilibrium:

$$SF_6 + 4H_2O \rightarrow H_2SO_4 + 6HF$$

lies on the right. The resistance of SF_6 to nucleophilic reaction arises from the steric screening of the sulfur atom, and the filling of all orbitals by fluorine electrons, giving no point of attack for a Lewis base. Sulfur tetrafluoride, however, hydrolyzes readily.

Organic derivatives, RSF_5, cannot be prepared directly from SF_6, but from SF_4 or $SClF_5$.

Sulfur tetrafluoride is a colorless, reactive gas, prepared:

$$\tfrac{1}{8}S_8 + 4CoF_3 \xrightarrow{130^\circ} SF_4 + CoF_2$$

$$3SCl_2 + 4NaF \xrightarrow[75\,^\circ\text{C}]{CH_3CN} SF_4 + S_2Cl_2 + 4NaCl$$

The tetrafluoride reacts with water to give SO_2 and HF, with SOF_2 as intermediate. With fluorides having large cations pentafluorosulfates(IV) are formed:

$$CsF + SF_4 \rightarrow Cs[SF_5]$$

In presence of NO_2, SF_4 is oxidized by O_2 to thionyl tetrafluoride, SOF_4. SF_4 fluorinates carbonyl and thiocarbonyl groups:

$$\text{>C=O} + SF_4 \rightarrow \text{>CF}_2 + SOF_2$$

and oxides, sulfides, and carbonyls:

$$I_2O_5 + 5SF_4 \rightarrow 2IF_5 + 5SOF_2$$

or adds to certain bonds, sometimes with insertion:

$$R—C{\equiv}N + SF_4 \rightarrow R—CF_2—N{=}SF_2$$

$$F_5S—O—O—SF_5 + SF_4 \rightarrow F_5S—O—SF_4—O—SF_5$$

Sulfur tetrafluoride is fluorinated to SF_6, and chlorinated to $SClF_5$:

$$SF_4 + ClF \xrightarrow{380\,°C} SClF_5$$

$$SF_4 + Cl_2 + CsF \xrightarrow{110\,°C} SClF_5 + CsCl$$

Sulfur chloropentafluoride is a colorless gas, reactive in radical mechanisms as the SCl bond is weaker, and since chlorine is open to attack by nucleophiles, the compound hydrolyzes and is a strong oxidizing agent. At $> 400\,°C$ or photolytically $SClF_5$ decomposes via SF_5 radicals to SF_4 and SF_6:

$$2SClF_5 \rightarrow SF_4 + SF_6 + Cl_2$$

Therefore, $SClF_5$ can be used to introduce SF_5 groups:

$$SClF_5 + C_2F_4 \rightarrow F_5S—CF_2—CF_2—Cl$$

$$2SClF_5 + O_2 \xrightarrow{h \cdot v} F_5S—O—O—SF_5 + Cl_2$$

With hydrogen, the toxic disulfur decafluoride is formed photolytically:

$$2SClF_5 + H_2 \rightarrow F_5S—SF_5 + 2HCl$$

The S_2F_{10} molecule (b.p. $30\,°C$) with a long (2.21 Å), weak S—S bond, disproportionates at $150°$ into SF_4 and SF_6. It cleaves easily into SF_5 radicals and, hence, reacts with chlorine or bromine to yield $SClF_5$ or $SBrF_5$.

The lower sulfur fluorides, FSSF and SSF_2, are prepared by fluorination of S_8 with AgF, or by reacting S_2Cl_2 with potassium fluorosulfite:

$$\tfrac{3}{8}S_8 + 2AgF \xrightarrow{125\,°C} FSSF + Ag_2S$$

$$S_2Cl_2 + 2KSO_2F \rightarrow SSF_2 + 2KCl + 2SO_2$$

Difluorodisulfane, FSSF, is a colorless gas (b.p. $15\,°C$) which isomerizes in presence of alkali metals to thiothionyl fluoride, SSF_2 (b.p. $-11\,°C$), which disproportionates slowly to S_8 and SF_4. Both fluorides have S—S double bonds:

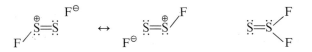

Chlorides and Bromides:
Sulfur reacts with chlorine with cleavage of the S—S bonds:

$$\text{—S—S—} + Cl_2 \rightarrow \text{—S—Cl} + \text{Cl—S—}$$

to produce various dichlorosulfanes, S_nCl_2, depending on the ratio of reactants. Excess chlorine with $FeCl_3$ as a catalyst yields a dark red colored liquid, SCl_2 (b.p. 60 °C) whose decomposition:

$$2SCl_2 \rightleftharpoons S_2Cl_2 + Cl_2$$

can be retarded by PCl_3.

With less chlorine, S_8 reacts to form S_2Cl_2, a straw-yellow liquid with a pungent odor (b.p. 138 °C), the most stable sulfur chloride, which like all sulfur halides, hydrolyzes in a complex reaction to HCl, H_2S, S, SO_2 and sulfoxo acids. The orange-red colored oily sulfur chlorides, S_nCl_2 ($n \leq 8$) are products of the condensation of SCl_2 or S_2Cl_2 with sulfur hydrides:

$$\text{Cl—S}_2\text{—Cl} + \text{H—S}_n\text{—H} + \text{Cl—S}_2\text{—Cl} \rightarrow \text{Cl—S}_{n+4}\text{—Cl} + 2HCl$$

Sulfur dichloride is oxidized by oxygen or SO_3 to $SOCl_2$ and SO_2Cl_2. Sulfur chlorides and bromides react with hydrogen compounds to liberate hydrogen halides and form element-sulfur bonds. For example, the reaction with HBr results in bromosulfanes:

$$S_nCl_2 + 2HBr \rightarrow S_nBr_2 + 2HCl$$

Sulfur dichloride is angular (symmetry C_{2v}) with single bonds. The S_2Cl_2 molecule has a structure analogous to H_2O_2 (symmetry C_2). Higher chlorosulfanes presumably exist in non-planar, spiral chains.

At -78 °C SCl_2 reacts with liquid chlorine to yield white crystals of SCl_4 which decompose above $-30°$ into SCl_2 and Cl_2. Sulfur hexachloride is not known. Solid SCl_4 is presumably a trichlorosulfonium salt: $[SCl_3^-]Cl^+$, and the cation is found in, e.g., $[SCl_3^-]$ $[AlCl_4^+]$ and $[SCl_3^-]$ $[AsF_6^-]$, which are formed from SCl_2, Cl_2 and $AlCl_3$ or AsF_3, respectively. The organic derivatives, RSCl, (R = organic radical), prepared from the disulfides, RSSR, and chlorine, react similarly, e.g., with chlorine to from the trichlorides $RSCl_3$. The corresponding bromides and iodides are unknown, but organic sulfur trifluorides, RSF_3, exist.

8.12.3. Sulfoxo dihalides

Thionyl dihalides, SOX_2 ($X=F$, Cl, Br), are pyramidal with sp^3 hybridized sulfur(IV). Thionyl chloride is prepared by oxidation of SCl_2 with SO_3 or SO_2Cl_2, or in the laboratory:

$$SO_2 + PCl_5 \rightarrow SOCl_2 + POCl_3$$

and isolated by fractional distillation. Thionyl chloride, $SOCl_2$ (b.p. 76 °C), hydrolyzes to SO_2 and HCl, and reacts with —OH, —NH, and —SH groups with HCl elimination. The other thionyl halides are made by halogen exchange:

$$SOCl_2 \xrightarrow[CH_3CN]{+NaF, -NaCl} SOClF \xrightarrow[CH_3CN]{+NaF, -NaCl} SOF_2$$

$$SOCl_2 \xrightarrow[\text{liqu. } SO_2]{+KBr, -KCl} SOBrCl \xrightarrow[\text{liqu. } SO_2]{+KBr, -KCl} SOBr_2$$

Sulfuryl halides, SO_2X_2, disulfuryl halides, $S_2O_5X_2$, thionyl tetrafluoride SOF_4, and SF_6 derivatives contain sulfur(VI). Sulfuryl fluoride SO_2F_2, the difluoride of sulfuric acid, is generated by heating solid $Ba(SO_3F)_2$, which is accessible from $BaCl_2$ and HSO_3F:

$$Ba(SO_3F)_2 \rightarrow BaSO_4 + SO_2F_2$$

Sulfuryl fluoride is also obtained from SO_2Cl_2 by exchange with NaF. The chloride is a colorless liquid, produced from chlorine and SO_2, with activated carbon as catalyst, and is utilized to introduce —Cl and —SO_2Cl groups into organic compounds. Sulfuryl halides hydrolyze to HX and sulfuric acid and react with NH_3 to form sulfuryl diamide, $SO_2(NH_2)_2$.

Polysulfuryl halides, $X-SO_2-(O-SO_2)_n-X$, derivatives of polysulfuric acids, are obtained as mixtures from the action of liquid sulfur trioxide on halides like BF_3 or CCl_4:

$$CCl_4 + S_3O_9 \rightarrow S_3O_8Cl_2 + COCl_2$$

Peroxodisulfuryl fluoride, $S_2O_6F_2$ (b.p. $67\,^\circ C$) derived from $S_2O_5F_2$ by replacement of an oxygen bridge atom by a peroxo group, hence is the fluoride of peroxodisulfuric acid, $H_2S_2O_8$, formed in the reaction of fluorine with SO_3 in presence of AgF_2 with SO_3F_2 (b.p. $-31\,^\circ C$) as an intermediate:

$$SO_3 + F_2 \rightarrow F-SO_2-O-F \xrightarrow{+SO_3} F-SO_2-O-O-SO_2-F$$

The other SO_3X_2 compounds are obtained from the elements:

$$Cl_2 + S_2O_6F_2 \rightarrow 2Cl(SO_3F)$$
$$Br_2 + 2S_2O_6F_2 \rightarrow Br(SO_3F) + Br(SO_3F)_3$$
$$I_2 + S_2O_6F_2 \rightarrow I(SO_3F), \ I_3(SO_3F), \ I(SO_3F)_3$$

Fluorination of SOF_2 at $150\,^\circ C$ with a platinum metal catalyst furnishes thionyl tetrafluoride which is further fluorinated by CsF and F_2 to the colorless, reactive gas OSF_6:

$$O{=}SF_2 + F_2 \xrightarrow{Pt} O{=}SF_4 \xrightarrow{CsF, \ +F_2} FO-SF_5$$

8.12.4. Selenium and Tellurium Halides

The fluorides of selenium and tellurium are like those of sulfur, but the colorless gaseous hexafluorides, formed by burning the elements in fluorine, are more reactive than SF_6. Tellurium hexafluoride, a Lewis acid, hydrolyzes with liquid water already at room temperature within several hours and is converted by CsF to Cs_2TeF_8, and by tertiary amines to adducts, e.g., $(R_3N)_2TeF_6$.

Selenium tetrafluoride is a colorless liquid, formed by fluorination of selenium in an F_2/N_2 mixture.

Selenium reacts with chlorine to give pale-yellow crystals of $SeCl_4$ which sublime upon heating to ca. 195 °C with substantial dissociation into $SeCl_2$ and Cl_2. $SeCl_4$ is both a donor and an acceptor of chloride ions. It is formed from SeO_2 and concentrated HCl, and reacts with NH_4Cl to give the yellow ammonium hexachloroselenate(IV), which is crystallized by bubbling HCl into the solution:

$$SeO_2 + 4HCl \xrightarrow{-2H_2O} SeCl_4 \xrightarrow{+2NH_4Cl} (NH_4)_2SeCl_6$$

but $SeCl_4$ also reacts with $AlCl_3$ by heating in SO_2Cl_2 to give yellow crystals of $AlSeCl_7$, consisting of the ions $SeCl_3^+$ and $AlCl_4^-$ in which three $d(SeCl) = 2.11$ Å, and three $= 3.04$ Å, to form a distorted octahedron about selenium. The cation and anion are bound:

$$SeCl_3^+ \ AlCl_4^- \ \leftrightarrow \ Cl_3Se\overset{\oplus}{-}Cl\overset{\ominus}{-}AlCl_3$$

Crystalline $TeCl_4$ is similar, consisting of $TeCl_3^+$ ions bridged via chlorine to Te_4Cl_{16} molecules, in which each tellurium is surrounded by six chlorine atoms in a distorted octahedron.

The formation of the hexachloro- and bromo-anions, EX_6^{2-} and the cations EX_3^+ is typical of these elements, unlike sulfur. Sulfur halide analogues, such as the oxohalides $SeOCl_2$, $RSeOCl$, and SeO_2Cl_2, are known as well as the reaction:

$$TeCl_4 + H_2S_7 \rightarrow S_7TeCl_2 + 2HCl,$$

in which the orange-colored heptasulfur tellurium dichloride containing an eight-membered S_7Te ring is formed.

Bonding in the Selenium and Tellurium Halides:

The bonds in octahedral SeF_6 and TeF_6 are like those in SF_6, since the electronegativities of the sp^3d^2 hybridized central atoms are similar. In this configuration unoccupied orbitals remain, and additional atoms can be bound if the center atom is sufficiently large (as in TeF_8^{2-}).

The ions SeX_6^{2-} and TeX_6^{2-} have 14 electrons in the valence shell. Their structures are octahedral, although complete hybridization of all occupied valence orbitals would give a less symmetrical sp^3d^3 hybrid. The VSEPR theory also demands a lower than O_h symmetry. d-Orbital participation, however, is improbable, since chlorine and bromine are not sufficiently contracting. The free electron pair is apparently in an s-orbital of the valence shell, and only three p-orbitals are used to bind the ligands. Each orthogonal p-orbital can bind two *trans*-ligands in a three-center bond similar to that discussed for HF_2^- (p. 172) and I_3^- (p. 252) to give O_h geometry.

The ions $SeCl_3^+$ and $TeCl_3^+$ are isoelectronic with the fifth group chlorides, $AsCl_3$ and $SbCl_3$, and have the same symmetry, C_{3v}. The bridging of $TeCl_3^+$ ions via chlorine in Te_4Cl_{16} can be also rationalized by three-center bonds.

8.13. Sulfur-Nitrogen Compounds

Sulfur and nitrogen can be bound in three ways:

$$-\ddot{S}-\ddot{N}< \qquad >\ddot{S}=\ddot{N}- \qquad -\ddot{S}\equiv N:$$

Higher oxidation states of sulfur are possible after promotion of valence electrons. Tetrasulfur tetranitride, S_4N_4, formed in many reactions of sulfur compounds with ammonia is best produced by passing S_2Cl_2 over heated NH_4Cl in a complex redox reaction:

$$2S_2Cl_2 + 4NH_4Cl \rightarrow S_4N_4, \; HCl + \text{other compounds}$$

The orange-colored crystals of S_4N_4 consist of cage-like molecules in which the SN distances are almost identical:

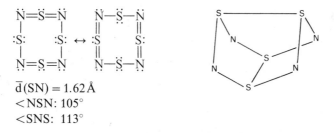

$$\bar{d}(SN) = 1.62 \, \text{Å}$$
$$<NSN: 105°$$
$$<SNS: 113°$$

Fig. 90.   Molecular structure of tetrasulfur tetranitride, S_4N_4

The spherical cage (see Figure 90) has sulfur atom distances, $d(SS) = 2.58$ Å, even larger than those in the dithionite ion, and much larger than the single bond distance (2.06 Å), but less than the van der Waals distance (3.5 Å), facing each other across the rings, to give partial bonds between the sulfur atoms. MO calculations indicate that these S—S bonds play an important role in the S_4N_4 cage, *vs.* in the open crown of the S_8 ring. Tetrasulfur tetranitride is thermodynamically unstable to decomposition to the elements, and it can detonate or pyrolyze. Figure 91 shows some of its reactions, including the formation of other rings and SN-multiple bonded compounds with delocalized π-electrons.

Disulfur dinitride, S_2N_2, forms colorless crystals containing planar rings with D_{2h} symmetry. The colorless crystals of $S_4(NH)_4$ contain rings isostructural with S_8 [$d(SN) = 1.67$ Å]. The colorless gas, NSF_3 (symmetry C_{3v}), contains SN triple bonds [$d(SN) = 1.42$ Å]. The ring of $(NSF)_4$, in which the fluorine atoms are bonded to sulfur, is puckered, containing localized SN double- [$d(SN) = 1.54$ Å] and single [$d(SN) = 1.66$ Å] -bonds, and bears no similarity with the ring in $S_4(NH)_4$. Sulfur nitride fluoride, NSF, is a colorless gas, which polymerizes to colorless crystals of $(NSF)_3$. Chlorination of S_4N_4 produces NSCl, which then polymerizes to solid $(NSCl)_3$.

Sulfur-nitrogen bonds can be synthesized via condensation and other reactions as,

$$2(CH_3)_2NH + Cl_2SO \rightarrow (CH_3)_2N\!-\!SO\!-\!N(CH_3)_2 + 2HCl$$
$$C_6H_5\!-\!NH_2 + Cl_2SO \rightarrow C_6H_5\!-\!N\!=\!S\!=\!O + 2HCl$$
$$3R\!-\!NH_2 + SF_4 \rightarrow R\!-\!N\!=\!SF_2 + 2[RNH_3]F$$
$$R\!-\!N\!=\!SF_2 + 3H_2N\!-\!R \rightarrow R\!-\!N\!=\!S\!=\!N\!-\!R + 2[RNH_3]F$$

Sulfur diimides are obtained by SO_2 elimination from N-sulfinyl compounds:

$$2R\!-\!N\!=\!S\!=\!O \rightarrow R\!-\!N\!=\!S\!=\!N\!-\!R + SO_2$$

The parent N-sulfinyl compound is thionyl imide, HNSO, generated in the stoichiometric, vapor phase reaction of NH_3 with $SOCl_2$ under reduced pressure:

$$3NH_3 + SOCl_2 \rightarrow HNSO + 2NH_4Cl$$

The ammonolysis of SOF_4 yields the imido derivative of sulfuryl chloride, indicating that the $=O$ and $=NH$ groups are exchangeable:

$$SOF_4 + 3NH_3 \rightarrow F_2S\!\!\begin{array}{c} \diagup O \\ \diagdown NH \end{array} + 2NH_4F$$

The reactions of sulfur halides in liquid ammonia are complex; for example, the reaction of S_2Cl_2 with NH_3, aside from S_8, yields the eight-membered ring sulfur imides S_7NH, $S_6(NH)_2$, and $S_5(NH)_3$, derived from S_8 by alternate substitution with imido groups, which can be separated by column chromatography. The final product is $S_4(NH)_4$. While there is only one heptasulfur imide, $S_7(NH)$, there are three separable solid isomeric hexasulfur diimides and two isomeric pentasulfur triimides, depending on the substitution pattern. All isomers have been isolated as colorless crystals.

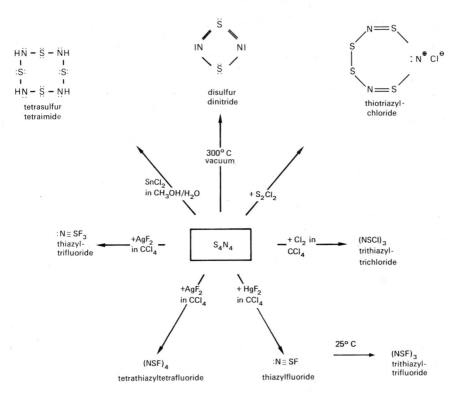

Fig. 91.   Some reactions of tetrasulfur tetranitride.

9. The Halogens

9.1. General

Of the five elements of the VIIth main group, only the first four, fluorine, chlorine, bromine and iodine, are of interest here. All the isotopes of astatine, discovered in 1940, are radioactive; its most stable isotope $^{210}_{85}At$ has a half-life of 8.3 hours. Some of the properties of the halogens are listed in Table 42.

Table 42   Properties of the Halogens (The dissociation energy, D^0_0, and internuclear distance, d, refer to the vapor phase.)

Element	Valence Electron Configuration	m.p. [°C]	b.p. [°C]	D^0_0 [kJ/mol]	d [Å]	Isotopes [Mol-%]
F	$2s^2p^5$	-220	-188	158	1.44	^{19}F: 100
Cl	$3s^2p^5$	-101	-34.6	244	1.99	^{35}Cl: 75.53;
						^{37}Cl: 24.47
Br	$4s^2p^5$	-7	$+59$	193	2.28	^{79}Br: 50.54;
						^{81}Br: 49.46
I	$5s^2p^5$	113.5	184.4	151	2.67	^{127}I: 100

Halogen Molecules:
The halogens form diatomic molecules in all states of aggregation. At low temperatures they are ordered in molecular lattices bound by van der Waals forces, explaining their low melting and boiling points. Fluorine and chlorine are yellow-green gases, bromine a red-brown liquid, and iodine forms glossy, gray-black crystals.

The rise in melting and boiling points from fluorine to iodine stems from the dispersion effect, based upon the polarizability of the atoms (p. 114) which increases with increasing atomic radius. Other forces come into play in the staggered, layered chlorine, bromine and iodine lattices in which molecules in neighboring layers are above and below holes. The van der Waals forces are reflected in the large interlayer distances (for iodine: d = 4.35 to 4.50 Å), and the cleavage of crystals parallel to the layers. Within the layers, however, smaller than van der Waals distances are noted, so that weak, partial covalent bonds may be assumed. The effect becomes more pronounced from chlorine to iodine. The smallest distances, e.g., in iodine are 3.57 and 4.05 Å (single bond distance 2.67 Å, van der Waals distance 4.4 Å), at which distances considerable orbital overlap must exist, to form intermolecular, many-center σ-bonds, spread through the layer and populated with delocalized electrons, reflected in the properties of iodine (lustre, color, moderate electric conductivity). Orbital overlap can take place in the iodine lattice:

$$
\begin{array}{ccccc}
& \text{I} & & & \\
& | & & & \\
\cdots \text{I---I} & \cdots\cdots \text{I} & \cdots\cdots \text{I} & \cdots\cdots \text{I---I} \\
\uparrow & \uparrow & \uparrow & | \\
d_1 & d_2 & d_3 & \text{I}
\end{array}
\qquad
\begin{array}{l}
d_1 = 2.67\,\text{Å} \\
d_2 = 3.57\,\text{Å} \\
d_3 = 4.05\,\text{Å}
\end{array}
$$

The many-center bonds which are important in halogen chemistry, are described for the polyhalides on p. 253.

Halogen Atoms:

The halogens dissociate at high temperatures. The degree of thermal dissociation, α, depends on pressure, temperature and dissociation energy (Table 42), reaching a maximum for fluorine, owing to repulsion of free electron pairs. At 1000 K and 1 bar, α is ca. 4% for F_2, 0,03% for Cl_2, 0.2% for Br_2, and 3% for I_2.

Reactivity:

Fluorine is the most reactive element. It reacts at $> 25\,^\circ\text{C}$ with all other elements, except with O_2, He, Ne, Ar, and Kr, and with many inorganic and with most organic compounds. The reactions are frequently vigorous or explosive. Chlorine, bromine and iodine are much less reactive.

The high reactivity is related to the small dissociation energy of the F_2-molecule leading to a low activation energy, and to the greater strength of the bonds fluorine forms. The heat liberated accelerates the reactions. The mean bond energies of CX bonds in the carbon tetrahalides, CX_4, are:

C—F	C—Cl	C—Br	C—I
486	327	276	239 kJ/mol

The high bond energies in covalent fluorides are associated with the ionic-covalent resonance energy, which is a function of the electronegativity difference (p. 140). For fluorine, the difference in X_E-values is particulary large, but the O—F and N—F bonds have low bond energies, owing to electron pair repulsion.

Halide Ions:

Negative halide anions with rare gas configurations and spherical symmetry form the salts to which halogens (= salt former) owe their name.

The tendency for ionization:

$$X_2 + 2e^- \rightleftharpoons 2X^-$$

decreases markedly from F_2 to I_2. The equilibrium lies to the right the more electro-negative the halogen. The free energy, $\Delta G°$, decreases markedly form F_2 to I_2, as can be seen in the corresponding reduction potentials, $E°$, defined:

$$E° = \frac{-\Delta G°}{n \cdot \mathfrak{F}}$$

n = number of electrons (here, two)
$\mathfrak{F}$ = Faraday constant, 96,487 C/mol

The $E°$ values are:

F_2	Cl_2	Br_2	I_2	At_2
+2.87	+1.36	+1.07	+0.54	+0.3 Volt

As shown by the thermodynamic relations, the halogens displace their higher homologues from salts, e.g., fluorine reacts with $CaCl_2$:

$$F_2 + CaCl_2 \rightarrow Cl_2 + CaF_2$$

The strengths of the hydrogen halides as reducing agents likewise increases from HF to HI. While HF, or F^- can only be oxidized to F_2 electrochemically, weak oxidizing agents liberate I_2 from HI or the iodides.

Covalent Compounds:
The monovalent halogens have oxidation states -1 or $+1$ in covalent compounds, for example, in the binary interhalogen compounds ClF, BrF, and ICl. The oxida-tion state of fluorine in all compounds, except F_2, is -1, since it is always negative. The valence shell of the fluorine atom lacks d-orbitals, and hence in F^- and $-$ F fluorine is electronically saturated. Activation of the 3s and 3p orbitals for covalent bonding is excluded, owing to their high orbital energies. Promotion of a 2p-electron to the 3s or 3p level requires 1227 and 1390 kJ/mole respectively, energies which prohibit bond formation.
Singly-bonded fluorine, being a Lewis base can enter into additional coordinate bonds with coordination number 2, e.g., in hydrogen-bonded fluoro-compounds (p. 167), and in fluorides such as polymeric beryllium difluoride, $(BeF_2)_n$:

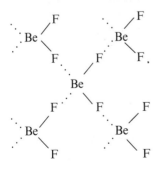

The beryllium atoms are tetrahedrally coordinated, and all atoms acquire rare gas configurations.

Excepting hydrogen bonds fluorine compounds with coordination higher than two are unknown. Owing to the high ionization energy, compounds with a fluoro-cation, F^+, or positively polarized fluorine cannot exist but this state is known for the remaining halogens, whose ionization energies decrease markedly with increasing atomic weight. Fluorine and oxygen form compounds with positively polarized Cl-, Br-, and I-atoms, as in interhalogens and in bromine nitrate, $BrONO_2$, chlorine hydroxide, ClOH, iodine trisfluorosulfate, $I(SO_3F)_3$, etc.

The cations F^+, Cl^+, Br^+ and I^+ are such strong Lewis acids, i.e., so electrophilic, that they are only known coordinated, for example in pyridine, C_5H_5N, salts[9]:

$$[F(py)]F \quad [Cl(py)_2]NO_3 \quad [Br(py)_2]NO_3 \quad [I(py)]NO_3 \quad [I(py)_2]NO_3$$

In the heavier halogens, higher coordination numbers are possible with fluorine, oxygen and chlorine ligand atoms using d-orbitals. Chlorine and bromine reach a maximum coordination number of five (in ClF_5, BrF_5) and iodine seven (in IF_7), derived from the atomic states:

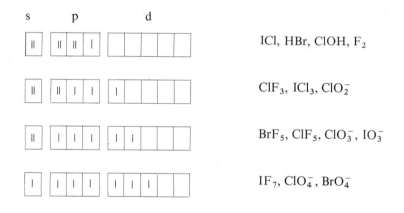

ICl, HBr, ClOH, F_2

ClF_3, ICl_3, ClO_2^-

BrF_5, ClF_5, ClO_3^-, IO_3^-

IF_7, ClO_4^-, BrO_4^-

Fluorine has a special position among the halogens similar to oxygen in the chalcogen group, and will be dealt with separately.

9 In the fluorine compound, prepared from fluorine and pyridine at $-78\,°C$, the positive charge is most likely associated with the pyridine; the others may have positively polarized halogen atoms.

9.2. Fluorine

9.2.1. Preparation

Fluorine occurs as fluorspar, CaF_2, cryolite, Na_3AlF_6, and fluoroapatite, $Ca_5(PO_4)_3(F, Cl)$. Fluorspar is reacted with dilute sulfuric acid to give aqueous hydrofluoric acid (HF aq):

$$CaF_2 + H_2SO_4(aq) \rightarrow 2HF(aq) + CaSO_4$$

Anhydrous hydrogen fluoride is obtained by fractional distillation (m.p. $-83\,°C$, b.p. $19.5\,°C$).
Elemental fluorine, owing to its high standard potential can only be prepared by anodic oxidation in electrolytes which lack other anions like, e.g., OH^-. Anhydrous HF is only weakly dissociated.

$$3HF \rightleftharpoons H_2F^+ + HF_2^- \qquad [H_2F^+][HF_2^-] \approx 10^{-10}$$

For the electrolysis, $KF \cdot xHF$ melts are used, where x is 1 to 13, and the melting points between $-100°$ and $250\,°C$ (the higher the HF content, the lower the m.p.). The fluoride ions in the electrolyte are strongly solvated as $H_nF_{n+1}^-$ ions.

The electrolysis cells and cathodes are steel or copper; the anodes are nickel or graphite-free carbon. The electrolyte composition is kept constant by addition of HF. The operating voltage is 10 V. The fluorine generated is contaminated with HF and CF_4, and is purified by cooling to $-140\,°C$. Fluorine is commercially available in cylinders.

9.2.2. Properties of Fluorine

Elemental fluorine and HF are corrosive, etching chemicals; F_2 is recognized even at 0.01 ppm by its odor which is similar to O_3 or chlorine. Fluorine and some co-valent fluorides such as S_2F_{10}, PF_3, etc., are toxic.

Few materials withstand fluorine, but some metals and alloys, whose surfaces form an impermeable film of metal fluoride, such as Fe, Al, Mg, Ni, brass, bronze, and a Ni—Cu alloy (Monel), are resistant (passivation). Glass and ceramics cannot be used, since SiO_2 is fluorinated to SiF_4 in the presence of trace amounts of HF:

$$SiO_2 + 4HF \rightarrow SiF_4 + 2H_2O$$

$$2H_2O + 2F_2 \rightarrow 4HF + O_2$$

Polytetrafluoroethylene (Teflon) is used for gaskets, condensers and tubing.
To destroy small amounts of F_2, $CaCl_2$, Al_2O_3 or 30% KOH is used:

$$2Al_2O_3 + 6F_2 \rightarrow 4AlF_3 + 3O_2$$

$$2KOH + F_2 \rightarrow 2KF + H_2O + \tfrac{1}{2}O_2$$

9.2.3. Preparation of Fluorides

Fluorides are prepared from HF, either through elemental fluorine or directly by reaction of hydrogen fluoride.

Hydrogen fluoride is associated (cf. p. 167) in all phases, and resembles water as a solvent. Salts with the anions F^-, HF_2^-, $H_2F_3^-$, and $H_3F_4^-$ are derived from it. Pure HF is made on a laboratory scale by thermal decomposition of KHF_2. Fluoride ion-acceptors behave as acids in HF.

$$2\,HF + SbF_5 \rightleftharpoons H_2F^+ + SbF_6^-$$

and soluble ionic fluorides behave as bases. In water HF is weakly acidic (p. 160), but etches glass and is stored in polyethylene, platinum, lead or paraffinized glass vessels.

Covalent and ionic fluorides are prepared by:

a) Reaction of anhydrous or aqueous HF with an oxide, hydroxide or carbonate to prepare KF, KHF_2, NH_4F, NH_4HF_2, BaF_2, AlF_3, Na_3AlF_6, BF_3, KBF_4, SiF_4, etc.

b) Halogen exchange with HF:

$$PCl_3 + 3\,HF \xrightarrow{50°} PF_3 + 3\,HCl$$

c) Fluorination with F_2:

$$\tfrac{1}{8}S_8 + 3\,F_2 \rightarrow SF_6$$

$$I_2 + 5\,F_2 \rightarrow 2\,IF_5$$

$$U_3O_8 + 9\,F_2 \rightarrow 3\,UF_6 + 4\,O_2$$

$$AgCl + F_2 \rightarrow AgF_2 + \tfrac{1}{2}Cl_2$$

Many fluorine reactions are accompanied by ignition and strong evolution of heat, for example, solid F_2 and liquid H_2 react vigorously even at 20K, and rocket engines have been fueled with F_2 in combination with H_2, B_2H_6, N_2H_4, C_2H_5OH, Li, LiH, and BeH_2 yielding reaction temperatures of 4000 to 5600 K.

d) Fluorination with fluorinating agents (AgF, AgF_2, CoF_3, MnF_3, ClF_3, IF_5, AsF_5, KSO_2F):

$$3\,PCl_5 + 5\,AsF_3 \xrightarrow{25\,°C} 3\,PF_5 + 5\,AsCl_3$$

$$\tfrac{3}{8}S_8 + 2\,AgF \longrightarrow S_2F_2 + Ag_2S$$

e) Electrolytic fluorination:

The starting material is dissolved in anhydrous HF and KF is added to increase the conductivity. Electrolysis is carried out using a large Ni anode with voltages selected (<8 V) to inhibit F_2-evolution:

$$H_2O \rightarrow OF_2, O_3 \qquad\qquad H_2S \rightarrow SF_6$$

$$NH_4^+ \rightarrow NF_3, N_2, N_2F_2, NH_2F \qquad CS_2 \rightarrow CF_3SF_5, SF_6$$

$$OC(NH_2)_2 \rightarrow NF_3, COF_2 \qquad\qquad HSO_3F \rightarrow SO_2F_2, OF_2$$

$$NaClO_4 \rightarrow FClO_3$$

Hydrogen fluoride dissolves inorganic and organic compounds, making the electrolytic procedure widely applicable but the products depend strongly on the reaction conditions.

9.2.4. Fluorinated Hydrocarbons

About 80% of the HF produced from fluorspar is used in the manufacture of C—Cl—F compounds:

$$CCl_3F \qquad CHCl_2F \qquad CCl_2F—CCl_2F$$

$$CCl_2F_2 \qquad CHClF_2 \qquad CClF_2—CClF_2$$

The Freons are colorless, non-toxic, and are characterized by low boiling points, high chemical resistance, non-flammability, and good solvent properties. They are prepared from the corresponding C—Cl or C—Cl—H compounds by partial halogen exchange with HF in the presence of SbF_5 as catalyst, and are used as coolants, propellants for aerosols, in foam production, and as solvents for active compounds such as SO_3.

Thermolysis of $CHClF_2$ with loss of HCl yields tetrafluoroethylene, which is polymerized in a radical process to polytetrafluoroethylene (Teflon), $(—CF_2—CF_2—)_n$, which has high chemical resistance, being attacked only by F_2, ClF_3 and other fluorinating agents and by molten alkali metals, and has a use range of -200 to $+260\,°C$. Similarly resistant are polytrifluorochloroethylene $(—CF_2—CFCl—)_n$, (Kel–F), as well as Viton $(—CHF—CF_2—CF_2—)_n$ and Teflon FEP $(—CH_2—$ $—CF_2—CF(CF_3)—)_n$, available as oils, greases, waxes, or solids depending on the degree of polymerization.

9.2.5. Bond Relationships in the Fluorides

The non-metal fluorides are discussed with their respective elements.
Fluorine is closer related to oxygen than to the other halogens in electronegativity:

	F	O	Cl	Br	I
x_E(Allred-Rochow):	4.1	3.5	2.8	2.7	2.2

and in atomic and ionic radii:

	F	O	Cl	F$^-$	O^{2-}	Cl$^-$
r[Å]:	0.64	0.66	0.99	1.36	1.40	1.81

This results in certain analogies in the crystal chemistry of fluorides and oxides and in the ability of fluorine to promote elements to their highest oxidation states, as in AgF_2, K_2NiF_6 and SF_6.

Higher coordination numbers are attained with fluorine than with other halogens, and analogous chlorides, bromides and iodides do not exist for:

$$NF_4^+ \quad SF_6 \quad XeF_6 \quad UF_6 \quad IF_7 \quad ReF_7 \quad OsF_7 \quad XeF_8^{2-} \quad TeF_8^{2-}$$

Higher bond energies or steric reasons may be responsible.

Despite its high electronegativity, fluorine forms with all non-metals polar compounds, and not salts. Fluorine exerts a strongly inductive effect in these covalent bonds, however, as illustrated by:

a) The SO-bond in SOF_2 is stronger than in any other thionyl compound, and likewise for the sulfuryl, seleninyl, selenyl, and phosphoryl compounds;
b) $(CF_3)_3N$ is a weaker Lewis base than $(CH_3)_3N$, since the electronegativity of the CF_3-group is ca. 3.5, vs. 2.5 for the CH_3-group.
c) CF_3COOH is a much stronger acid than CH_3COOH; and
d) NF_3 is less basic than NH_3 and thus forms ammonium compounds (NF_4^+) only with extremely strong Lewis acids.

Fluoride Ion Acceptors:

Fluorides, such as BF_3, AsF_5, SbF_5 PF_5, PtF_5, AlF_3, and SiF_4 can add fluoride ions to become complex anions, e.g.:

$$AlF_6^{3-} \quad SiF_6^{2-} \quad PF_6^- \quad SF_6$$

which are isoelectronic with SF_6. The central atoms are octahedral (sp^3d^2-hybrids), and owing to d-orbital occupation and steric reasons, resist nucleophilic (H_2O) attack. Whereas SiF_4 and PF_5 hydrolyze immediately, the hexafluoride ions dissolve in water without decomposition, and survive in melts. The corresponding acids H_3AlF_6, H_2SiF_6, and HPF_6, which are completely dissociated in aqueous solution, cannot be isolated, but decompose into HF and the corresponding fluorides, and upon cooling H_2SiF_6-solutions, the oxonium salt $(H_3O)_2SiF_6$ crystallizes.

Covalent fluorides accept fluoride ions to give salts, for which no other halogen analogues exist:

$$IF_7 + AsF_5 \rightarrow [IF_6][AsF_6]$$

$$2XeF_2 + PtF_5 \rightarrow [Xe_2F_3][PtF_6]$$

$$O_2F_2 + PF_5 \rightarrow [O_2][PF_6] + \tfrac{1}{2}F_2$$

$$\left\{ \begin{array}{l} 3ONF + 2IrF_6 \rightarrow 2[NO][IrF_6] + ONF_3 \\ ONF_3 + AsF_5 \rightarrow [ONF_2][AsF_6] \end{array} \right\}$$

$$NF_3 + F_2 + SbF_5 \xrightarrow[185\ bar]{200\,°C} [NF_4][SbF_6]$$

This Lewis acid behavior arises from the unoccupied p- or d-orbitals of the valence shell, and the positive charge generated by the inductive effect of the fluorine ligands. In the absence of Lewis base, intra- or intermolecular coordinate bonds are formed with the central atom. Intramolecular π-bonds are present in BF_3, SiF_4, and PF_3 (p. 111), while the fluorine bridging in SbF_5 corresponds to $(BeF_2)_n$, with intermolecular coordinate bonds. Gaseous SbF_5 has a trigonal bipyramidal structure, but in the liquid phase below 15 °C (m.p. 7 °C), *cis*-connected SbF_6 octahedra are formed, giving rise to chains or rings:

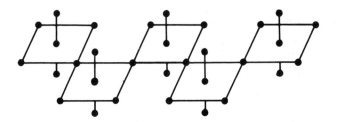

The structure in solution is determined from the ^{19}F NMR spectrum. Fluorine consists only of ^{19}F atoms with nuclear spin, $I = \tfrac{1}{2}$, and the position of the NMR signal depends on the state of the fluorine atoms. In polymeric SbF_5, there are three kinds of fluorines:
a) the two axial fluorine atoms
b) the two singly-coordinated fluorine atoms in the equatorial plane, and
c) the bridge fluorine atom.
Since the intensity of the resonance signal is proportional to the number of fluorine atoms, the lines in the expected ^{19}F NMR spectrum for SbF_5 should have an intensity ratio of 2 : 2 : 1, which is observed. If the SbF_6 octahedra were connected in *trans*-orientation, only two lines with intensity ratio 4 : 1 would be observed.

SbF_5 is a tetramer in crystalline state, with the $(SbF_5)_4$ units made up of slightly distorted *cis*-joined SbF_6 octahedra.

9.2.6. Stabilization of Lower Oxidation States

Subvalent compounds, in which non-metals are found in unusual oxidation states, are formed preferentially with fluorine and oxygen.

Carbon monoxide is the most common suboxide. It is thermodynamically unstable with respect to CO_2 and graphite, but owing to the high activation enthalpy and entropy, CO is metastable at room temperature. The stability of other subvalent compounds is similarly kinetically based.

Boron monofluoride, BF, is an isostere of CO, and can be prepared by reducing BF_3 with crystalline boron:

$$BF_3(g.) + 2\,B(s.) \xrightarrow[\text{1 torr}]{2000\,°C} 3\,BF(g.) \xrightarrow{+BF_3} F_2B{-}BF_2, F_2B{-}BF{-}BF_2$$

but reacts on cooling immediately with BF_3 by insertion into the B—F bond to give B_2F_4 and with the latter to give B_3F_5. With CO, $(F_2B)_3BCO$ is obtained and with PF_3, the crystalline $(F_2B)_3BPF_3$. The BF/BF_3 mixture is reacted after the heated zone, and immediately quenched with liquid nitrogen. In contrast to CO, BF is single bonded as evident from the internuclear distance and the force constant. The free electron pair and the unoccupied orbitals at boron give carbene-like properties, and BF reacts by addition and insertion reactions with formation of two new single bonds. It disproportionates at high temperatures or in the presence of boron to boron and BF_3, and polymerizes at low temperatures.

The subvalent fluorides CF_2 and SiF_2 are also carbene analogues, but are more stable than BF; CF_2 is formed in the thermolysis and photolysis reactions:

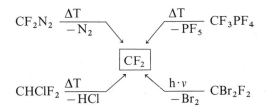

At the low pressures of the pyrolysis reaction, the half-life for dimerization to C_2F_4 is ca. 1 sec, and C_2F_4 reacts further with additional CF_2 to form cyclo-C_3F_6, cyclo-C_4F_8, etc., which are intermediates in the thermal polymerization yielding polytetrafluoroethylene (p. 245). The CF_2 and SiF_2 molecules are angular, diamagnetic, and single bonded. Silicon difluoride, which has a half-life of ca. 150 sec in the vapor phase (0.1 torr), is formed in high yield in the reduction of SiF_4 with silicon, and in the thermal disproportionation of Si_2F_6:

$$SiF_4(g.) + Si(s.) \xrightarrow[\text{0.1–0.2 torr}]{1150\,°C} SiF_2(g.) \xleftarrow[-\,SiF_4]{700\,°C} Si_2F_6(g.)$$

In the absence of other reactants, SiF_2 inserts into available SiF_4 to form perfluoro-polysilanes, Si_nF_{n+2} (n = 1 to 14), or forms biradical chain polymers, $(SiF_2)_n$, which can be quenched and identified in the polymerization. Silicon difluoride also inserts into BF_3 as with SiF_4, i.e., to form $F_2B(SiF_2)_nF$ (n = 2, 3 colorless liquids). Hexafluorobenzene forms $C_6F_5SiF_3$ and $C_6F_4(SiF_3)_2$, while with benzene:

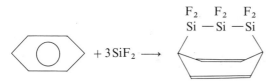

Subvalent compounds are important as reaction intermediates, and for prepara-tive purposes. The subfluorides NF_2 (p. 297) and O_2F (p. 192) are free radicals, in equilibrium with their dimers, N_2F_4 and O_4F_2.

9.3. Chlorine, Bromine, Iodine

9.3.1. Preparation and Properties of the Elements

Chlorine:
Owing to its high reactivity, chlorine occurs only bound, mainly as chlorides of sodium, potassium and magnesium. Sodium chloride, found in sea water, inland lakes, and in underground deposits created by evaporation, is the chief starting material for chlorine production which is carried out by electrolysis of aqueous NaCl solution (anodic oxidation of chloride ions):

$$2\,Na^+ + 2\,Cl^- + 2\,H_2O \rightarrow 2\,Na^+ + 2\,OH^- + H_2\uparrow + Cl_2\uparrow$$

Electrolysis of hydrochloric acid is also feasible. In the laboratory the oxidation of chloride ion with oxidizing agents in acid solution is used, e.g., by reacting concentrated HCl with MnO_2 or $KMnO_4$:

$$4\,HCl + MnO_2 \rightarrow Cl_2 + MnCl_2 + 2\,H_2O$$

Chlorine is a heavy, greenish-yellow, toxic gas with a pungent odor which attacks the mucosa. It is readily liquified by pressure (vapor pressure at $20\,^\circ C$: 6.5 bar), and reacts with metals often with spontaneous ignition, e.g., with alkali and alkaline earth metals, Cu, Fe, As, Sb, and Bi. Among non-metals, H_2 reacts after ignition explosively in a chain reaction. Many hydrogen compounds react vigorously:

$$C_2H_2 + Cl_2 \rightarrow 2\,C(s.) + 2\,HCl$$

$$NH_3 + 3\,Cl_2 \rightarrow NCl_3 + 3\,HCl$$

$$H_2S + 2\,Cl_2 \rightarrow SCl_2 + 2\,HCl$$

Cl_2 is soluble in water to give a concentration-dependent equilibrium (p. 258):

$$Cl_2 + H_2O \rightleftharpoons HOCl + HCl$$

Bromine:
Bromine occurs in nature bound in salts analogous to the chlorides and is obtained by bubbling chlorine into the acidic bromide solutions from sea water and from enriched solutions from the production of KCl:

$$2\,Br^- + Cl_2 \rightarrow Br_2 + 2\,Cl^-$$

In the laboratory, bromine is generated by oxidation of KBr in acid solution, using, e.g., MnO_2 and H_2SO_4.
Bromine is a deep red-brown liquid at room temperature. It is less soluble in water than chlorine, but completely miscible with non-polar solvents. Bromine is less reactive than chlorine but in general both elements react in a similar way.

Iodine:
Iodine is found in nature in iodides, iodates and bound in organic compounds. The burning of seaweed, which concentrates iodine, gives an ash rich in iodide salts. Chilean nitre ($NaNO_3$) contains iodine as $Ca(IO_3)_2$, which is recovered as $NaIO_3$. Iodine and iodides are produced by reduction. Conversion of iodides to I_2 requires only weak oxidizing agents or electrolysis. Iodine forms gray-black, lustrous, scaly, sublimable crystals. Its vapor as well as solutions in CCl_4, $CHCl_3$, or CS_2 are violet colored, but in water, ether, dioxane, etc., iodine is brown and with aromatic hydrocarbons red solutions are obtained. The brown and red colors are ascribed to intermolecular interactions in charge transfer complexes in which electronic transition from solvent molecules (donors) to I_2-molecules take place. Such complexes, also known for Br_2 and Cl_2, are characterized by strong light absorption.

9.3.2. Halides

Almost all elements form binary halides in which the halogen is the more electronegative, which may crystallize in an ionic lattice (KCl, AlF_3), form polymers (graphite fluoride, $[SiCl_2]_n$), or exist in small molecules (SCl_2, PF_5). More complex compounds are also known, e.g., the oxyhalides (SOF_4), hydroxyhalides (HSO_3Cl), complex anions (SiF_6^{2-}), and cations (PCl_4^+), etc.

The fluorides of hydrogen and some non-metals were treated in Section 9.2. and the chlorides, bromides and iodides are treated with their respective elements. Interhalogen compounds are discussed in Sections 9.3.3 and 9.3.4.

Hydrogen Halides HCl, HBr, and HI:
Hydrogen halides can be synthesized from the elements. The free energy, $\Delta G°$, (negative), of the reaction:

$$H_2 + X_2 \rightleftharpoons 2\,HX \qquad X:\text{ Halogen (F, Cl, Br, I)}$$

decreases from fluorine to iodine, and the heavier the halogen, the less the equilibrium lies to the right. Fluorine and chlorine can be quantitatively converted in a hydrogen atmosphere, but HBr is only obtained in good yield at 150 to 300 °C, in the presence of a catalyst (active carbon or platinum). The formation of HI from the gaseous elements is only exothermic by 4 kJ/mol and HI dissociates to the elements on heating. The homogeneous vapor phase reactions proceed by radical mechanisms.

Hydrogen chloride is made like HF by protonation of chloride ions, i.e., by reacting NaCl with conc. H_2SO_4 to yield HCl and $NaHSO_4$. The $NaHSO_4$ reacts with additional NaCl to give Na_2SO_4 only at red heat. Also, HCl is formed in the chlorination of organic compounds:

$$\gtrless C{-}H + Cl_2 \;\rightarrow\; \gtrless C{-}Cl + HCl$$

HBr can likewise be made from KBr and nonvolatile, nonoxidizing acids (e.g., H_3PO_4), but the hydrolysis of bromides is easier:

$$PBr_3 + 3\,H_2O \;\rightarrow\; 3\,HBr{\uparrow} + H_3PO_3$$

Hydrogen iodide is obtained analogously from PI_3.

The hydrogen halides are colorless, readily liquified gases which form molecular lattices. Hydrogen fluoride is strongly associated by hydrogen bonding in all phases, while in the other compounds dipole association is only present in condensed phases. The dipole moments decrease from HF (1.9 D) to HI (0.4). Like HF, the other hydrogen halides form acid salts, $M(HX_2)$, where the anions are HX complexes of X^-.

Hydrogen halides dissolve in water in which they react to form unstable oxonium and hydronium salts (p. 156). The acid strengths increase from HF to HI (p. 160). Owing to strong hydrogen bonding, HCl is also very soluble in alcohols and ethers.

A characteristic property of HBr and HI and their anions is easy oxidation, as predicted from their redox potential values (p. 241). Aqueous iodide solutions turn yellow and finally brown upon exposure to air (formation of I_3^-).

9.3.3. Polyhalide Ions

More elemental iodine dissolves in aqueous potassium iodide than in pure water owing to complex formation between I_2 molecules and iodide ions:

$$I^- + I_2 \rightleftharpoons I_3^- \qquad\qquad K_B = \frac{[I_3^-]}{[I^-][I_2]}$$

The black salt $KI_3 \cdot H_2O$ can be isolated from the dark brown solution. Larger cations permit isolation of anhydrous triiodides, as in RbI_3, CsI_3, and $[Co(NH_3)_6] [I_3]_3$.

The tendency to form polyhalides is less pronounced for the lighter homologs, but crystalline NH_4Br_3 and $[(CH_3)_4N]Cl_3$ have been prepared. Mixed polyhalide ions, e.g., ICl_2^-, I_2Br^-, $IBrF^-$, and $BrCl_2^-$, etc., are generated from a halide (or X^- donor) and a halogen molecule, or an interhalogen compound:

$$PCl_5 + ICl \rightarrow [PCl_4] [ICl_2]$$
$$Cs[ICl_2] + ICl \rightarrow Cs[I_2Cl_3]$$

Large monovalent cations (p. 69) are used in the isolation of such salts.

The complex formation constant, K_B, defined above, has values in water at $25\,^\circ C$: I_3^- : 725; ICl_2^- : 167; Br_3^- : 18; Cl_3^- : 0.01. The triiodides are the most stable, and the same holds for the anhydrous salts, which dissociate upon heating. With CsI_3, the iodine equilibrium pressure reaches 1 bar only at $250\,^\circ C$.

The trihalide ions are linear and in solution the internuclear distances are equal (symmetry $D_{\infty h}$). In the crystalline state, however, asymmetrical interactions with cations reduces the symmetry. Thus, the d(I-I) internuclear distances in NH_4I_3 are 2.79 and 3.11 Å (symmetry $C_{\infty v}$). In the mixed ions, the most electropositive halogen is at the center.

The internuclear distances and valence force constants show the bonds to be weaker in the anions than in the molecules X_2 (see Table 43).

Table 43   Force Constants of Trihalide Ions and the Corresponding Halogen Molecules (in mdyn/Å)

Br_2: 2.46	BrCl: 2.67	I_2: 1.72	ICl: 2.38
Br_3^-: 0.94	$BrCl_2^-$: 1.08	I_3^-: 0.70	ICl_2^-: 1.06

These findings can be described within the framework of VB-theory as follows:

$$:\ddot{B}r:^{\ominus} \quad :\ddot{B}r\!-\!\ddot{B}r: \quad \leftrightarrow \quad :\ddot{B}r\!-\!\ddot{B}r: \quad :\ddot{B}r:^{\ominus}$$

which gives a mean bond order of 0.5, without stipulating d-orbital participation at the center atom.

A more elegant interpretation is furnished by MO theory. If the z axis is made to coincide with the molecular axis, only the three $5p_z$-orbitals of the iodine atoms overlap in the linear I_3^- ion. The interaction of these three σ-orbitals leads to three molecular orbitals, as in the linear HF_2^- ion (p. 172). Of these three molecular orbitals, one is bonding, one non-bonding, and one anti-bonding (Figure 92).

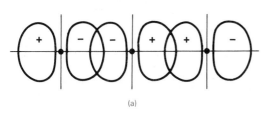
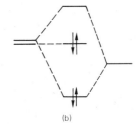

(a)   (b)

Fig. 92.   Three-center bond in the I_3^- ion. a) Overlap of three p-orbitals forming the bonding molecular orbital, b) Splitting of atomic orbitals to molecular orbitals

There are four valence electrons in these molecular orbitals, but since only two are bonding, an MO bond order of 0.5 results. A valence angle of 180° is favored for the 3-center-4-electron bond, since smaller angles would reduce the overlap.

Iodine atoms tend to form many-center bonds as in crystalline iodine and the anions I_5^-, I_7^-, I_9^-, and I_8^{2-}, which have been isolated from aqueous solution as I_2-molecular adducts of I^- and I_3^-. The I_5^- ion in $[(CH_3)_4N]I_5$ has a symmetry C_{2v}:

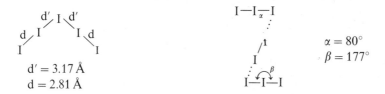

$$d' = 3.17 \text{ Å}$$
$$d = 2.81 \text{ Å}$$

$$\alpha = 80°$$
$$\beta = 177°$$

The planar I_5^- ion consists of a central iodide ion bonded to two I_2 molecules via two p-orbitals. Internuclear distance data for the ions I_7^- and I_9^- indicate addition of two and three I_2 molecules, respectively, to the I_3^- ion. The anions are further interconnected by weak, many-center bonds in the solid state. The diamagnetic salt, Cs_2I_8, contains the Z-shaped I_8^{2-} anion (see above) made up of two asymmetric I_3^- ions connected to a central I_2-molecule.

Of possible bromine and chlorine analogues of the polyiodides, only a pentabromide, Br_5^-, is known.

9.3.4. Positive Halogen Ions

Positive ions exist for the chalcogens (O_2^+, S_8^{2+}, Se_4^{2+}, etc.), more stable for the less electronegative elements. Strong oxidizing agents are required to produce positive ions of the non-metals. Owing to the nucleophilic character of such ions, they can only be prepared with weakly nucleophilic anions.

Iodine is the easiest halogen to oxidize, and molten iodine has a conductivity based on self-dissociation:

$$3I_2 \rightleftharpoons I_3^+ + I_3^-$$

The I_3^+ ion is also accessible chemically and bromo- and chloro-cations are synthesized by oxidation of the elements with interhalogen compounds or other strong oxidizing agents:

$$Cl_2 + ClF + AsF_5 \xrightarrow{-78\,°C} [Cl_3][AsF_6] \quad \text{yellow crystals}$$

$$2\,ClF + BF_3 \xrightarrow{-130\,°C} [Cl_2F][BF_4] \quad \text{colorless crystals}$$

These two salts revert to the starting materials at 25 °C.

$$4\tfrac{1}{2}\,Br_2 + BrF_5 + 15\,SbF_5 \;\rightarrow\; 5\,[Br_2][Sb_3F_{16}] \quad \text{red, paramagnetic crystals, m.p. 69 °C}$$

$$4\,Br_2 + BrF_3 + 3\,AsF_5 \;\rightarrow\; 3\,[Br_3][AsF_6] \quad \text{brown, subliming crystals, decomp. 70 °C}$$

$$1\tfrac{1}{2}\,Br_2 + [O_2][AsF_6] \;\xrightarrow{\;-O_2\;}$$

The iodine ions I_2^+, I_3^+, I_5^+, I_4^{2+}, and the mixed cations I_2Cl^+ and I_2Br^+ have been identified. The blue, paramagnetic I_2^+, is generated by dissolving I_2 in fuming sulfuric acid (65 % SO_3 in H_2SO_4), in which SO_3 is reduced to SO_2. In sulfuric acid solution, I_2 is oxidized by $S_2O_6F_2$ to I_2^+ which on cooling presumably dimerizes to the diamagnetic I_4^{2+}.

Dissolving I_2 in oleum, which only contains 25 % SO_3, or oxidizing an I_2—H_2SO_4 mixture with HIO_3 results in the brown I_3^+ ion, which reacts with additional I_2 to give the brown I_5^+.

The polyhalo cations are strong electrophiles and are only stable in solutions of very low Lewis basicity. They disproportionate in water.

The covalent bonds in the cations Cl_2^+, Br_2^+ and I_2^+ are stronger than in the corresponding halogens Cl_2, Br_2 and I_2, as seen in the shorter internuclear distances and increased force constants (Table 44).

Table 44   Internuclear Distances and Force Constants in Halogen Compounds X_2 and X_2^+

	Cl_2	Cl_2^+	Br_2	Br_2^+	I_2	I_2^+
f[mdyn/Å]	3.38	4.42	2.41	3.05	1.70	2.12
d[Å]	1.98	1.89	2.27	2.15	2.66	2.56

This strengthening is understandable from the MO diagrams in which the uppermost occupied levels are the antibonding π^*-orbitals. Oxidation of X_2 to X_2^+ removes one antibonding electron, and the MO bond order increases from 1.0 to 1.5.

Cations of type X_3^+ are angular; Cl_3^+ is isoelectronic with SCl_2, and like the latter has single bonds. The ion Cl_2F^+ has symmetry C_s, i.e., the atomic arrangement is ClClF. In neither cation do d-orbitals participate in the bonds. The cations ClF_2^+, BrF_2^+, and ICl_2^+ are covered in Section 9.3.5.

The ions I_3^+ and I_5^+ may be considered as complexes of an I^+-ion with I_2 molecules, i.e., as coordinated I^+ cations. This ion also exists as $[I(pyridine)_2]^+$, formed by disproportionation of I_2:

$$I_2 + 2\,py + AgNO_3 \xrightarrow{\text{CHCl}_3} [I(py)_2]NO_3 + AgI$$

$$py: C_5H_5N$$

The iodo salt gives conducting solutions in $CHCl_3$ and acetone, which deposit iodine at the cathode. Analogous, but less stable complexes have been prepared with chlorine and bromine.

9.3.5. Interhalogen Compounds

Interhalogen compounds comprise a class, XY_n, where X and Y are different halogens, and n ranges from 1 to 7. Ternary compounds are also known.
The most simple XY molecules correspond to the elemental halogens, X_2 and Y_2:

ClF	BrF	BrCl	IF	ICl	IBr
colorless	light red	gas	solid brown	red	red-brown
gas	gas		decomp.	crystals	crystals
			$>0\,^\circ C$		

and can be synthesized directly from the elements. Different compounds result depending on reaction conditions. In the excess of one halogen (Y_2), XY_3, XY_5, and XY_7 result:

$$\tfrac{1}{2}X_2 + \tfrac{1}{2}Y_2 \ \rightarrow \ XY \ \xrightarrow{+Y_2} \ XY_3 \ \xrightarrow{Y_2} \ XY_5 \ \xrightarrow{Y_2} \ XY_7$$

for example:

ClF_3	BrF_3	IF_3	ICl_3	ClF_5	BrF_5	IF_5	IF_7
colorless	colorless	yellow	yellow	colorless	colorless	colorless	colorless
gas	liquid	solid	crystals	gas	liquid	liquid	gas
		decomp.					
		$>-35^\circ$					

Excepting iodine trichloride, all others are fluorides of chlorine, bromine and iodine. Neither fluoro-halides of type FY_n nor bromides or iodides, XBr_n or XI_n, exist. Only fluorine, and in one case chlorine, are suitable substituents, and only Cl, Br, and I can assume the central position. One explanation is furnished by the VB description where promotion of s- and p-electrons to the d-level is postulated. Since the promotion energies decrease Cl > Br > I, it is understandable that only iodine forms a heptafluoride, and that the thermodynamic stability of the pentafluorides increases $ClF_5 < BrF_5 < IF_5$. Fluorine is the best substituent for stabilizing higher oxidation states (p. 246).

The structures of the interhalogen compounds are consistent with d-orbital participation and hybridization of all occupied valence orbitals (cf. table 12 p. 88). The halides XY_3 are T-shaped (pseudo-trigonal bipyramids), the pentahalides are square pyramids (pseudo-octahedra), and IF_7 forms a distorted pentagonal bipyramid (p. 89). The structures also follow from the principles of electron pair repulsion (p. 117). They hold, however, only for the vapor state. In condensed phases, association via bridging halogens is observed; for example, ClF_3, BrF_3, and ICl_3 are dimers in the solid state. The planar I_2Cl_6 molecule is formed by two pseudo-octahedra with a common edge:

$$d(I-Cl) = 2.4 \, \text{Å} \qquad \alpha = 94°$$
$$d(I \cdots Cl) = 2.7 \, \text{Å} \qquad \beta = 84°$$

Dichloroiodobenzene, $C_6H_5ICl_2$, readily obtained from iodobenzene and chlorine, forms yellow needles.

Halogen fluorides are reactive, and ClF_3 and BrF_3 are used commercially as fluorinating agents.

Interhalogen compounds such as ICl, I_2Cl_6, BrF_3 and IF_5 form electrically conducting liquids, arising from self-dissociation:

$$2\,BrF_3 \rightleftharpoons BrF_2^+ + BrF_4^-$$
$$2\,IF_5 \rightleftharpoons IF_4^+ + IF_6^-$$

Dissociation then arises from a slight shift in the bridging halogen atoms:

These dissociation equilibria, which are reminiscent of hydrogen-bonded solvents, allow the halide ions to be isolated as salts with strong Lewis acids acting as halogen ion acceptors:

$$BrF_3 + SbF_5 \rightarrow [BrF_2][SbF_6]$$
$$ICl_3 + AlCl_3 \rightarrow [ICl_2][AlCl_4]$$
$$IF_5 + SO_3 \rightarrow [IF_4][SO_3F]$$
$$IF_7 + AsF_5 \rightarrow [IF_6][AsF_6]$$

The anions are generated by reaction with ionic halides:

$$NaF + BrF_3 \rightarrow Na[BrF_4]$$
$$KF + ICl_3 \rightarrow K[ICl_3F]$$
$$KF + IF_5 \rightarrow K[IF_6]$$

and halide and complex halide ions can sometimes be further halogenated:

$$KCl + 2F_2 \rightarrow K[ClF_4]$$
$$[ICl_2]^- + Cl_2 \rightarrow [ICl_4]^-$$

The ionic nature of these salts is confirmed by their structure and by, e.g., the solubility of $[BrF_2][SbF_6]$ in BrF_3 to yield a conducting solution. The ion, BrF_2^+, is an acid neutralizable by the base, BrF_4^-. The end point of the neutralization reaction:

$$[BrF_2][SbF_6] + Ag[BrF_4] \rightarrow Ag[SbF_6] + 2BrF_3$$

is recognized by a minimum in electric conductance. The structures of the ions $[XY_{n-1}]^+$ and $[XY_{n+1}]^-$ are consistent with VB and electron pair repulsion principles.

BrF_2^+	BrF_4^-	IF_4^+	IF_6^+	IF_6^-
angular	square	pseudo-trigonal	octahedral	presumably
(C_{2v})	planar	bipyramidal	(O_h)	pseudo-pentagonal
	(D_{4h})	(C_{2v})		bipyramidal

The tendency of the heavier halogens to achieve higher coordination numbers is reflected in the formation of the halogen fluorides, and in their association by fluorine bridges in condensed phases as well as between anions and cations in such salts as $[BrF_2][SbF_6]$ or $[ClF_2][AsF_6]$. For example, the Lewis acid BrF_2^+ extends two weak coordinate bonds to two fluorine atoms of the octahedral anions:

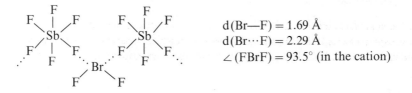

$$d(Br—F) = 1.69 \text{ Å}$$
$$d(Br\cdots F) = 2.29 \text{ Å}$$
$$\angle(FBrF) = 93.5° \text{ (in the cation)}$$

The four fluorine atoms about bromine lie in a plane with the latter (pseudo-octahedral coordination), with internuclear distances, $d(Br—F)$, markedly smaller than the van der Waals distance of 3.25 Å. This cation-anion interaction of partial ionic and covalent character finds its expression in a distortion of the SbF_6 octahedra, absent in a purely ionic, non-directional interaction.

9.3.6. Halogen Oxygen Compounds

The oxides of chlorine, bromine and iodine listed in Table 45 contain the halogens in positive oxidation states [unlike the binary fluorine-oxygen compounds which are oxygen fluorides and not halogen oxides, and for this reason are discussed in the chapter on oxygen (see Section 7.4)], since oxygen is the more electronegative element here.

Table 45   Oxides of the Halogens

Oxidation state	Cl	Br	I
+1	Cl_2O	Br_2O	
+2			
+3	Cl_2O_3		
+4	ClO_2	BrO_2	
+5			I_2O_5
+6	Cl_2O_6		
+7	Cl_2O_7		$I_2O_7(?)$

Halogen oxides are endothermic compounds, with the exception of I_2O_5. They are strong oxidizing agents which decompose into the elements on heating, sometimes explosively.

Formally, $ClO(ClO_3)$ (chlorine perchlorate), $IO(IO_3)$ and $I(IO_3)_3$ should also be classified among the halogen oxides, but they are discussed later (p. 266).

Chlorine oxides:

Dichlorine oxide is formed as a yellow-red gas in the reaction of chlorine gas with dry HgO at 20 °C:

$$2Cl_2 + 2HgO \rightarrow Cl_2O + HgO \cdot HgCl_2$$

Chlorine oxide is liquified at 2 °C. It explodes on warming or in contact with oxidizable substances, and hydrolyses to hypochlorous acid so that it may be considered as the anhydride of that acid:

$$Cl_2O + H_2O \rightleftharpoons 2HOCl$$

The Cl_2O molecule is angular (symmetry C_{2v}). The Cl—O internuclear distances and valence force constant argue for a bond order <1.

Chlorine dioxide, a yellow gas (b.p. 10 °C) is obtained from chlorates, for example, in the laboratory by adding concentrated sulfuric acid to $KClO_3$ at 0 °C:

$$3KClO_3 + 3H_2SO_4 \rightarrow 3HClO_3 + 3KHSO_4$$
$$3HClO_3 \rightarrow 2ClO_2 + H_3OClO_4$$

On a technical scale, $NaClO_3$ in sulfuric acid solution is reduced with SO_2:

$$2\,NaClO_3 + SO_2 + H_2SO_4 \;\rightarrow\; 2\,ClO_2 + 2\,NaHSO_4$$

Chlorine dioxide is explosive, decomposing into the elements. It can be handled with safety as the pyridine adduct, $C_5H_5N \cdot ClO_2$, or after dilution in CO_2. Chlorine dioxide is moderately soluble in water, but in alkaline solution disproportionation to chlorite and chlorate ions takes place:

$$2\,ClO_2 + 2\,OH^- \;\rightarrow\; ClO_2^- + ClO_3^- + H_2O$$

The ClO_2 molecule, isoelectronic with the radical anion SO_2^- (p. 229) but with no tendency to dimerize, is angular with $\angle OClO = 117.5°$. Internuclear distances and valence force constants indicate a bond order of about 1.5 based on resonance:

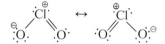

In the photolysis of ClO_2 at low temperatures, dichlorine trioxide, Cl_2O_3, is formed as a dark-brown solid of unknown structure which decomposes into the elements above $-45\,°C$. Oxidation of ClO_2 at $0\,°C$ with ozone produces the dimeric chlorine trioxide, Cl_2O_6:

$$2\,ClO_2 + 2\,O_3 \;\rightarrow\; Cl_2O_6 + 2\,O_2$$

Another route to the dichlorine hexoxide is the reaction:

$$ClO_2F + HClO_4 \;\rightarrow\; ClO_2ClO_4 + HF$$

Cl_2O_6 is a deep-red liquid, m.p. $3.5\,°C$, soluble in CCl_4 but which reacts with water to form chlorate and perchlorate ions, so that it can be considered a mixed anhydride:

$$Cl_2O_6 + 3\,H_2O \;\rightarrow\; 2\,H_3O^+ + ClO_3^- + ClO_4^-$$

The probable structure of Cl_2O_6 is that of a chloryl perchlorate, $O_2ClOClO_3$. Dichlorine heptoxide, Cl_2O_7, which is isoelectronic with the disulfate ion, $S_2O_7^{2-}$, with similar structure is the most stable oxide of chlorine. It is generated by dehydration of cold perchloric acid with P_2O_5:

$$2\,HClO_4 + P_2O_5 \;\rightarrow\; Cl_2O_7 + 2\,HPO_3$$

The oxide is distilled from the polymeric metaphosphoric acid as a colorless liquid. The product hydrolyzes to regenerate $HClO_4$.

Bromine Oxides:

The unstable oxygen-rich bromine oxides of unknown structure are polymeric. The oxide BrO_2, a yellow solid which decomposes at $0\,°C$ into the elements, is produced in the ozonization of bromine in $CFCl_3$ at $-50\,°C$. Decomposition yields Br_2O, Br_3O_8 and BrO_3; Br_2O is a volatile, brown solid which decomposes to the elements above $-40\,°C$. It reacts with NaOH to form sodium hypobromite, NaOBr.

Iodine Oxides:

Diiodine pentoxide, I_2O_5, can be prepared from iodic acid by dehydration at $250\,°C$:

$$2\,HIO_3 \rightleftharpoons I_2O_5 + H_2O$$

The product is a colorless, crystalline solid, unstable to decomposition to the elements at $>300\,°C$, which consists of I_2O_5 molecules linked via weak coordinate bonds.

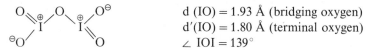

d (IO) = 1.93 Å (bridging oxygen)
d'(IO) = 1.80 Å (terminal oxygen)
∠ IOI = 139°

At $170\,°C$ I_2O_5 reacts with CO to form CO_2 and I_2 quantitatively, which is utilized in the iodimetric determination of CO.

A heptoxide, I_2O_7, generated in the dehydration of periodic acid with oleum has not been authenticated.

Oxo Acids:

Of the oxygen acids of chlorine, bromine, and iodine listed in Table 46, only three are obtainable in anhydrous form, namely $HClO_4$, HIO_3 and H_5IO_6. The other oxo acids have only been identified in aqueous solution or in the vapor phase, or as salts.

Table 46   Oxo Acids of the Halogens

Oxidation state	Cl	Br	I
+1	HClO	HBrO	HIO
+3	$HClO_2$	$HBrO_2$	
+5	$HClO_3$	$HBrO_3$	HIO_3
+7	$HClO_4$	$HBrO_4$	HIO_4, H_5IO_6

The molecular formulae in Table 46 do not reflect molecular structure since hydrogen is always present in —OH groups. The nomenclature is as follows: HOCl, hypochlorous acid; HOClO chlorous acid; $HOClO_2$, chloric acid; $HOClO_3$, perchloric acid, and the bromine and iodine acids are named analogously. The structures of these acids and their corresponding anions are discussed on p. 242.

Polynuclear acids, which are common among other heavier non-metals (e.g., $H_2S_2O_7$) are known only for iodine in the di and tri-periodates.

Chlorine Oxygen Acids:

Hypochlorous acid HOCl:

Chlorine gas reacts with water to give HOCl:

$$Cl_2 + H_2O \rightleftharpoons HOCl + HCl$$

Hypochlorous acid oxidizes Cl^- ions to Cl_2 in water, shifting the equilibrium to the left. Concentrated HOCl solutions result when chloride ions, e.g., are trapped with suspended HgO in the form of $HgO \cdot HgCl_2$, but slowly decompose to hydrochloric acid and O_2. Dichlorine oxide, the anhydride of HOCl, can be extracted into CCl_4 from concentrated solutions where it is in equilibrium with HOCl (p. 258). The solutions contain Cl_2, HOCl, and Cl_2O. Hypochlorous acid is a weak acid in water but a very powerful oxidizing agent.

In the vapor phase, the HOCl molecule is angular ($\angle HOCl = 102°$). Like Cl_2O and the ClO^- anion it has a weak Cl—O single bond.

Bubbling chlorine into strong alkali solution at $0°C$ produces hypochlorites and chlorides:

$$Cl_2 + 2\,NaOH \rightarrow NaOCl + NaCl + H_2O$$

Hypochlorite salts, which are markedly hydrolyzed, are bleaches and disinfectants. The are weaker oxidizing agents than HOCl.

Chlorous Acid, $HClO_2$:

The unstable $HClO_2$ is formed in the disproportionation of ClO_2 in water (p. 259). The more stable salts find use as bleaches. Sodium chlorite is formed by bubbling ClO_2 into Na_2O_2 or $NaOH/H_2O_2$ solutions, or by reduction of $NaClO_3$ with oxalic acid. In aqueous solution $NaClO_2$ is strongly oxidizing, and when dry forms explosive mixtures with oxidizable material.

The ClO_2^- anion is angular with some bond strengthening by resonance:

compensated by repulsions of free electron pairs on adjacent atoms. The anion resembles the isoelectronic cation, $ClF_2{}^+$, and is best represented by the first of the above structures.

Chloric Acid, $HClO_3$:

Chloric acid is unstable in aqueous solution of $>40\%$ concentration; dilute solutions can be prepared:

$$Ba(ClO_3)_2 + H_2SO_4(aq) \rightarrow 2\,HClO_3(aq) + BaSO_4\downarrow$$

Chlorate salts are obtained by disproportionation of hot aqueous solution of hypochlorites:

$$3\,ClO^- \rightarrow ClO_3^- + 2\,Cl^-$$

In this reaction the ClO^- anion is oxidized by the free acid HOCl. Chlorine is bubbled into hot NaOH, or a hot NaCl solution is electrolyzed and the anodically generated chlorine allowed to react with the cathodically formed NaOH.

Aqueous chloric acid and solid chlorates are strong oxidizing agents; $KClO_3$ is used for the manufacture of matches and explosives.

Like the isoelectronic sulfite ion, SO_3^{2-}, the chlorate anion has a trigonal-pyramidal structure with $f(Cl-O)$ corresponding to a bond order of about 1.2. The free acid, $HClO_3$, however, is not isostructural with hydrogen sulfite ion, HSO_3^-, but has an OH group:

Perchloric Acid, $HClO_4$:

Perchloric acid, the most stable oxychlorine acid, can be distilled in vacuo as a colorless liquid from a mixture of $KClO_4$ and H_2SO_4:

$$KClO_4 + H_2SO_4 \rightarrow HClO_4 + KHSO_4$$

but explodes on heating or with combustible matter. The strongly acidic aqueous solutions are stable and ionize completely on dilution.

Perchlorates are prepared by anodic oxidation of chlorates and, hence from chlorine:

$$\tfrac{1}{2}Cl_2 + 4H_2O \rightarrow ClO_4^- + 8H + 7e^-$$

or by thermal disproportionation of alkali chlorates:

$$4KClO_3 \rightarrow 3KClO_4 + KCl$$

On heating, $KClO_4$ decomposes to KCl and O_2. The perchlorate ion is isoelectronic with the ions SO_4^{2-}, PO_4^{3-}, and SiO_4^{4-}, and like them is tetrahedral. The $Cl-O$ bond order is 1.5. In the free acid, the OH group is connected to the pyramidal ClO_3 with bond order of 1.6:

The $Cl-O$ bond strength increases with the oxidation state of chlorine, as shown by the valence force constants (in mdyn/Å):

$$ClO^- : 3.3 \qquad ClO_2^- : 4.2 \qquad ClO_3^- : 5.6 \qquad ClO_4^- : 7.2$$

d-Orbital contraction at chlorine increases with oxygen attachment and improves the overlap; loss of free electron pairs on chlorine decreases replusion with those of oxygen. As a consequence many disproportionation reactions are observed which create chloride and anions richer in oxygen, with $HClO_4$ and ClO_4^- the most stable, and analogous behavior is found with other non-metals (S, Se, Br, I).

The dissociation constants of chlorine oxo acids in aqueous solution increase strongly with the oxidation state of chlorine. The increasing bond order, $b(Cl-O)$, in the anions, from ClO^- to ClO_4^- reduces the negative charge at oxygen, enhancing the heterolytic cleavage of the OH bond

$$-\ddot{O}-H + H_2O \; \rightarrow \; -\ddot{O}{:}^- + H_3O^+$$

The increase in dissociation constants in the series $H_4SiO_4 < H_3PO_4 < H_2SO_4 < < HClO_4$ is rationalized similarly.

Bromine Oxo Acids

The bromine analogues of the chlorine oxo acids are known only in aqueous solution or as salts. The disproportionation of bromine in alkali produces hypobromites, bromites or bromates depending on the temperature:

$$Br_2 + 2OH^- \; \rightarrow \; BrO^- + Br^- + H_2O$$
$$2\,BrO^- \; \rightarrow \; BrO_2^- + Br^-$$
$$BrO^- + BrO_2^- \; \rightarrow \; BrO_3^- + Br^-$$

However, perbromates cannot be obtained by disproportionation, and their existence was in doubt until 1968. Perbromate is produced by oxidation of bromate with fluorine gas, or XeF_2, or by anodic oxidation in aqueous solution:

$$BrO_3^- + F_2 + 2OH^- \; \rightarrow \; BrO_4^- + 2F^- + H_2O$$
$$BrO_3^- + XeF_2 + H_2O \; \rightarrow \; BrO_4^- + Xe + 2HF$$

and $RbBrO_4$ precipitated with RbF, from which perbromic acid can be prepared by ion exchange.

Iodine Oxo Acids

Hypoiodite ions are formed when iodine is dissolved in sodium hydroxide, but they disproportionate to iodate and iodide ions:

$$I_2 + 2NaOH \; \rightarrow \; NaIO + NaI + H_2O$$
$$3\,NaIO \; \rightarrow \; NaIO_3 + 2\,NaI$$

The acid HOI, a reaction intermediate in aqueous iodine solution with HgO, as well as HIO_2 and the iodites are unknown.

Iodic acid, HIO_3, precipitates as colorless crystals in the oxidation of iodine with HNO_3, Cl_2, H_2O_2, or $HClO_3$ in aqueous solution, and is a strong oxidizing agent. Dehydration yields I_2O_5.

The iodates, MIO_3, contain the pyramidal IO_3^-. In the acid salts $MIO_3 \cdot HIO_3$ or $MIO_3 \cdot 2HIO_3$, the iodic acid molecules are hydrogen bonded to the iodate ions, as in the nitric acid analogues.

Strong oxidizing agents like NaOCl yield periodates:

$$IO_3^- + ClO^- \xrightarrow[100\,°C]{OH^-} IO_4^- + Cl^-$$

The salt $Na_3H_2IO_6$ which precipitates upon cooling is a derivative of the only isolatable periodic acid, the hexaoxoiodine (VII) acid, H_5IO_6, prepared by treating the barium salt with concentrated nitric acid. Recrystallization of the sodium salt from dilute nitric acid produces $NaIO_4$.

Periodic acid, H_5IO_6, and the ions $H_4IO_6^-$, $H_3IO_6^{2-}$, and $H_2IO_6^{3-}$ are octahedral. Thus iodine behaves like its neighbors in the periodic system, forming compounds for which no lighter homologs are known (p. 225). The acid decomposes at $>130\,°C$ to I_2O_5, water and O_2.

Acid Halides:

Acid halides are formally derived from the halogen oxo acids by replacing OH-groups by halogen. The halogen, like the OH group must be more electronegative than the central atom, and only fluorine can qualify. Most important are the halogenyl fluorides FXO_2, the perhalogenyl fluorides FXO_3 (X = Cl, Br, I), and the iodooxypentafluoride, IOF_5, which is a derivative of periodic acid, $IO(OH)_5$. The oxohalide $ClOF_3$ cannot be derived from an acid.

Halogenyl Fluorides $FClO_2$, $FBrO_2$, FIO_2:

These compounds of symmetry C_s are derivatives of the acids HXO_3:

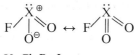

X: Cl, Br, I

$FClO_2$ occurs as a colorless gas in the fluorination of ClO_2 with elemental fluorine or a fluorinating agent:

$$ClO_2 + AgF_2(CoF_3, BrF_3) \xrightarrow{25\,°C} FClO_2 + AgF$$

Chloryl fluoride hydrolyzes:

$$FClO_2 + H_2O \rightarrow HClO_3 + HF$$

and forms chloryl salts:

$$FClO_2 + BF_3 \rightleftharpoons ClO_2[BF_4] \quad 225 \text{ torr dissociation pressure at } 25\,^\circ C$$

$$FClO_2 + AsF_5 \rightarrow ClO_2[AsF_6] \quad \text{stable at } 25\,^\circ C$$

$$2FClO_2 + SnF_4 \rightarrow (ClO_2)_2[SnF_6]$$

which consist of isolated ions identified by IR spectroscopy. Like the isoelectronic SO_2, the chloryl cation is angular with bond order b(ClO), in ClO_2^+ and $FClO_2$ 1.8. Chloryl fluoride is a Lewis acid towards fluoride ion forming difluorochlorate, $K[ClO_2F_2]$, from KF which can also be prepared from $KClO_3$ and F_2 by exchanging one oxygen for two fluorine atoms. Crystalline bromyl fluoride is prepared by fluorination of BrO_2 with BrF_5, and iodyl fluoride is likewise formed when I_2O_5 is dissolved in anhydrous hydrogen fluoride:

$$I_2O_5 + HF \rightarrow FIO_2 + HIO_3$$

The pseudo-trigonal bipyramidal liquid chlorine oxotrifluoride, $ClOF_3$, forms by reaction of fluorine gas with Cl_2O, $NaClO_2$, or $ClNO_3$, and reacts with BF_3, AsF_5 and SbF_5 to yield salts of the pseudo-tetrahedral cation $ClOF_2^+$.

Perhalogenyl Fluorides $FClO_3$, $FBrO_3$, FIO_3:

The pseudotetrahedral fluorides FXO_3 (symmetry C_{3v}) are derivatives of the perhalogen acids HXO_4; $FClO_3$, isoelectronic with the perchlorate ion, forms in the fluorination of $KClO_3$ with fluorine:

$$KClO_3 + F_2 \rightarrow FClO_3\uparrow + KF$$

or by reaction of $KClO_4$ with fluorosulfuric acid at $50\text{--}85\,^\circ C$:

$$KClO_4 + HSO_3F \rightarrow FClO_3\uparrow + KHSO_4$$

Perchloryl fluoride is a colorless gas (b.p. -46.7°) which resists hydrolysis like SF_6 owing to kinetic inhibition of nucleophilic attack, so that, e.g., reaction with concentrated sodium hydroxide only takes place at $200\text{--}300\,^\circ C$, although the equilibrium shown below lies completely to the right:

$$FClO_3 + 2OH^- \rightleftharpoons F^- + ClO_4^- + H_2O$$

$FClO_3$ does not form perchloryl salts with fluoride ion acceptors.

Perbromyl fluoride, prepared from $KBrO_4$ and SbF_5 in liquid HF, reacts with alkali at $25\,^\circ C$ to give perbromate and fluoride. Periodyl fluoride forms in the fluorination of periodates with fluorine in HF:

$$IO_4^- + F_2 \rightarrow FIO_3 + F^- + \tfrac{1}{2}O_2$$

The fluorides IO_2F_3, $HOIOF_4$, and IOF_5 are also derived from iodine(VII) acids. The pseudo-octahedral, liquid IOF_5 (symmetry C_{4v}), prepared by reacting IF_7 and SiO_2 at $100\,^\circ C$, contains an IO bond of order 2.0.

Halogen Derivatives of Oxyacids of the Non-Metals:
Since chlorine, bromine, and iodine can occur in oxidation state $+1$, the X^+ cations can replace hydrogen in the oxo acids as, e.g., in chlorine nitrate, $ClNO_3$. Bromine and iodine in their $+3$ oxidation can replace several hydrogen atoms as in iodine nitrate $I(NO_3)_3$ and iodine phosphate IPO_4, which are not acid halides, but contain positive halogen bound to oxygen by covalent bonds. Examples include HNO_3, H_3PO_4, H_2SO_4, HSO_3F, HIO_3, $HClO_4$ and CH_3COOH. The properties of such compounds will be discussed below.

Nitrates:
Chlorine nitrate, $ClNO_3$, has the structure $Cl\!-\!O\!-\!NO_2$, representing the mixed anhydride of the acids $HOCl$ and HNO_3. It differs from nitrosyl chloride, $ClNO$, and from nitryl chloride, $ClNO_2$, by the $Cl\!-\!O$ bond. It is generated as a pale yellow liquid (m.p. $-107\,^\circ C$) by the reaction of dichlorine monoxide with dinitrogen pentoxide:

$$Cl_2O + N_2O_5 \xrightarrow{\;-20\,^\circ C\;} 2\,ClNO_3$$

Alkaline hydrolysis forms OCl^- and NO_3^-:

$$ClNO_3 + 2\,OH^- \;\rightarrow\; ClO^- + NO_3^- + H_2O$$

The positive chlorine of $ClNO_3$ reacts with negative polarized chlorine to form Cl_2. Other halogen nitrates can be prepared in this manner:

$$BrCl + ClNO_3 \;\rightarrow\; BrNO_3 + Cl_2$$
$$ICl_3 + 3\,ClNO_3 \;\rightarrow\; I(NO_3)_3 + 3\,Cl_2$$

and ICl, HCl, $TiCl_4$ and CrO_2Cl_2, etc., react in an analogous way.

Fluorosulfates:
Chlorine fluorosulfate, $ClSO_3F$, a yellow liquid, is prepared from ClF and SO_3:

$$ClF + SO_3 \;\rightarrow\; ClSO_3F$$

and reacts at $-45\,^\circ C$ with $CsClO_4$ or nitronium perchlorate to form chlorine perchlorate:

$$ClSO_3F + NO_2ClO_4 \;\rightarrow\; ClClO_4 + NO_2SO_3F$$

a yellow liquid which is formally a chlorine oxide and reacts with HCl to Cl_2 and $HClO_4$, and with $AgCl$ to form $AgClO_4$. From bromine and $ClClO_4$ at $-45\,^\circ C$, bromine perchlorate, $BrClO_4$, a red liquid which decomposes above $-20\,^\circ C$ is formed:

$$Br_2 + 2\,ClClO_4 \;\rightarrow\; Cl_2 + 2\,BrClO_4$$

which like $HClO_4$ and $ClOClO_3$ has symmetry C_s.

Iodine trisfluorosulfate, $I(SO_3F)_3$, a yellow polymeric solid is formed in the oxidation of iodine with peroxodisulfuryl fluoride (p. 233) and hydrolyzed to $IO(SO_3F)$ and HSO_3F. $I(SO_3F)_3$ acts on KSO_3F in liquid disulfuryl fluoride to give the square planar tetrakis-fluorosulfatoiodate(III):

$$KSO_3F + I(SO_3F)_3 \xrightarrow{S_2O_5F_2} K[I(SO_3F)_4]$$

which is also prepared from $KICl_4$ and $S_2O_6F_2$. The analogous bromine salt is prepared from $KBrO_3$ and $S_2O_6F_2$ with oxygen elimination. The bisfluorosulfatobromate(I) is also known:

$$CsBr + S_2O_6F_2 \xrightarrow{Br_2} Cs[Br(SO_3F)_2]$$

Iodates:

Ozonization of iodine at $-78°C$ in CCl_3F yields I_4O_9 as a yellow, hygroscopic powder:

$$2I_2 + 9O_3 \rightarrow I_4O_9 + 9O_2$$

which is a polymeric iodine(III) iodate, $I(IO_3)_3$. No iodine(III) oxo acid exists, and the hydrolysis products are iodine and HIO_3. Yellow iodine(III) oxo iodate $IOIO_3$, formally an iodine oxide (I_2O_4), is formed in treating iodic acid with hot concentrated sulfuric acid:

$$3HIO_3 \rightarrow IO(IO_3) + HIO_4 + H_2O$$

I_2O_4 decomposes at $135°C$ into iodine and I_2O_5, and in alkaline hydrolysis yields iodide and iodate, or with water iodine and HIO_3.

9.4. Pseudohalogens

Certain monovalent atoms or groups behave like halogens:

—CN	—SCN	—N_3	—OCN
cyanide	thiocyanate	azide	cyanate

These groups, X, form hydrogen compounds, HX, which dissociate in water like the hydrogen halides. The silver salts, AgX, derived from these acids are insoluble like AgI. The ions like I^- can also be oxidized by mild oxidizing agents to free pseudohalogens as, for example, dicyanogen, $(CN)_2$, and $(SCN)_2$. Halogens and pseudohalogens are bonded in, e.g., cyanogen chloride, ClCN, and fluoroazide, FN_3. Pseudohalogens can replace the halogens in, for example, PCl_3 to give $P(CN)_3$, and $AgCl_2^-$ to give $Ag(CN)_2^-$.
The pseudohalogens are similar to iodine in their electronegativities.

10. The Rare Gases

10.1. General

The elements of the VIIIth main group differ from the rest of the non-metals in their inertness, a consequence of their electron configuration. The rare gases with exception of helium which fills its outer shell with only two electrons, have a s^2p^6 outer configuration, without unpaired electrons. The stability of this configuration is borne out by the high ionization energies, which are larger than any of the elements within the same period (p. 60). Slater effective nuclear charge values, Z^*, are likewise larger, and the rare gas atoms are smaller and more electronegative than their neighboring non-metals. Their electron affinities are negative, hence these elements do not form negative ions (p. 61).

The spherically symmetrical rare gas atoms with no unpaired electrons are restricted to van der Waals interactions, as reflected in their low boiling points (Table 17, p. 115).

Rare gas atoms are isoelectronic with the corresponding halide and chalcogenide ions:

$$N^{3-} \qquad O^{2-} \qquad F^- \qquad Ne$$

but no chemical similarites exist. For example, neon does not react like F^- to form complexes with strong Lewis acids, nor is it protonated in aqueous solutions like the analogous anions. Compound formation on the part of the rare gases is according to VB theory only possible after promotion of valence electrons[10].

The lowest unoccupied AO's of the rare gases are the s-orbitals of the $(n + 1)$ shells. From argon on, the empty nd-orbitals are at higher energy than the $(n + 1)$s-orbital. The promotion energies of an np-electron to the $(n + 1)$s levels are:

$$Ne: 16.6 \qquad Ar: 11.5 \qquad Kr: 9.9 \qquad Xe: 8.3 \qquad Rn: 6.8 \text{ eV}$$

suggesting that formation of covalent bonds after valence electron promotion might be favored for the heavier rare gases and stable compounds are only found for krypton, xenon, and radon. Few kryton compounds are known, but a large number of xenon compounds have been reported, the first binary compound in 1962. The inherent radioactivity of radon is an obstacle to experimental investiga-

---

10 The diatomic ions He_2^+ and XeF^+ which are identified spectroscopically, are excepted here.

tion, and only qualitative data on radon compounds are available. The chemistry of the rare gases is, therefore, practically identical with the chemistry of xenon.

The $(n+1)$s orbital does not significantly participate in the bonding of rare gas compounds like XeF_2 and XeF_4. The energies of the AO's change in bound rare gas atoms, particularly the nd-level which is stabilized in the substituent field as in SF_6 (p. 199) in which the valence d-orbitals participate in the bonding. The stepwise electron excitation process gives the rare gases the oxidation states $+2$, $+4$, $+6$, and $+8$:

$$s^2p^6 \rightarrow s^2p^5d \rightarrow s^2p^4d^2, \quad \text{etc.}$$

Bonding relationships are discussed on p. 278.

10.2. Occurrence, Production, and Utilization

All the rare gases are present in air, where apart from 78.09 vol % N_2, 20.95% O_2, and 0.03% CO_2, they amount to 0.935%, which is almost exclusively argon:

> He 0.0005 Vol.-%
> Ne 0.0016 Vol.-%
> Ar 0.9327 Vol.-%
> Kr 0.0001 Vol.-%
> Xe $8 \cdot 10^{-6}$ Vol.-%
> Rn $6 \cdot 10^{-18}$ Vol.-%

Production of the rare gases is by fractional distillation or condensation of air. In addition helium occurs in North America in natural gas in a few per cent.

Helium is also a product of radioactive decay of some minerals (uranium ore, monazite sand, etc.) and is obtained by grinding and heating. Radon is collected over radium salts in closed vessels. The lighter gases are available in cylinders. Helium is used in lighter-than-air aircraft and ballons, and in the laboratory as the carrier gas in vapor-phase chromatography. Neon is used in electric discharge signs; argon is used as inert blanket in welding and chemistry and as a filling gas in light bulbs.

10.3. Xenon Compounds

Xenon bonds with electronegative elements (e.g., F, O and Cl) which contract its d-orbitals. With one fluorine bonded, other non-metals (e.g., N and B) will also bond. The most stable xenon compounds are the fluorides which are stable thermodynamically, and can be synthesized directly from the elements, and some fluoro-xenates derived from them. Xenon-oxygen compounds are thermodynamically unstable (endothermic) and decompose, often explosively. They are prepared from the fluorides.

10.3.1. Xenon Fluorides

Xenon reacts on heating, photon excitation, or electric discharge with elemental fluorine to yield XeF_2, XeF_4, and XeF_6 involving the equilibria:

$$Xe + F_2 \rightleftharpoons XeF_2$$
$$XeF_2 + F_2 \rightleftharpoons XeF_4$$
$$XeF_4 + F_2 \rightleftharpoons XeF_6$$

Table 47   Preparation and Properties of the Xenon Fluorides

	XeF_2	XeF_4	XeF_6
Preparation: Molar ratio Xe: F_2	1:1–3	1:5	1:20
Activation:	400° or irradiation or discharge	as for XeF_2	300° and 60 bar or discharge
Properties: melting point (°C) enthalpy of formation, (gas, 25 °C, kJ/mol)	130–140° −109	117° −218	49° −293
average bond energy (kJ/mol)	130	130	126
molecular symmetry	$D_{\infty h}$	D_{4h}	
internuclear distance (cryst.; Å)	2.00	1.95	1.89
valence force constant, (gas, mdyne/Å)	2.8	3.0	2.8

Larger ratios of F_2:Xe increase the yield of the higher fluorides; XeF_6 requires higher pressure. Table 47 summarizes the conditions for formation and the properties of the fluorides. The solids are sublimable *in vacuo* and except for XeF_6 in which there are fluorine bridges between XeF_6 moieties as in SbF_5 (p. 247) consist of isolated molecules.

The fluoride XeF_8, which is theoretically possible, has not been prepared.

The structures of the di- and tetrafluorides which are monomeric in all phases correspond to the expectations of the theory of electron pair repulsion, and of the VB theory, assuming complete hybridization of all occupied xenon valence orbitals:

XeF_2 symmetrically linear (pseudo-trigonal-bipyramidal) sp^3d-hybrid.

XeF_4 square, planar (pseudo-octahedral) sp^3d^2-hybrid.

Xenon hexafluoride is monomeric in the gas phase, but as a solid and yellow melt the molecules are associated via fluorine bridges. Crystalline XeF_6 exists in four modifications. The isolated XeF_6 molecule whose structure is unknown, can take several forms according to the theory of electron pair repulsion, differing little in energy or relative stability, all, however, of lower symmetry than O_h. The experimental data (UV, IR, Raman spectra, electron diffraction) are best explained by assuming three isomers, differing in stability by 5 to 15 kJ/mol, present simultaneously in gaseous XeF_6 at 25 to 100 °C. The most stable isomer would have O_h symmetry, with others derived by elongation or compression along a C_3 axis, thus lowering the symmetry to D_{3d}.

10.3.2. Xenon Chlorides and Bromides

Xenon chlorides have not been isolated but there is proof of their existence. The solid from a xenon-chlorine mixture passed through a microwave discharge and condensed at 20K has an IR absorption spectrum ascribable to a linear $XeCl_2$ molecule. The fragment $XeCl^+$ was identified by mass spectroscopy from a mixture of xenon, fluorine and carbon tetrachloride exposed to a discharge. Traces of the unstable xenon compounds $XeCl_4$ and $XeBr_2$, as well as $XeCl_2$ have been demonstrated by the Mössbauer effect from the β-decay of the isoelectronic ions $^{129}ICl_4^-$, $^{129}IBr_2^-$, and $^{129}ICl_2^-$. $XeBr_2$ has a half life of only 10^{-10}s.

10.3.3. Reactions of the Xenon Fluorides

The xenon fluorides which decompose to the elements upon heating are oxidizing and fluorinating agents. In redox reactions, xenon returns to the zero oxidation state.

Xenon fluorides react with hydroxides with condensation:

$$—Xe—F + H—O— \rightarrow —Xe—O— + HF$$

Xenon oxides are synthesized in this manner (See Sections 10.3.4. to 10.3.6). In addition, XeF_2, XeF_4, and XeF_6 act as F^--donors and -acceptors to give salts with XeF_n-cations or -anions by analogy with the interhalogen compounds.

Redox Reactions:

The di- and tetrafluoride react with hydrogen at 300–400 °C (XeF_6 at 25 °C):

$$XeF_2 + H_2 \rightarrow Xe + 2HF$$

Shaking with mercury produces xenon and HgF_2 or Hg_2F_2:

$$XeF_2 + Hg \rightarrow Xe + HgF_2$$

These reactions serve in the assay of xenon fluorides. Aqueous iodide solution is oxidized to I_2.

Oxidative power increases from XeF_2 to XeF_6. The solvents HF, SO_2, CH_3NO_2, CH_3CN, CCl_4, dioxane, etc., form stable solutions of XeF_2, while XeF_4 and XeF_6 give stable solutions only in liquid HF.

Fluoride Ion Exchange Reactions:
Xenon difluoride is a F^- donor for strong Lewis acids, as with the pentafluorides of arsenic, antimony, ruthenium, iridium, and platinum:

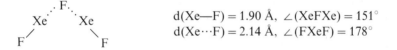

$$XeF_2 + MF_5 \begin{cases} \xrightarrow{2:1} [Xe_2F_3][MF_6] \\ \xrightarrow{1:1} [XeF][MF_6] \\ \xrightarrow{1:2} [XeF][M_2F_{11}] \end{cases}$$

The mostly colored products which melt at 50 to 150 °C are ionic. The planar cation, $Xe_2F_3^+$ in the hexafluoroarsenate has the geometry:

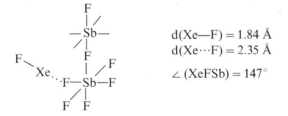

$$d(Xe—F) = 1.90 \text{ Å}, \quad \angle(XeFXe) = 151°$$
$$d(Xe \cdots F) = 2.14 \text{ Å}, \quad \angle(FXeF) = 178°$$

In $[XeF][Sb_2F_{11}]$, the fluorine bridges are similar to liquid SbF_5:

$$d(Xe—F) = 1.84 \text{ Å}$$
$$d(Xe \cdots F) = 2.35 \text{ Å}$$
$$\angle(XeFSb) = 147°$$

Similar short distances between the cation, XeF^+, and fluorine in the anion are found in the $[XeF][MF_6]$ salts, since XeF^+ is a strong Lewis base. By contrast, crystals of $XeF_2 \cdot IF_5$ and $XeF_2 \cdot XeF_4$ are molecular with only van der Waals interactions.

Xenon tetrafluoride reacts with SbF_5 forming $[XeF_3][SbF_6]$ and $[XeF_3][Sb_2F_{11}]$. Xenon hexafluoride can accept or donate fluoride ion, as with AsF_5, SbF_5, and PtF_5 to form $[XeF_5][MF_6]$ salts. The yellow platinum compound has square pyramidal XeF_5^+ cations connected loosely via asymmetrical fluorine bridges with four O_h PtF_6-anions each, so that the xenon atom is coordinated with a lone electron pair, 5 nearest and 4 next-nearest fluorine atoms.

Similarly high coordination numbers occur in the fluoro-xenates, which are formed in the reaction of XeF_6 with alkali metal fluorides (excepting LiF). The yellow salt $Cs^+XeF_7^-$ from CsF and liquid XeF_6 at $50\,^\circ C$ decomposes into the stable $Cs_2^+XeF_8^{2-}$ and XeF_6:

$$CsF + XeF_6 \xrightarrow{50\,^\circ C} CsXeF_7 \xrightarrow[-XeF_6]{\gg 50\,^\circ C} Cs_2XeF_8 \xrightarrow{400\,^\circ C} CsF, Xe, F_2$$

10.3.4. Oxides and Oxo Salts of Xenon

Water dissolves XeF_2 at $0\,^\circ C$ to give a yellow molecular solution which slowly hydrolyzes with base catalysis to oxidize the water:

$$XeF_2 + H_2O \rightarrow Xe + 2HF + \tfrac{1}{2}O_2$$

Xenon tetrafluoride hydrolyzes immediately to give Xe, O_2, HF, and xenon trioxide, which is also formed by hydrolysis of XeF_6. Evaporation yields colorless, explosive crystals; XeO_3 is hygroscopic and a strong oxidizing agent. It dissolves as molecules in water, yet the solution is weakly acidic:

$$2H_2O + XeO_3 \rightleftharpoons H_3O^+ + HXeO_4^-$$

Therefore, XeO_3 is a Lewis acid. Colorless salts of the hypothetical acid, H_2XeO_4, e.g., $NaHXeO_4 \cdot 1.5\,H_2O$ and $CsHXeO_4$, have been prepared with base.
Strong bases bring about disproportionation of XeO_3 and XeF_6 into Xe(0) and Xe(VIII):

$$2HXeO_4^- + 4Na^+ + 2OH^- \rightarrow Na_4XeO_6 + Xe + O_2 + 2H_2O$$

Colorless, thermally stable perxenates like $Na_4XeO_6 \cdot nH_2O$ and Ba_2XeO_6 are also obtained by ozonolysis of XeO_3 in the metal hydroxide. The sodium salt consists of O_h XeO_6^{4-} ions which hydrolyze:

$$XeO_6^{4-} + H_2O \rightleftharpoons HXeO_6^{3-} + OH^-$$

A perxenon acid is not known. Concentrated sulfuric acid evolves XeO_4 from Ba_2XeO_6:

$$Ba_2XeO_6 + 4H_2SO_4 \rightarrow 2BaSO_4\downarrow + XeO_4\uparrow + 2H_3OHSO_4$$

The yellow, explosive xenon tetroxide, isoelectronic with IO_4^-, decomposes $>0\,^\circ C$ to the elements. The less volatile XeO_3 forms a molecular lattice with distorted trigonal pyramidal XeO_3 units bridged via oxygen atoms:

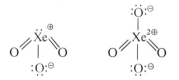

The strong tendency of XeO_3 to increase coordination number is also shown with fluoride ions:

$$MF + XeO_3 \xrightarrow[0\,°C]{H_2O} MXeO_3F \qquad M:K, Rb, Cs$$

The fluoroxenates (VI) are polymeric chain-like anions consisting of XeO_3 pyramids linked via F^--bridges resulting in distorted octahedral (2 fluorine, 3 oxygen atoms and one electron pair) coordination.

10.3.5. Oxofluorides of Xenon

XeF_6 reacts with water and certain oxides to form the oxofluorides $XeOF_4$ and XeO_2F_2:

$$2XeF_6(g.) + SiO_2 \xrightarrow{50\,°C} 2XeOF_4 + SiF_4$$

$$XeF_6(g.) + H_2O(g.) \xrightarrow{20\,°C} XeOF_4 + 2HF$$

The former is a colorless liquid containing molecules of C_{4v}, square pyramidal symmetry. Hydrolysis yields XeO_3.
Other oxofluorides occur in exchange reactions:

$$XeF_6 + 2XeO_3 \xrightarrow{25\,°C} 3XeO_2F_2 \text{ (colorless crystals, m.p. 31\,°C)}$$

$$XeF_6 + XeO_4 \longrightarrow XeO_3F_2 + XeOF_4$$

$$XeF_6 + XeO_3F_2 \longrightarrow XeO_2F_4 + XeOF_4$$

The IR spectrum of XeO_3F_2 suggests D_{3h} symmetry, while XeO_2F_2 is pseudo-trigonal bi-pyramidal with C_{2v} symmetry.

10.3.6. Other Xenon Compounds

Xenon fluorides undergo hydrolysis or alcoholysis with HF formation. Oxygen acids undergo condensation reactions, e.g., XeF_2 condenses with $HClO_4$ at $< -75\,°C$:

$$XeF_2 + HOClO_3 \xrightarrow{-HF} FXeOClO_3 \xrightarrow[-HF]{+HClO_4} Xe(OClO_3)_2 \xrightarrow{25°} Xe, O_2, Cl_2O_7$$

$$\qquad\qquad\qquad\quad \text{colorless} \qquad\qquad \text{pale}$$
$$\qquad\qquad\qquad\quad \text{crystals} \qquad\qquad\; \text{yellow}$$
$$\qquad\qquad\qquad\quad \text{m.p. } 16.5\,°C$$

The xenon fluoride perchlorate is more stable than the bisperchlorate, which contains only xenon-oxygen bonds.

Fluorosulfuric acid, HSO_3F and pentafluoroorthotelluric acid, $HOTeF_5$, react with XeF_2 to form:

$XeF(SO_3F)$	$Xe(SO_3F)_2$	$XeF(OTeF_5)$	$Xe(OTeF_5)_2$
colorless	yellow	yellow	colorless
crystals	m.p. 44 °C	liquid	crystals
m.p. 37 °C			m.p. 36 °C

The two tellurium compounds are stable $\leq 120\,°C$, but the fluorosulfates decompose slowly into xenon, $S_2O_6F_2$ and possibly XeF_2.

X-ray analysis of $XeF(SO_3F)$ indicates a molecular structure:

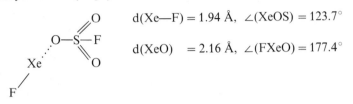

$d(Xe\!-\!F) = 1.94\ Å,\quad \angle(XeOS) = 123.7°$

$d(XeO)\ \ = 2.16\ Å,\quad \angle(FXeO) = 177.4°$

in which the internuclear distance, $d(XeO)$, is larger than the single bond distance of 1.96 Å but smaller than the van der Waals distance, so that the interaction between XeF and SO_3F groups must be partially covalent and partially ionic:

$$F\!-\!Xe\!-\!OSO_2F \leftrightarrow [FXe]^+[SO_3F]^-$$

Fluorosulfuric acid reacts similarly with XeF_4 and XeF_6 at low temperatures. With the former, $XeF_2(SO_3F)_2$ is isolated as a yellow-green liquid which slowly decomposes into xenon, XeF_4 and $S_2O_6F_2$. From XeF_6, $XeF_4(SO_3F)_2$, likewise a liquid is obtained which decomposes into XeF_4 and $S_2O_6F_2$. (See the analogous halogen derivatives on p. 266).

A xenon-nitrogen bond is present in the colorless, solid $FXeN(SO_2F)_2$ which is obtained:

$$XeF_2 + HN(SO_2F)_2 \xrightarrow[0\,°C]{CCl_2F_2} FXeN(SO_2F)_2 + HF$$

and decomposes at 70 °C to Xe, XeF_2 and $[N(SO_2F)_2]_2$.

The recently synthesized $FXeBF_2$, prepared from Xe and O_2BF_4, decomposes into xenon and BF_3 upon warming:

$$Xe + O_2BF_4 \xrightarrow[-O_2,\,\frac{1}{2}F_2]{-100\,°C} F\!-\!Xe\!-\!BF_2 \xrightarrow{-30\,°C} Xe + BF_3$$

According to its vibrational spectra this colorless crystalline compound is not a $Xe\!-\!BF_3$-adduct.

10.4. Krypton and Radon Compounds

Only one binary krypton compound is known with certainty; the sublimable solid difluoride, KrF_2, which is produced in a $1:1$ krypton-fluorine gas mixture at 20 torr and $-196\,°C$ in a high voltage discharge; or by reacting krypton and fluorine or OF_2 in sun light. It decomposes into the elements at $> -10\,°C$.

Endothermic KrF_2 forms colorless crystals, consisting of linear molecules of symmetry $D_{\infty h}$, soluble in hydrogen fluoride. The mean $Kr - F$ bond energy is 50 kJ/mol, the force constant, $f(KrF)$, is 2.46 mdyn/Å. The covalent radius of the krypton atom is 1.24 Å from the internuclear distance, $d(KrF)$, of 1.89 Å. It reacts with mercury to give HgF_2 and krypton, and hydrolyzes with base catalysis to give oxygen, krypton and HF. From SbF_5 and KrF_2 at $-20\,°C$, colorless crystals of $KrF_2 \cdot 2\,SbF_5$ are produced which decompose to krypton, fluorine and SbF_5 above its melting point of $ca.\ 50\,°C$.
The most stable radon isotope, ^{222}Rn, has a half-life of 3.8 days which makes investigation of the chemistry of radon difficult. No pure compounds have been isolated, but radon and fluorine react at 400 °C to give a stable product of low volatility which is reduced by hydrogen at 500 °C.

10.5. Electronegativities of the Rare Gases

Electronegativity values can be calculated by the Allred-Rochow Method, if the effective nuclear charge number, Z^*, has been obtained from the Slater rules (p. 47).

$$x_E = 0.359 \cdot \frac{Z^*}{r^2} + 0.744$$

For a 5p-electron of xenon or a 4p-electron of krypton, $Z^* = 8.25$.
The covalent radii, r, for xenon (II) and krypton (II) are determined from the KrF internuclear distances in gaseous XeF_2 (1.98 Å) and KrF_2 [$d(KrF) = 1.89$ Å], by subtracting the fluorine radius (0.64 Å). These radii correspond to those of the adjacent halogen atoms:

$$r_1(Xe) = 1.34 \text{ Å} \qquad r_1(I) = 1.33 \text{ Å}$$
$$r_1(Kr) = 1.25 \text{ Å} \qquad r_1(Br) = 1.14 \text{ Å}$$

For Xe(IV), 1.31 Å, and for Xe(VI) 1.25 Å are obtained. Thus the x_E values for Xe(II) and Kr(II) are:

Xe: 2.4   İ: 2.2
Kr: 2.6   Br: 2.7

Deriving the single bond radius of Xe from the bond distance of XeF^+ in $XeFSb_2F_{11}$ gives a somewhat higher x_E value, and the same holds for Kr if the atomic radius is taken from KrF^+.

The electronegativity of xenon is similar to that of sulfur, and the bonding in XeF_6 should be comparable with that of SF_6.

10.6. Bond Theory of the Rare Gas Compounds

Using the XeF_2 molecule as an example, the descriptions offered by the VB- and MO-theories will be tested.

The VB description includes the 5d-orbitals in the bonding of the xenon compounds. The large electronegativity difference is reflected in the structures:

$$:\!\overset{..}{\underset{..}{F}}\!-\!\overset{..}{\underset{..}{Xe}}\!-\!\overset{..}{\underset{..}{F}}\!: \quad \leftrightarrow \quad :\!\overset{..}{\underset{..}{F}}\!-\!\overset{\oplus}{\underset{..}{Xe}}\!: \quad :\!\overset{\ominus}{\underset{..}{F}}\!: \quad \leftrightarrow \quad :\!\overset{..}{\underset{..}{F}}\!: \quad :\!\overset{\oplus}{\underset{..}{Xe}}\!-\!\overset{..}{\underset{..}{F}}\!:$$

$$\text{(I)} \qquad\qquad \text{(II)} \qquad\qquad \text{(III)}$$

A linear pd- or pseudotrigonal bipyramidal sp^3d-hybrid at xenon is the basis for (I), but d-orbitals are not required for (II) and (III) in which there are only 8 valence electrons at the xenon atom. Quantum mechanical calculations suggest that (I) contributes 70% and (II) and (III) 15% each. In KrF_2 the $4d_{z^2}$ orbital is used along with the $4p_z$ orbital.

Promotion of a p-electron to a d-orbital in xenon requires 955 kJ/mol. The d-orbital is diffuse and unsuited for hybridization, but contracts in the field of fluorine or oxygen atoms. The low Xe—F bond energy is related to the large promotion energy.

The MO-description of the XeF_2 molecule corresponds to that for the I_3^- ion, which also has a linear structure. The p_z-orbital of xenon overlaps with the p_z orbitals of the fluorine atoms to give a 3-center 4-electron bond (cf. Fig. 92, p. 253). The bonding in XeF_4 and XeF_6 involves the other 5p orbitals of the xenon atom in bonding with each pair of fluorine atoms. Thus the square planar XeF_4 and the O_h XeF_6 molecules are formed. Yet the MO-treatment is unsatisfactory. It is not clear, for example, why XeI_2, isoelectronic with I_3^-, does not exist. Furthermore, the 3-center bonds of order 0.5 exhibit valence force constants f(Xe—F), 2.8 to 3.0 mdyn/Å comparable with 3.6 mdyn/Å for the 2-center bond in IF whose bond is stronger owing to the larger difference in electronegativities.

With the exception of XeF_6, the molecular structures of the rare gas compounds are in agreement with the VB-theory postulates. Participation of 5d-orbitals follows from quantum mechanical calculations on XeF_2 and $XeCl_2$, as well as from the existence of the ions XeF_8^{2-}, XeO_3F^-, and XeO_6^{4-}.

11. Nitrogen

Nitrogen, phosphorus, arsenic, antimony and bismuth form the Vth main group of the periodic system. However, the chemistry of nitrogen differs from that of its higher homologs, just as the chemistry of oxygen differs from that of sulfur. The reasons are the same: an abrupt increase in atomic radius, a higher electronegativity, and the absence of d-orbitals in the valence shell of the nitrogen atom.

The chemistry of nitrogen will be treated separately; the chemistry of its lower congeners will be covered in Section 12.

11.1. Elemental Nitrogen

Dinitrogen, N_2, is the main constituent of air (78.09 vol%). Most nitrogen compounds are prepared indirectly from atmospheric nitrogen.

Nitrogen is produced by fractional distillation of liquid air, or by converting atmospheric oxygen into CO_2 with glowing coke and removing it by scrubbing. The rare gases are obtained as well. Nitrogen free of rare gases is prepared in the laboratory by thermal decomposition of sodium azide at 275 °C:

$$2\,NaN_3 \;\rightarrow\; 2\,Na + 3\,N_2$$

Nitrogen recovered from air is reduced with hydrogen to ammonia (Haber-Bosch process). From ammonia NO is manufactured by catalytic oxidation (Ostwald process) which is reacted with oxygen to yield NO_2 which is injected into water and with additional air oxidation nitric acid is produced:

$$N_2 \xrightarrow[\text{Fe}]{H_2} NH_3 \xrightarrow[\text{Pt/Rh}]{O_2} NO \xrightarrow{O_2} NO_2 \xrightarrow{H_2O,\, O_2} HNO_3$$

Other nitrogen compounds are prepared from either NH_3 or HNO_3.

Molecular nitrogen is an inert gas (m.p. 63.2 K; b.p. 77.4 K) which requires high temperatures or a catalyst for NN-bond cleavage. Ammonia synthesis, for example, is carried out catalytically at 400–500 °C and 10^2–10^3 bar.

The N_2-molecule has a triple bond according to both VB- and MO-theory (p. 82 and 105) and an extremely high dissociation energy:

$$:N{\equiv}N: \;\rightarrow\; 2\,N \qquad D° = 945 \text{ kJ/mol}$$

Electric discharges produce reactive nitrogen atoms.

Despite its inertness there are microorganisms and algae which have the ability to assimilate atmospheric nitrogen and to utilize it via NH_3 as an intermediate   the synthesis of amino acids. This remarkable process plays an important role in life, since the soluble nitrogen compounds in soil are insufficient to provide for the nitrogen demand of plants. Enzymatic N_2-assimilation is not yet understood, but coordinate bonding of N_2 to heavy metals (iron, molybdenum etc.) present in all N_2-assimilating microorganisms forms the first step. This hypothesis has gained credibility after the synthesis of inorganic complexes with N_2 ligands was carried out successfully in 1965.

11.2. N_2 as Complex Ligand

The nitrogen molecule is an isostere of CO and isoelectronic with the NO^+ and CN^- ions, of which numerous transition metal complexes are known:

$$:N{\equiv}N: \qquad :\overset{\ominus}{C}{\equiv}\overset{\oplus}{O}: \qquad :\overset{\oplus}{N}{\equiv}\overset{\ominus}{O}: \qquad :C{\equiv}N:$$

Thus N_2-complexes can be expected. The two free electron pairs are localized in sp-hybridized orbitals on opposite sides of the triple bond, well-suited for coordinate bond formation (p. 82) and > 50 dinitrogen complexes, some containing two and three N_2 molecules, have now been prepared by:

a) Addition of N_2 or ligand exchange with N_2:

$$[Ru(H_2O)(NH_3)_5]Cl_2 + N_2 \xrightarrow{-H_2O} [Ru(N_2)(NH_3)_5]Cl_2 \quad \text{(yellow crystals)}$$

b) Oxidation of hydrazino complexes with H_2O_2:

$$(C_5H_5)Mn(CO)_2(N_2H_4) \xrightarrow[+H_2O_2/Cu^{2+}]{-40\,°C} (C_5H_5)Mn(CO)_2N_2 \quad \text{(red-brown crystals)}$$

c) Reaction of hydrazine with a metal salt (catalytic·disproportionation of N_2H_4 into nitrogen and ammonia:

$$(NH_4)_2OsCl_6 + N_2H_4(aq) \xrightarrow[H_2O]{100\,°C} [Os(N_2)(NH_3)_5]Cl \quad \text{(yellow crystals)}$$

d) Thermolysis of an azido complex:

$$[Ru(H_2O)(NH_3)_5]^{2+} \xrightarrow[-H_2O]{+N_3^-} [Ru(N_3)(NH_3)_5]^+ \xrightarrow[-\frac{1}{2}N_2]{\Delta T} [Ru(N_2)(NH_3)_5]^{2+}$$

e) Reaction of an ammine complex with HNO_2:

$$[Os(N_2)(NH_3)_5]Cl_2 + HNO_2(aq) \rightarrow [Os(N_2)_2(NH_3)_4]Cl_2$$

Bonding in the N_2 complexes has been clarified by structural evidence as well as by vibrational spectroscopy:

M—N≡N: M—N≡N—M

end-on coordination   cis-positions   linear bridges in
in mono-nuclear   of two N_2 ligands   binuclear
complexes   complexes

The linear M—N—N bond of the N_2 complexes corresponds to the CO bond in metal carbonyls, however, the M—N—N—M bridge bonds have no CO analogue. The metal-N_2 σ-bond is reinforced by a π-bond:

$$M \underset{\sigma}{\leftarrow} N \equiv N: \quad \leftrightarrow \quad M \underset{\sigma}{\overset{\pi}{\rightleftharpoons}} N = N:$$

The σ-bond is formed by overlap of the uppermost occupied σ-molecular orbital of the N_2-molecule with an unoccupied σ-orbital of the center atom.
This orbital will in general be an sp-hybrid (Fig. 93a) through which negative charge is transferred to the center atom. Electrons are returned to the ligands by the overlap of a filled d-orbital of the metal which has π-symmetry, with an unoccupied π^*-orbital of the N_2-molecule (Fig. 93b).
This π-(back) bond reinforces the metal-ligand interaction. However, since anti-

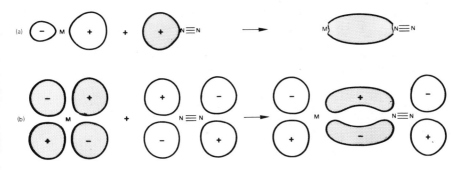

Fig. 93. Bonding of the molecule N_2 in dinitrogen complexes. The occupied orbitals are shaded.   a) σ-bond   b) π-(back)bond.

bonding orbitals are occupied, the NN-bond is weakened, and the internuclear distances $d(NN)$ are always larger in N_2-complexes while the valence force constants are smaller than in N_2. The bond order $b(NN)$ is 2.9 to 2.5, $vs.$ 3.0 for the N_2 molecule.

Back bonding presupposes a low oxidation state for the metal with filled d-orbitals. Oxidation of the metal leads to elimination of N_2. The N_2-complexes also liberate nitrogen on treatment with CO (irreversible ligand exchange) reflecting the greater stability of the carbonyl complexes. So far, complexes of type $M(N_2)_x$, corresponding to the carbonyls $Ni(CO)_4$ or $Fe(CO)_5$ have been obtained only at low temperatures.

Some N_2-complexes are stable at $> 300°$, but attempts to reduce the complexed N_2-molecule to N_2H_4 or NH_3 have been unsuccessful so far.

Spectroscopic data indicate that metal surfaces also bind N_2 as in the technical synthesis of ammonia, where a complex bond is probably first formed with the iron catalyst. The NN-bond is thus weakened, facilitating reaction with hydrogen.

11.3. Bond Relationships in Nitrogen Compounds

Nitrogen, the first element in the Vth main group, has five valence electrons of ground state configuration $2s^2p^3$. The valence shell has no d-orbitals, and the higher unoccupied 3s and 3p levels are of no importance in nitrogen compounds.

Like the other non-metals of the first octad, the nitrogen atom often achieves the neon configuration by forming three covalent bonds, or by the formation of the corresponding ions.

covalent:	NH_3, NCl_3
covalent and ionic:	NH_2^- in KNH_2, NH^{2-} in Li_2NH
ionic:	N^{3-} in Ba_3N_2

In NH_3 and most of its derivatives, the nitrogen atom is approximately sp^3-hybridized.

Like carbon and oxygen, the nitrogen atom uses multiple bonding with bond order ≤ 3.0 to complete its electron octet. Suitable partners for π-bonding include C, N, O, P and S. The following examples illustrate this bond type:

$R{-}C{\equiv}N:$	$:N{\equiv}N:$	$:N{\equiv}O:$	$:N{\equiv}SF_3$
covalent cyanide and cyano complexes	nitrogen	nitrosyl ion	sulfur nitride trifluoride

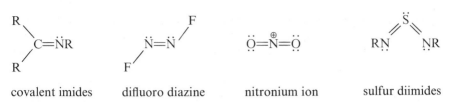

| covalent imides | difluoro diazine | nitronium ion | sulfur diimides |

The free electron pair in the 2s orbital of the nitrogen atoms can also be employed in coordinate bond formation, in the ions NH_4^+, NF_4^+, $N_2H_6^{2+}$ and in the following compounds:

    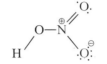

| boron trifluoride alkylamine | trifluoro amine oxide | nitric acid | copper(II) tetraammine ion |

These examples contain coordinate σ-bonds, but the nitrogen can also form coordinate π-bonds. Compounds of the type R_3N (R = any radical) are generally pyramidal Lewis bases in which ($\angle R-N-R$) is close to tetrahedral, and the free electron pair is in an sp^3-hybrid orbital. In contrast, trisilylamine contains a planar framework Si_3N:

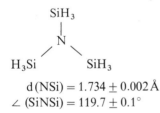

$$d(NSi) = 1.734 \pm 0.002\,\text{Å}$$
$$\angle\,(SiNSi) = 119.7 \pm 0.1°$$

with an sp^2-hybridized nitrogen atom. One explanation involves a delocalized coordinate π-bond formed by overlap of the doubly occupied p-orbital of the nitrogen atom perpendicular to the molecular plane with the 3d-orbitals of the silicon atoms which have π-symmetry (Fig. 94a). The consequences of this 4-center, 2-electron-π-bond are:

1) increase of the valence angle at the nitrogen
2) strengthening of the Si—N bonds
3) loss of Lewis basicity of the nitrogen atom toward weaker Lewis acids than the adjacent silicon atoms.

The rotation of the silyl groups about the Si—N bond is not restricted, however, since silicon has two perpendicular d-orbitals available which can alternate with rotation. The situation corresponds to that in acetylene.

π-Bonds of this kind are possible when an electron donor, $\equiv$N:, is adjacent to an acceptor, such as Be, B, Al, and S. Delocalization need not always be complete, and the relative weight of the structure on the right varies over wide limits:

$$\underset{\diagdown}{\overset{\diagup}{>}}\ddot{\text{N}}\underset{\sigma}{-}\text{X} \leftrightarrow \underset{\diagdown}{\overset{\diagup}{>}}\overset{\oplus}{\text{N}}\underset{\sigma}{\overset{\pi\;\ominus}{\equiv}}\text{X} \qquad \text{X: Be, B, Al, Si, S}$$

Further compounds with π-bonds of this type are, for example:

$\text{H}_3\text{Si}{=}\text{N}{=}\text{C}{=}\text{S}$

$\angle\,(\text{SiNC}) = 180°$
(cf. figure 94b)

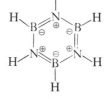

$\angle\,(\text{BNB}) \approx 120°$
$\angle\,(\text{NBN}) \approx 120°$

$\angle\,(\text{SNS}) = 120°$

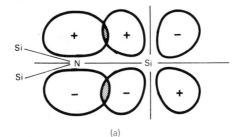

(a)

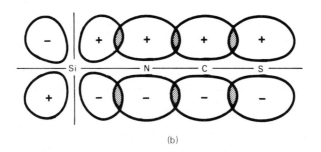

(b)

Fig. 94. Schematic Representation of Orbital Overlap in Many-Center π-Bonds
(a) in $(\text{SiH}_3)_3\text{N}$ (b) in SiH_3NCS

The coordination at the nitrogen atom is planar also in $[(H_3Si)_2N]_2$ and $\{[(CH_3)_3Si]_2N\}_2$.

Similar coordinative π-bonds can also originate from oxygen atoms (p. 184) and fluorine atoms (p. 247) but not in the higher homologs of nitrogen to the same extent. The molecules $P(SiH_3)_3$ and $As(SiH_3)_3$, for example, are pyramidal showing that the $(p \rightarrow d)\pi$-overlap between these groups if present, does not lead to planarity.

Energies and Enthalpies of Formation:
Owing to the large dissociation energy of the N_2-molecule, considerable energy is required in the formation of nitrogen compounds from the elements, e.g.:

$$\tfrac{1}{2}N_2 + \tfrac{1}{2}O_2 \rightleftharpoons NO \qquad \Delta H^{\circ}_{298} = 90 \text{ kJ/mol NO}$$

The nitrogen oxides, the azides, hydrazine, S_4N_4 and NCl_3 are endothermic (metastable) at room temperature and decompose spontaneously if the activation energy is small (e.g., in NCl_3) or if it is lowered catalytically as in N_2H_4 by Cu^{2+} ions. The product of decomposition is generally N_2, or a stable, exothermic nitrogen compound, for example, NH_3 or NF_3.

Thus an extraordinary large energy is liberated in the formation of N_2 from the reaction of two endothermic nitrogen compounds. Such reactions are suitable for rocket propulsion, for example, the third stage of the latest European rocket and the engines of the lunar excursion module of the Apollo program used liquid N_2O_4 as oxidizer and a mixture of N_2H_4 and $(CH_3)_2NNH_2$ as fuel. Mixing produces self-ignition, and combustion with a red flame takes place:

$$2\,N_2H_4 + N_2O_4 \rightarrow 3\,N_2 + 4\,H_2O \qquad \Delta H^{\circ}_{298} = -1252 \text{ kJ/mol } N_2O_4$$

Single bonds between nitrogen atoms are quite weak owing to the mutual repulsion of free electron pairs in the structural element $>\ddot{N}-\ddot{N}<$ (cf. p. 131) or in $>\ddot{N}-\ddot{O}\diagdown$ as well.

As a consequence, hydrazine, N_2H_4, and hydroxylamine, NH_2OH, have low thermal stabilities and higher homologs of these compounds, e.g., triazane, N_3H_5, or hydroxylhydrazine, N_2H_3OH, have not been isolated. The higher homologs of hydrogen peroxide, H_2O_3 and H_2O_4, are similarly unstable.

Salts of hydrazine containing the cation $N_2H_6^{2+}$ are much more stable. Since the repulsion of the free electron pairs is now eliminated (cf. p. 138). The positive charges are delocalized over six hydrogen atoms. Likewise, salts with the cation $H_2N-\overset{+}{N}H_2-NH_2$, or better the organic derivatives $H_2N-\overset{+}{N}R_2-NH_2$ can be prepared, while triazane itself, N_3H_5, is unknown. However, tetrazene, N_4H_4, has been isolated.

Certain compounds exist as free radicals at room temperature, for example the oxides NO and NO_2, the fluoride NF_2, and Frémy's salt, $K_2[ON(SO_3)_2]$. The nitrogen lacks an octet in these compounds, even though the unpaired electron is delocalized. These radicals are in equilibrium with their diamagnetic dimers, which predominate in the condensed phase and at low temperatures. The dissociation energies are small compared to normal single bonds:

$$N_2O_2 \rightleftharpoons 2\,NO \qquad\qquad \Delta H°: 10.5 \text{ kJ/mol dimer}$$

$$N_2O_4 \rightleftharpoons 2\,NO_2 \qquad\qquad 65 \quad \text{kJ/mol dimer}$$

$$N_2O_3 \rightleftharpoons NO + NO_2 \qquad\quad 40 \quad \text{kJ/mol dimer}$$

$$N_2F_4 \rightleftharpoons 2\,NF_2 \qquad\qquad 85 \quad \text{kJ/mol dimer}$$

$$[ON(SO_3)_2]_2^{4-} \rightleftharpoons 2[ON(SO_3)_2]^{2-}$$

(crystalline, yellow) (violet in H_2O)

Nitrogen-nitrogen bonds[11] are present in N_2O_2, N_2O_4, N_2O_3, and N_2F_4, but Fremy's salt may contain an O—O bond.

11.4. Hydrides

11.4.1. General

Five volatile hydrides of nitrogen are known: ammonia, NH_3, hydrazine, H_2NNH_2, diimine, $HN{=}NH$, tetrazene, N_4H_4, and hydrogen azide, N_3H. Additional binary NH-compounds are ammonium azide, NH_4N_3, and hydrazonium azide, $N_2H_5N_3$. Ammonia is the most important nitrogen hydride, and many other nitrogen compounds are derived from it. In liquid form it is a water-like solvent (p. 157 and 291).

11.4.2. Ammonia, NH_3

Ammonia is formed in an exothermic reaction from the elements if sufficiently high temperatures (400–500°) and a catalyst (α-iron with Al_2O_3 and K_2O) for activation of the N_2 molecule are provided:

$$N_2 + 3\,H_2 \rightleftharpoons 2\,NH_3 \qquad \Delta H° = -92 \text{ kJ/mol } N_2$$

High pressure (100–1000 bar) shifts the equilibrium in favor of NH_3 formation. The technical product is mainly utilized in fertilizers.

Liquid NH_3 is available in cylinders (b.p. $-33.4°$). Ammonia can be generated from ammonium salts in the laboratory:

$$NH_4Cl + NaOH \rightarrow NH_3 + NaCl + H_2O$$

---

11 One reason for the low bond energies probably may be the repulsion of the two partially positively charged nitrogen or oxygen atoms.

Upon heating, NH_3 partially decomposes into the elements, or is oxidized to N_2 and H_2O in presence of air or O_2. Passing NH_3-air mixtures over certain catalysts gives NO:

$$4\,NH_3 + 5\,O_2 \;\rightarrow\; 4\,NO + 6\,H_2O, \qquad \Delta H^\circ = -904 \text{ kJ/formula weight}$$

which is formed at 600 to 800°, and upon cooling reacts with additional O_2 to give NO_2 which is bubbled into water to yield nitric acid:

$$2\,NO_2 + \tfrac{1}{2}O_2 + H_2O \;\rightarrow\; 2\,HNO_3$$

Production of nitric acid is based on this process.
Ammonia is a pungent gas which, owing to the structural similarity with H_2O and hydrogen bonding, is very soluble in water. Concentrated solutions contain up to 35 wt-% NH_3. Most of the NH_3 is physically dissolved, but basicity and electrolytic conductance argue for a equilibrium:

$$NH_3(aq) + H_2O \;\rightleftharpoons\; NH_4^+ + OH^-$$

which lies almost entirely to the left $(K = [NH_4^+][OH^-]/[NH_3]$, at 25°C is 1.8×10^{-5} mol $\cdot$ l^{-1}). Thus at 25°C a 0.1 M NH_3 solution is less than 1% dissociated, i.e., ammonia is a weak base. No compound NH_4OH exists.
Ammonia reacts with stronger proton donors to form the corresponding ammonium salts. Owing to their similar cationic radii, the solubilities are very similar to the corresponding potassium salts. The regular tetrahedral ion, NH_4^+, is isoelectronic with BH_4^- and CH_4. Ammonium salts are completely dissociated in aqueous solution, and owing to hydrolysis are weakly acidic, if the anion is derived from a strong acid:

$$NH_4^+ + H_2O \;\rightleftharpoons\; NH_3 + H_3O^+$$

The hydrogen atoms in NH_3 can be replaced by strongly electropositive metals under anhydrous conditions. Thus gaseous NH_3 reacts with alkali metals on heating to produce ionic amides:

$$Li + NH_3 \xrightarrow{\;400\,^\circ C\;} LiNH_2 + \tfrac{1}{2}H_2$$

On stronger heating certain amides split off NH_3 to form imides and finally nitrides:

$$2\,MNH_2 \;\rightarrow\; M_2NH + NH_3 \qquad 3\,M_2NH \;\rightarrow\; 2\,M_3N + NH_3$$

The alkali and alkaline earth salts which contain the ions NH_2^-, NH^{2-}, and N^{3-} hydrolyze to NH_3 and the corresponding hydroxides.
Almost all non-metals form bonds to nitrogen. Two reactions for forming element-N-bonds are used to synthesize many compounds from NH_3, amides and imides:

$$E-X + H-N\langle \;\rightarrow\; E-N\langle \;+\; HX$$

E: B, C, Si, N, P, As, S, Se, Cl
X: F, Cl(Br, I)

$$E-OH + H-N\langle \;\rightarrow\; E-N\langle \;+\; H_2O$$

E: B, N, P, S, Cl

11.4.3. Hydrazine, N_2H_4

Hydrazine is one of the few compounds having a N—N single bond. This bond has a small bond energy (p. 131), and hydrazine is an endothermic compound whose synthesis requires gentle conditions:

a) in the Raschig synthesis NH_3 is oxidized with aqueous NaOCl to NH_2Cl, which then reacts with HCl elimination to N_2H_4:

$$NH_3 \quad + NaOCl \;\rightarrow\; NH_2Cl + NaOH$$

$$NH_2Cl + HNH_2 + NaOH \;\rightarrow\; H_2N-NH_2 + NaCl + H_2O$$

Addition of glue or gelatin suppresses a competing reaction which is catalyzed by traces of heavy metals:

$$2\,NH_2Cl + N_2H_4 \;\rightarrow\; 2\,NH_4Cl + N_2$$

Hydrazine forms in 70% yield.

b) In a more recent synthesis, NH_3 is condensed with a ketone in presence of chlorine to form a ketazine via an intermediate diazacyclopropane derivative:

$$2\,R_2C{=}O + Cl_2 + 2\,NH_3 \;\rightarrow\; R_2C{=}N-N{=}CR_2 + 2\,H_2O + 2\,HCl\;(\rightarrow NH_4Cl)$$

The ketazine is hydrolyzed with water under pressure and the ketone is recovered:

$$R_2C{=}N-N{=}CR_2 + 2\,H_2O \;\rightarrow\; N_2H_4 + 2\,R_2CO$$

Hydrazine can be distilled from aqueous solution and concentrated to 95% by distilling over solid NaOH. Anhydrous hydrazine is obtained by dehydrating the concentrate with BaO $[\rightarrow Ba(OH)_2]$ or Ba_3N_4 (barium pernitride) the latter reacting with water quantitatively to give $Ba(OH)_2$, N_2H_4 and N_2.
Hydrazine is an oily, colorless liquid (m.p. 1.5°, b.p. 113,5 °C) which fumes strongly when exposed to air and explodes on heating or ignition. The N_2H_4 molecule is twisted similar to H_2O_2 with a dihedral $\angle H-N-N-H$ of 90° (symmetry C_2). The molecular dipole moment is 1.85 D. The valence angles at nitrogen correspond to sp^3-hybridization, but the derivative $[(CH_3)_3Si]_2N-N[Si(CH_3)_3]$ is planar as rationalized in the corresponding NH_3 derivatives by $(p \rightarrow d)\,\pi$-bonding. Hydrazine is completely miscible with water and the resulting solution which can be handled

safely reacts as a reducing agent, as a complexing agent, or as a weak base. Hydrazine reduces halogens to hydrogen halides, Cu(II) salts to Cu(I)-oxide, reduces selenites and tellurites and silver and mercury salts to the elements. In all cases N_2H_4 is oxidized to N_2.

Hydrazine is a bifunctional base from which salts of the cations $N_2H_5^+$ and $N_2H_6^{2+}$ (isoelectronic with CH_3NH_2 and C_2H_6, respectively) are derived. The former dissolve in water with simple dissociation, however, those containing $N_2H_6^{2+}$ suffer marked hydrolysis:

$$N_2H_6^{2+} + H_2O \rightleftharpoons N_2H_5^+ + H_3O^+$$

Hydrazinium dichloride, $N_2H_6Cl_2$, and sulfate, $(N_2H_6)SO_4$, are precipitated from aqueous N_2H_4 solution by excess acid. Hydrazine is a weaker base than NH_3, but like it forms complexes with metal ions, e.g., $M(N_2H_4)_2Cl_2$, where M = Mn, Fe, Co, Ni, Cu, Zn.

11.4.4. Diimine

Diimine is prepared as follows:

$$
\begin{array}{c}
R \\
\diagdown \\
\quad N-NH_2 + NaN[Si(CH_3)_3]_2 \rightarrow Na[RN-NH_2] + HN[Si(CH_3)_3]_2 \\
\diagup \\
H
\end{array}
$$

$$Na[RN-NH_2] \xrightarrow[10^{-5}\text{ torr}]{60\,^\circ C} HN=NH + NaR \qquad R = -SO_2-C_6H_4-CH_3$$

but disproportionates in the vapor phase into N_2 and N_2H_4. Diimine can only be prepared under small partial pressures, but condenses to a yellow solid at 77K. Upon warming it discolores and decomposes to N_2, N_2H_4 and NH_4N_3:

$$2N_2H_2 \diagdown \begin{array}{l} \rightarrow N_2 + N_2H_4 \\ \rightarrow NH_4N_3 \end{array}$$

11.4.5. Hydrogen Azide, HN_3

HN_3 is the parent compound of a number of covalent and ionic azides. In contrast to NH_3 and N_2H_4, hydrogen azide is an acid in water and thus should be termed nitrogen hydrogen acid.

Sodium azide is prepared by passing N_2O over molten $NaNH_2$:

$$2\,NaNH_2 + N_2O \rightarrow NaN_3 + NaOH + NH_3$$

The NaN_3 can be crystallized from the aqueous mixture. Sodium azide is also prepared from hydrazine and ethyl nitrite in ether:

$$N_2H_4 + C_2H_5ONO + CH_3ONa \xrightarrow{0\,°C} NaN_3 + C_2H_5OH + CH_3OH + H_2O$$

Aqueous HN_3 can be obtained from NaN_3 via ion exchange or distillation with dilute H_2SO_4.

Anhydrous HN_3 is a water-white, mobile liquid (b.p. 36 °C) which decomposes with explosion to the elements. Aqueous solutions with up to 20 wt% HN_3, however, are not dangerous to handle. In water, HN_3 is a weak acid, comparable to acetic acid. The salts, resembling the chlorides in appearance and solubilities, are obtained from NaN_3 by double decomposition. The anion N_3^- is a pseudohalide. In contrast to the HN_3 molecule, the azide ion has two identical NN-bonds:

$d(NN) = 1.15\,\text{Å}$
$\angle(NNN) = 180°$

$d(NN) = 1.24, 1.13\,\text{Å}$, respectivelly
$\angle(NNN) = 180°$
$\angle(HNN) \approx 110°$

The N_3^- ion is isoelectronic with CO_2, N_2O, NO_2^+ and OCN^- and the bonding is similar in having two three-center four-electron π-bonds superimposed over both two-center σ-bonds. The three-center bonds arise from three p-orbitals available at each sp-hybridized nitrogen atom in two mutually perpendicular planes. Their overlap results in bonding, non-bonding, and antibonding orbitals. Each of the two π-bonds contains four electrons, two in a bonding and two in a non-bonding state. The total bond order amounts to 2.0. Protonation of the ion at one of the terminal N-atoms perturbs the symmetry, since a former π-orbital is required for a σ-bond to the hydrogen. Therefore, a 3-center 4-electron π-bond and a 2-center π-bond remain, and HN_3 and other covalent azides have two unlike NN-bonds (bond orders of 1.5 and 2.5). Thus the N_2 molecule which arises through decomposition is already preformed.

While ionic azides, upon heating, decompose smoothly to N_2 and metal, the covalent compounds and the heavy metal azides explode. Lead(II) azide is suitable as primer for explosives.

11.4.6. Hydroxylamine, NH_2OH

Hydroxylamine, formally a derivative of ammonia, is generated in the electrolytic reduction of nitric acid or nitrites on a cathode of lead amalgam, as well as in the reduction of NO or NO_2 with hydrogen by introducing the gases into hydrochloric acid solution in which platinum on charcoal is suspended as catalyst:

$$NO_2 + 2\tfrac{1}{2}H_2 \xrightarrow{Pt} NH_2OH + H_2O$$

Hydroxylamine is a weaker base in water than ammonia, but forms stable hydroxylammonium salts, $(NH_3OH)X$, with acids, from which it can be liberated by reaction with sodium methylate:

$$[NH_3OH]X + CH_3ONa \rightarrow NH_2OH + CH_3OH + NaX$$

Hydroxylamine forms colorless crystals (m.p. $33°C$) which decompose slowly at room temperature. The aqueous solution is unstable, disproportionating, depending upon the pH value, into NH_3 and N_2 (pH > 7) or NH_3 and N_2O (pH < 7). Pure NH_2OH explodes at $> 100°$ into NH_3, N_2 and H_2O.

Hydroxylamine is a strong reducing agent, e.g., towards Cu^{2+}, Hg^{2+} and Ag^+, becoming oxidized to N_2. With other oxidizing agents, however, N_2O, NO, NO_2^- and NO_3^- can be generated. Only very strong reducing agents (Sn^{II}, V^{II}, Cr^{II}) reduce NH_2OH to NH_3.

11.4.7. Non-aqueous Solvents

Many low viscosity liquids dissolve inorganic and organic compounds and have practical utility as solvents. The following anhydrous liquids are examples:

a) Protic solvents:

$$NH_3, HF, H_2SO_4, HSO_3F, CH_3COOH, HCl, H_2S, HCN$$

b) Aprotic solvents:

$$SO_2, N_2O_4, BrF_3, SeOCl_2, POCl_3, NOCl, COCl_2, \text{ and halides of As, Sb, Bi, and Hg.}$$

These solvents often have low electrical conductivity, yet their solutions are often good conductors indicating ionization. Many reactions can be followed by conductometric and potentiometric measurements.

In the following, only liquid ammonia will be discussed. Ammonia melts at $-77.7°C$ (triple point) and boils at $-33.4°C$.

Solubilities in Liquid Ammonia:

The solubility of an ionic compound is a function of the lattice energy and the sum of the solvation energies of the ions. Thus considerable differences in solubilities between water and ammonia are possible.

The solubilities of the potassium halides in NH_3 increase with increasing anionic radius, $KF < KCl < KBr < KI$, similar to their behavior in water. However, silver halides are opposite to their behavior in water ($AgF < AgCl < AgBr < AgI$). At $25°C$, 207 g AgI dissolve in 100 g of liquid NH_3. This high solubility is related to the high solvation energy of the cation which dissolves in NH_3 as $[Ag(NH_3)_2]^+$. The special solubility relations in the ammonia system permit the following double decomposition:

$$2AgCl + Ba(NO_3)_2 \xrightleftharpoons[H_2O]{NH_3} 2AgNO_3 + BaCl_2$$

Since $BaCl_2$ is the least soluble component in NH_3, the reaction proceeds to the right, unlike the situation in H_2O, owing to the near insolubility of $AgCl$.

Self-dissociation of Liquid Ammonia

Even at its boiling point, liquid ammonia is only very weakly dissociated:

$$2NH_3 \rightleftharpoons NH_4^+ + NH_2^- \qquad [NH_4^+][NH_2^-] = 10^{-29}(-33\,°C)$$

The ion product is even smaller than of ethanol (ca. 10^{-20}). Substances which enhance the concentration of ammonium ions are acids, e.g., soluble NH_4^+ salts of I^-, CN^-, SCN^-, NO_3^-, NO_2^-, N_3^- and BF_4^- anions which correspond to the oxonium salts in the aquo system.

Solutions of the ammonium salts can dissolve metals like Mg or Al with hydrogen evolution:

$$Mg + 2NH_4^+ \rightarrow Mg^{2+} + 2NH_3 + H_2\uparrow$$

and decompose magnesium silicide with generation of silane:

$$Mg_2Si + 4NH_4^+ \rightarrow 2Mg^{2+} + 4NH_3 + SiH_4\uparrow$$

Compounds which increase the concentration of amide ions in ammonia, e.g., KNH_2 and $Ba(NH_2)_2$ behave as bases. $NaNH_2$ is insoluble, and $LiNH_2$ and $Ca(NH_2)_2$ are only sparingly soluble. These ionic amides correspond to the hydroxides of the aquo system.

Neutralization reactions between acids and bases can be followed as in water, with indicators, or electrometrically. For example, in the titration of the strong base KNH_2 with NH_4Cl solution, the insoluble salt KCl is formed:

$$NH_4^+ + Cl^- + K^+ + NH_2^- \rightarrow KCl\downarrow + 2NH_3$$

The equivalence point consequently has a minimum of conductivity.

Ammonolysis Reactions:

Many non-metal halides, e.g., BCl_3 react with ammonia:

$$BCl_3 + 6NH_3 \rightarrow B(NH_2)_3 + 3NH_4Cl$$

Boron triamide is a barely dissociated base in the ammonia system, the analog of orthoboric acid, $B(OH)_3$, in the aquo system, which is formed in the hydrolysis of BCl_3. Rhombic sulfur reacts with liquid ammonia to give heptasulfur imide:

$$S_8 + NH_3 \rightarrow S_7NH + H_2S$$

Solvated Electrons in Liquid Ammonia:

Certain alkali and alkaline earth metals dissolve reversibly with good (Li, Na, K, Ca) to excellent (Cs) solubility in anhydrous ammonia to give blue, or in higher concentrations, bronze-colored solutions. The alkaline earth metals crystallize as hexaammoniates, e.g., $Ca(NH_3)_6$, on evaporation.

Blue colors originate at the cathode in the electrolysis of ammoniacal metal salt solutions.

The metal solutions have common properties which to a first approximation are independent of the dissolved metal.
Both the blue- and bronze-colored solutions have excellent electric conductivity, indicating ions or electrons or both. The conductivity of concentrated solutions corresponds to that of pure metals like sodium or mercury. These experimental findings are rationalized by a reversible dissociation:

$$M + xNH_3 \rightleftharpoons M^+_{am} + e^-_{am}$$

The common properties of the metal solutions are a consequence of solvated electrons which give rise to a molar susceptibility of one Bohr magneton in the very dilute solutions. With increasing concentration the molar susceptibility decreases markedly owing to dimerization of the electrons to give diamagnetic electron pairs:

$$2e^-_{am} \rightleftharpoons e^{2-}_{2(am)}$$

Information regarding the structure of the solvated electrons comes from the density which decreases from 0.68 g/ml at the boiling point of NH_3 to 0.477 g/ml

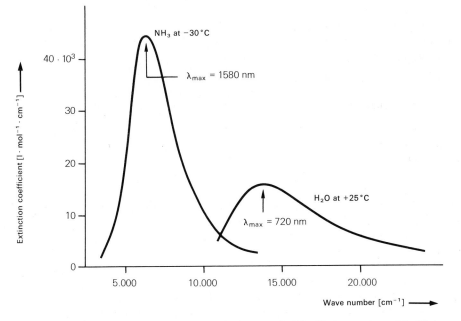

Fig. 95.   Absorption Spectrum of Solvated Electrons in Liquid Ammonia and in Water

for a saturated Li-solution at 19 °C. Since the metal cations behave normally in the solutions, the decrease in density, i.e., the volume expansion must be caused by the electrons whose volume requirement has been determined as 70 to 90 ml/mol, 10 to 40 times larger than for simple ions.

Coulombic repulsion between dissolved electrons and the electrons of the NH_3-molecules creates holes which can be occupied, without substantial influence on their radius, by one or two electrons, i.e., the holes behave like orbitals. The radius of the hole can be estimated from the density of the metal solutions as well as from the absorption spectrum[12] (Fig. 95) as about 3.5 Å.

The absorption spectrum consists of a single, broad band at 1580 nm, whose long wavelength branch extends into the visible range, producing the blue color.

Several models exist which explain the quasimetallic properties of the concentrated solutions which have metallic luster and are produced from the dilute solutions by concentration or cooling to give a concentrated lighter phase together with a dilute blue one which is heavier than the former.

Reactions of the Electrons in Ammonia:

In the amide reaction, the metal ammonia solutions decompose with formation of metal amide:

$$NH_3 + e_{am}^- \rightleftharpoons NH_2^- + \tfrac{1}{2}H_2 \qquad K = \frac{[NH_2^-] \cdot p_{H_2}^{1/2}}{[e_{am}^-]}$$

The metal solutions are metastable, and on standing or addition of a catalyst (Ni, Fe_3O_4, Pt) discolor with hydrogen evolution.

The amide reaction is reversible, i.e., a KNH_2-solution reacts with hydrogen under pressure (100 bar) to give e_{am}^- and NH_3. This reaction can be followed quantitatively by the production of blue color to yield $K = 5 \cdot 10^4 [bar^{1/2}]$ (25 °C), and $\Delta H_{298}^\circ = -67$ kJ/mol.

Solvated electrons are strong reducing agents, reducing non-metals to monomeric or polymeric anions (Table 48). Evaporation of solvent yields the salts.

Solvated Electrons in Water:

Solvated electrons can also be produced in water, ice, and other media, but reaction occurs more rapidly in liquid water than in ammonia to give hydroxyl ions and hydrogen atoms:

$$e_{aq}^- + H_2O \rightarrow OH^- + H$$

The half-life of e_{aq}^- is $0.8 \cdot 10^{-3}$ s, so that the electrons in water can only be observed for 1 ms. Hydrated electrons arise in α-, β-, or γ-bombardment of water:

12 The excitation energy of an electron, following the model of the electron in the box, is a function of the hole size.

Table 48   Reactions of Non-metals with Solvated Electrons in Liquid Ammonia

Non-Metal	Metal	Reaction Products
P	Li	Li_4P_2
	K	K_3P_x, K_6P_4
O_2	Li	Li_2O, Li_2O_2, LiO_2
	Na	Na_2O, Na_2O_2
	K	KO_2, K_2O_2
	Rb	Rb_2O_2, RbO_2
O_3	Li	$LiO_3 \cdot 4NH_3$
S_8	K	$K_2S, K_2S_2, K_2S_4, K_2S_x$
Se	K	$K_2Se, K_2Se_2, K_2Se_3, K_2Se_4$
Te	K	K_2Te, K_2Te_2

$$H_2O \rightarrow H_2O^+ + e_{aq}^-$$

or in photochemical ionization of anions:

$$I^- \xrightarrow[H_2O]{h\nu} I + e_{aq}^-$$

Pulse radiolysis with an intense beam of high energy electrons is used.
Hydrated electrons appear as intermediates in chemical reactions, e.g., in the reactions of hydrogen atoms with alkaline solutions:

$$H + OH^- \xrightarrow[H_2O]{} H_2O + e_{aq}^-$$

or the decomposition of sodium amalgam by water:

$$Na(Hg) \xrightarrow[H_2O]{} Na^+ + e_{aq}^-$$

In these cases, e_{aq}^- can be trapped by N_2O which is reduced to N_2.

The absorption spectrum of electrons in water is shifted from that of the blue metal NH_3-solutions, with the energy of the optical transition 167 kJ/mol vs. 84 kJ/mol in NH_3. The radius of the holes occupied by electrons in water is calculated from this to be 1.45 Å, corresponding to a requirement of 20 ml/mol. The effective ionic radius of the hydrated electrons is thus 2.5 to 3.0 Å.
The redox potential of hydrated electrons vs. the normal hydrogen electrode is -2.7 V, and hence nearly all inorganic compounds are reduced, excepting the alkalis and alkaline earths,

Table 49   Reactions of Hydrated Electrons with Several Simple Compounds

Starting Material	Primary Reaction Product	Subsequent Products
H	H^-	$H_2 + OH^-$
CO_2	CO_2^-	$C_2O_4^{2-}$
N_2O	N_2O^-	$N_2 + O^-$
NH_4^+	NH_4	$NH_3 + H$
NO_3^-	NO_3^{2-}	NO_2
Cu^{2+}	Cu^+	Cu^+
Zn^{2+}	Zn^+	$Zn + Zn^{2+}$
MnO_4^-	MnO_4^{2-}	MnO_4^{2-}

and the halide ions (see Table 49). Simple addition of electrons can be distinguished from reactions with dissociative electron capture. All reactions are extremely rapid, e.g., the rate of the bimolecular e_{aq}^- reactions with O_2 or N_2O is ca. $10^{10}\ \mathrm{l \cdot mol^{-1} \cdot s^{-1}}$. Hence even traces of these compounds take up e_{aq}^- markedly.

11.5. Halides and Oxohalides of Nitrogen

11.5.1. Halides

Nitrogen halides are formally derived from the hydrides by substitution of H by X (X = F, Cl, Br, I):

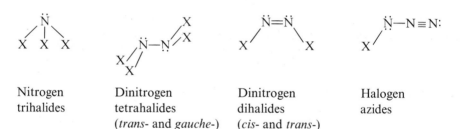

Nitrogen trihalides	Dinitrogen tetrahalides (*trans-* and *gauche-*)	Dinitrogen dihalides (*cis-* and *trans-*)	Halogen azides

Partially halogenated compounds are also known, e.g., haloamines, NH_2X, and dihaloamines, NHX_2, and mixed halides also exist. Stability decreases with increasing atomic weight of the halogen so that *all* the bond types are only realized for fluorine.

The trihalides result from complete halogenation of NH_3 or NH_4^+-compounds by the halogen:

$$NH_3 + 3X_2 \rightarrow NX_3 + 3HX$$

The trifluoride is produced as a colorless gas from the action of dilute fluorine in the presence of copper metal as well as in the electrofluorination of urea or NH_4F in hydrogen fluoride (p. 245). Nitrogen trifluoride is a weak Lewis base, forming complexes only with the strongest Lewis acids, e.g., to form NF_4SbF_6(p. 247), and reacts with O_2 in a glow discharge to give nitrogen oxotrifluoride ONF_3 (colorless gas, b.p. $-85\,°C$).

The trichloride, by contrast is an endothermic compound formed by bubbling Cl_2 into a acidified NH_4Cl solution. Pure NCl_3 is a yellow, highly explosive oil, soluble in organic solvents, which hydrolyzes to form NH_3 and $HOCl$.

Chlorination of NH_3 with insufficient halogen yields chloramine:

$$2\,NH_3 + Cl_2 \;\rightarrow\; NH_2Cl + NH_4Cl$$

Bromamine is prepared analogously in the reaction of bromine with ammonia in the vapor phase. Reaction of iodine with concentrated aqueous ammonia produces $NI_3 \cdot NH_3$, an explosive black solid.

Of the hydrazine tetrahalides only gaseous N_2F_4 is known. This exothermic compound (b.p. $-111\,°C$) is prepared in the reduction of NF_3 with copper at $375\,°C$, or with mercury vapor in a glow discharge:

$$2\,NF_3 + M \;\rightarrow\; N_2F_4 + MF_2$$
$$M: Cu, Hg$$

The molecular structure corresponds to hydrazine, but the gas contains both *trans*- and *gauche*-conformers.[13] In contrast to NF_3, N_2F_4 is extraordinarily reactive, owing to the small dissociation energy of the NN-bond which leads to NF_2 in equilibrium:

$$N_2F_4 \rightleftharpoons 2\,NF_2 \qquad \varDelta H° = 85 \text{ kJ/mol } N_2F_4$$

and many reactions of N_2F_4 lead to NF_2-containing compounds, e.g., with Cl_2 to $NClF_2$, with S_2F_{10} to F_2NSF_5, and with NO to F_2NNO. The NF_2 radical is isoelectronic with the ozonide ion and like it, is angular. Its stability is similar to that of the nitrogen radicals NO and NO_2.

Dinitrogen difluoride, N_2F_2, which occurs when NF radicals combine:

$$FN_3 \xrightarrow[-N_2]{70-90\,°C} FN$$
$$NHF_2 \xrightarrow[-KHF_2]{+KF} FN$$
$$\longrightarrow \tfrac{1}{2}N_2F_2 \text{ colorless gas}$$

consists of two isomers in rapid equilibrium at high temperature, of which the *trans*-form is more stable by about 12 kJ/mol, but allowing separation at room temperature or below by

13 In the *trans*-isomer, the NF_2 groups are twisted by 180° giving rise to an inversion center (zero dipole moment). In the *gauche*-isomer the twist is only 90°.

distillation or vapor phase chromatography. Dinitrogen difluoride is stable to O_2 and H_2O but decomposes at $300°$ into the elements. The salt $[N_2F][AsF_6]$ containing the linear cation N_2F^+ which is isoelectronic with CO_2, N_2O, and NO_2^+ forms from AsF_5.
Thermolysis of mixtures of FN_3 and ClN_3 yields N_2ClF:

$$FN_3 \xrightarrow[120°C]{-N_2} FN{-} \atop {} $$
$$\left.\begin{array}{}\\ \\ \\ \\\end{array}\right\} \rightarrow FN{=}NCl$$
$$ClN_3 \xrightarrow[120°C]{-N_2} ClN{-}$$

an explosive compound whose formation is evidence for the cleavage of covalent azides, RN_3, to N_2 and a carbene analogue RN, which can dimerize or further decompose. Halogens act on azides to form halogen azides:

$$Cl_2 + NaN_3 \xrightarrow{H_2O} ClN_3\uparrow + NaCl$$
$$I_2 + AgN_3 \longrightarrow IN_3 + AgI$$

The fluoride is an unstable, yellow-green gas which gives N_2F_2 and N_2. The other three halogen azides are explosive.

11.5.2. Oxohalides

Two series of oxohalides with nitrogen-halogen bonds are known:

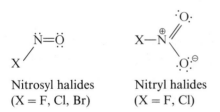

Nitrosyl halides	Nitryl halides
(X = F, Cl, Br)	(X = F, Cl)

The nitrosyl halides are acid halides of nitrous acid, HONO, where the OH group is replaced by X. Likewise the nitryl halides are derived from nitric acid, $HONO_2$. Halogen nitrates, $XONO_2$ (X = Cl, Br, I, p. 266), are derived by the replacement of hydrogen in nitric acid by X, but $FONO_2$ is called pernitryl fluoride.
The compounds ONOF, NOF_3 (p. 297) and $ONNF_2$ (p. 297) have no analogues with the other halogens.
The gaseous nitrosyl halides are prepared by reaction of NO with the halogen:

$$X_2 + 2NO \rightleftharpoons 2XNO$$

While FNO is stable, ClNO and BrNO revert partially to the starting materials. The fluoride is colorless (b.p. $-60°C$), ClNO is orange-yellow (b.p. $-6°C$) and BrNO is red (b.p. $0°C$). The molecules are angular with C_s symmetry.

The nitrosyl halides hydrolyze to X_2 and HNO_2, which disproportionates to HNO_3 and NO:

$$XNO + H_2O \rightarrow HNO_2 + HX$$

$$3 HNO_2 \rightarrow HNO_3 + 2 NO + H_2O$$

Nitryl fluoride and nitryl chloride are colorless gases consisting of planar molecules with C_{2v} symmetry and bonding as in nitric acid. The fluoride is prepared by fluorination of NO_2:

$$NO_2 + CoF_3 \xrightarrow{300\,°C} FNO_2 + CoF_2$$

$$2 NO_2 + F_2 \xrightarrow{25\,°C} 2 FNO_2$$

The chloride is obtained by oxidation of $ClNO$ with O_3, by chlorination of N_2O_5 with PCl_5, or best in the reaction of anhydrous nitric acid with chlorosulfuric acid:

$$HNO_3 + HSO_3Cl \xrightarrow{0\,°C} ClNO_2{\uparrow} + H_2SO_4$$

The poisonous, corrosive product decomposes at $100\,°C$ into Cl_2 and NO_2. In the alkaline hydrolysis nitrite and hypochlorite are formed.

11.6. Oxides of Nitrogen

11.6.1. General

The nitrogen oxides shown in Table 50 are endothermic compounds which de-compose into the elements on heating. The existence of the peroxides NO_3 or N_2O_6 is uncertain.

Table 50   Oxides of Nitrogen

Oxidation State of Nitrogen:	+1	+2	+3	+4	+5
	N_2O	NO	N_2O_3	NO_2	N_2O_5
		N_2O_2		N_2O_4	

Bond orders $b(NO)$ lie between 0.7 and 2.5.

11.6.2. Dinitrogen Oxide, N_2O

Thermal decomposition of ammonium nitrate produces gaseous N_2O:

$$NH_4NO_3 \xrightarrow{180–250\,°C} N_2O + 2 H_2O$$

At $>300\,°C$, explosion occurs. Condensation of amidosulfuric acid with concentrated nitric acid at $50-80°$ also gives N_2O:

$$H_2N-SO_3H + HNO_3 \rightarrow N_2O + H_3OHSO_4$$

The linear (symmetry $C_{\infty v}$) N_2O molecule is an isostere of CO_2 and isoelectronic with the ions NO_2^+ and N_3^-, and is characterized best by the first mesomeric structure:

$$\overset{\ominus}{N}=\overset{\oplus}{N}=\overset{}{\ddot{O}} \; \rightleftharpoons \; :N\equiv\overset{\oplus}{N}-\overset{\ominus}{\ddot{O}}: \qquad \mu(N_2O) = 0.16 \text{ D}$$

Dinitrogen oxide (nitrous oxide) which is soluble in water without hydrolysis has anesthetic properties.

11.6.3. Nitrogen Monoxide, NO, N_2O_2

Technical production of NO by catalytic oxidation of NH_3 is devoted to further oxidation to nitric acid (p. 279). In the laboratory sodium nitrite is reacted with $3M$ H_2SO_4:

$$3\,NaNO_2 + 3\,H_2SO_4 \rightarrow 2\,NO\uparrow + HNO_3 + H_2O + 3\,NaHSO_4$$

The nitrous acid formed disproportionates into NO and HNO_3. Reducing nitrites with iodide or Fe^{2+} also yields NO.

Paramagnetic NO is a colorless gas which in contrast with N_2O is reactive (b.p. $-152\,°C$, m.p. $-164\,°C$). Little dimerization occurs at room temperature:

$$2\,NO \rightleftharpoons N_2O_2 \quad \text{(cf. p. 286)}$$

In the liquid, and even more in the solid phase, however, nitrogen(II) oxide is largely dimerized, forming planar cis- and trans-N_2O_2, the trans-form being the more stable:

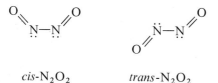

$$cis\text{-}N_2O_2 \qquad trans\text{-}N_2O_2$$

Formation of the NO-molecule is best understood by MO theory as with the simple scheme on p. 104 in which the π^*-level is occupied by one electron which is readily lost to give the nitrosyl cation NO^+, isoelectronic with the N_2-molecule, making NO a reducing agent. The MO bond order is 2.5 for NO, and 3.0 for NO^+.

Oxygen gas acts on NO via N_2O_2 in a reversible exothermic reaction to NO_2 and N_2O_4:

$$N_2O_2 + O_2 \rightleftharpoons 2\,NO_2 \rightleftharpoons N_2O_4$$

Since at very low NO concentrations, e.g., in air, only the monomer is available, oxidation does not take place. At high pressures NO decomposes on heating:

$$3\,NO \;\rightarrow\; N_2O + NO_2$$

The halogens produce the corresponding nitrosyl halides, XNO (p. 298), from which nitrosyl salts are obtained with suitable Lewis acids:

$$FNO + BF_3 \;\;\rightarrow\; NO[BF_4]$$

$$ClNO + SbCl_5 \;\rightarrow\; NO[SbCl_6]$$

As the nitrogen atom in the nitrosyl- or the nitrosonium cation, NO^+, is in the oxidation state $+3$, N_2O_3 can be used to obtain nitrosyl hydrogen sulfate (lead chamber crystals) from concentrated sulfuric acid:

$$N_2O_3 + 3\,H_2SO_4 \;\rightarrow\; 2\,NO[HSO_4] + H_3O[HSO_4]$$

The salt structure of $NOHSO_4$, $NOClO_4$, $(NO)_2PtCl_6$, and $NOAsF_6$ is confirmed by conductivity and cryoscopy. The stronger bond in the NO^+ cation is reflected in the force constant:

$$NO(gas.)\colon f = 15.5\ \text{mdyn/Å} \qquad NO^+(\text{in } NOBF_4)\colon f = 25.1\ \text{mdyn/Å}$$

Nitrosyl salts hydrolyze to yield nitrite ions or nitrous acid:

$$NO^+ + H_2O \;\rightarrow\; HNO_2 + H^+$$

Similar to CO, CN^-, and N_2, NO can act as a π-acid ligand in transition metal complexes, e.g., in $Cr(NO)_4$, $Co(CO)_3NO$, and $Na_2[(CN)_5NO]$, in which NO, a 3-electron ligand loses one electron in the center atom, and then is bonded as NO^+, like CO and N_2.

11.6.4. Dinitrogen Trioxide, N_2O_3

The anhydride of nitrous acid, N_2O_3 is obtained by saturation of liquid N_2O_4 with gaseous NO at $-80\,°C$:

$$NO + NO_2 \;\rightleftharpoons\; N_2O_3 \qquad \Delta H° = -40\ \text{kJ/mol}$$

The product is dissociated in the vapor phase to NO and NO_2. The diamagnetic molecules are planar, and like N_2O_2 and N_2O_4 have a long, weak N—N bond:

$$d(NN) = 1.84\,Å$$

Liquid N_2O_3 has a deep blue color; it solidifies at $-110°$ to light blue crystals. It dissolves in organic solvents with blue color.

With alkali N_2O_3 forms nitrites, and with water HNO_2:

$$N_2O_3 + 2OH^- \rightarrow 2NO_2^- + H_2O$$

An equimolar mixture of NO and NO_2 behaves like N_2O_3.

11.6.5. Nitrogen Dioxide, NO_2, N_2O_4

Technical production of NO_2 is devoted to nitric acid manufacture (p. 279). In the laboratory, NO_2 is obtained by mixing of NO and O_2, or by pyrolysis of heavy metal nitrates in an O_2-stream:

$$Pb(NO_3)_2 \xrightarrow{250-600\,°C} PbO + 2NO_2 + \tfrac{1}{2}O_2$$

Diamagnetic, colorless N_2O_4 is about 90% dissociated in the vapor phase at $100\,°C$ and 1 bar to brown, paramagnetic NO_2, which is a poisonous and very corrosive gas, decomposing above $150\,°C$ into NO and O_2.

$$2NO_2 \rightleftharpoons N_2O_4 \qquad \Delta H° = -65 \text{ kJ/mol } N_2O_4$$

Liquid N_2O_4 is brown colored near the boiling point ($21\,°C$), owing to the presence of NO_2. The color lightens on cooling and at $-11\,°C$ colorless N_2O_4 crystallizes. Nitrogen dioxide is easily reduced to nitrite ions, NO_2^-, or oxidized to nitronium ions NO_2^+. It is appropriate to compare the properties of these three species:

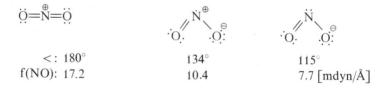

$<:\ 180°$	$134°$	$115°$
f(NO): 17.2	10.4	7.7 [mdyn/Å]

The nitrite anion is isoelectronic with ozone (p. 182). The sp^2-hybridized nitrogen atom forms two σ-bonds to oxygen on which is superimposed a 3-center 4-electron π-bond, to give a total MO bond order of 1.5. The electronic structure of NO_2 is analogous with the larger valence angle indicating a higher s-content in the hybrid orbitals employed for the σ-bonds (38% vs. 28% for NO_2^+), making the bonds stronger. The unpaired electron is located in a σ-orbital of sp-type on nitrogen. The repulsion between non-bonding electrons is less than in NO_2^- leading to further bond strengthening as borne out in the higher valence force constant. The three p-orbitals perpendicular to the molecular plane account for four electrons and form, as in O_3, a 3-center 4-electron π-bond, consisting of a bonding and a non-bonding electron pair.

In NO_2^+, like in N_3^- (p. 290) the strong σ-bonds originate from a linear sp-hybrid at the N-atom on which are superimposed two 3-center π-bonds formed by three p-orbitals with mutually perpendicular nodal planes. The total MO bond order is 2.0.

In the vapor phase N_2O_4 molecules are of symmetry D_{2h}:

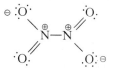

Less stable isomers, an N_2O_4 with two perpendicular NO_2 groups, or an $ONONO_2$ array, can be trapped at very low temperature as has been shown by vibrational spectroscopy.

Hydrolysis of NO_2 or N_2O_4 furnishes nitrite and nitrate:

$$N_2O_4 + 2OH^- \rightarrow NO_2^- + NO_3^- + H_2O$$

Hence, N_2O_4 is the mixed anhydride of nitrous and nitric acids.

In anhydrous acids like HNO_3 or H_2SO_4, N_2O_4 dissociates into NO^+ and NO_3^-, but the pure liquid is undissociated as shown by its low electric conductivity.

Monomeric and dimeric nitrogen dioxide, NO_2 and N_2O_4 are strong oxidizing agents which transform even noble metals like copper into their nitrates. Anhydrous metal nitrates can be prepared in this manner.

11.6.6. Dinitrogen Pentoxide, N_2O_5

The anhydride of nitric acid, N_2O_5, can be obtained by dehydration with P_2O_5:

$$2 HNO_3 + P_2O_5 \rightarrow N_2O_5 + \tfrac{2}{n}(HPO_3)_n$$

or by oxidation of NO_2/N_2O_4 with O_3. It forms colorless, sublimable crystals which decompose to NO_2 and O_2, and which react vigorously with water to HNO_3. In the vapor phase N_2O_5 has the structure:

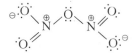

In the solid phase, however, N_2O_5 has an ionic nitronium nitrate structure, $NO_2^+NO_3^-$. Salts with NO_2^+ cations also occur when N_2O_5 dissolves in inorganic acids:

$$N_2O_5 + HClO_4 \rightarrow NO_2ClO_4 + HNO_3$$

$$N_2O_5 + HSO_3F \rightarrow NO_2SO_3F + HNO_3$$

Nitronium salts are also prepared from other nitrogen(V) compounds, e.g.:

$$HNO_3 + 2SO_3 \rightarrow NO_2HS_2O_7$$

$$ClNO_2 + SbCl_5 \rightarrow NO_2SbCl_6$$

These are colorless, hydrolyzable salts. The cation NO_2^+ is also present in nitrating acid, the mixture of concentrated HNO_3 and H_2SO_4 used to prepare nitro derivatives of aromatic hydrocarbons:

$$NO_2^+ + Aryl—H \rightarrow Aryl—NO_2 + H^+$$

11.7. Oxo Acids of Nitrogen

11.7.1. General

Nitrous acid, HNO_2, and nitric acid, HNO_3, and hyponitrous acid, HNO, or HON=NOH, can be isolated, but nitroxylic acid, $(HO)_2N—N(OH)_2$, peroxonitrous acid, HOONO, and peroxonitric acid, $HOONO_2$, are of doubtful existence, or exist only in solution or as salts.

11.7.2. Nitric Acid, HNO_3

Aqueous nitric acid is manufactured by introduction of NO_2 (from catalytic NH_3 oxidation) into water with air or oxygen for further oxidation:

$$2 NO_2 + \tfrac{1}{2}O_2 + H_2O \rightarrow 2 HNO_3$$

The aqueous acid is concentrated by evaporation to an azeotrope containing $69\% HNO_3$ and boiling at $121.8\,°C$. Concentrated HNO_3 is obtained by dehydration *in vacuo* or with P_2O_5. In the laboratory anhydrous nitric acid is best obtained by the action of concentrated H_2SO_4 on KNO_3, followed by vacuum distillation (b.p. $84\,°C/1$ bar). Distillation or exposure to light decomposes HNO_3 partly with brown discoloration:

$$2 HNO_3 \rightarrow 2 NO_2 + \tfrac{1}{2}O_2 + H_2O$$

Anhydrous nitric acid is a colorless, conducting liquid partially dissociated to NO_2^+, NO_3^-, and H_2O:

$$3 HNO_3 \rightleftharpoons NO_2^+ + H_3O^+ + 2 NO_3^-$$

The nitrate ion is trigonal-planar (symmetry D_{3h}), whereas the nitric acid molecule is planar with C_s symmetry in the vapor phase:

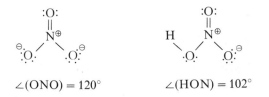

$$\angle(ONO) = 120° \qquad\qquad \angle(HON) = 102°$$

In both cases, the nitrogen atom is sp^2-hybridized with a delocalized 4-center 6-electron π-bond, formed by overlap of four p-orbitals perpendicular to the molecular plane (cf. BF_3, p. 111), superimposed on the σ-bond frame-work of the NO_3^-. All the three NO distances are equal, and the force constant of 8.0 mdyn/Å corresponds to an MO bond order of 1.3. The π-bond in the HNO_3 molecule is only delocalized over the nitro group, with the —OH connected to the nitrogen atom via a simple σ-bond. The $\angle ONO$ in the nitro group is $130°$.

Anhydrous nitric acid is a strong oxidizing agent. The aqueous solution is a strong acid. The concentrated acid dissolves copper and mercury, but not gold and platinum. In dissolving the metals, NO_3^- is reduced to NO or NO_2 depending on concentration:

$$NO_3^- + 4H^+ + 3e^- \rightleftharpoons NO + 2H_2O$$

Certain base metals (e.g., Al, Fe, Cr) are not dissolved by concentrated nitric acid because a dense adherent oxide layer is formed which prohibits further attack (passivity). Neutralization of HNO_3 with NH_3, metal hydroxides, or carbonates, produces the corresponding nitrates which are very soluble in water. Alkali metal nitrates become nitrites on heating. Heavy metal nitrates yield NO_2 and the metal oxide:

$$KNO_3 \rightarrow KNO_2 + \tfrac{1}{2}O_2$$
$$Cu(NO_3)_2 \rightarrow CuO + 2NO_2 + \tfrac{1}{2}O_2$$

Nitrates are good oxidizing agents particularly at higher temperatures. Nascent hydrogen reduces NO_3^- ions to NH_3.

The nitrate ion is a ligand (mostly monodentate) for transition metal ions, forming a Me-−O—NO_2 group. The anhydrous nitrates of Hg^{2+}, Mn^{2+}, Cu^{2+}, etc., contain coordinated not ionic nitrate groups.

11.7.3. Nitrous Acid, HNO_2

Pure HNO_2 cannot be obtained, but dilute solutions are stable. Nitrites, however, are stable as salts and can be obtained as follows:

$$NO + NO_2 \,(\text{or } N_2O_3) + 2NaOH \rightarrow 2NaNO_2 + H_2O$$
$$KNO_3 \rightarrow KNO_2 + \tfrac{1}{2}O_2$$

Nitrous acid is obtained from H_2SO_4 and $Ba(NO_2)_2$ prepared by double decomposition from $NaNO_2$ precipitating $BaSO_4$ and leaving a dilute HNO_2 solution which decomposes:

$$3HNO_2 \rightarrow HNO_3 + 2NO + H_2O$$

In the vapor phase HNO_2 decomposes:

$$2HNO_2 \rightleftharpoons NO_2 + NO + H_2O$$

Gaseous nitrous acid consist of planar *cis*- and *trans*-molecules:

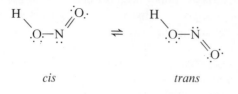

| cis | trans |

The *trans*-isomer is more stable by 2 kJ/mol. Aqueous HNO_2 is slightly stronger than acetic acid and reacts as reducing agent (towards MnO_4^-) and as oxidizing agent (towards I^- and Fe^{2+}).

The bonding in the anion has been discussed on p. 302. Organic derivatives. RNO_2. as well as nitrous acid esters, R—O—NO, are derived from it. The nitrite ion also behaves as a complex ligand, both as nitro group, $M \leftarrow NO_2$ and as nitrito group, $M \leftarrow ONO$. One of the best known nitro complexes is $Na_3[Co(NO_2)_6]$.

11.7.4. Hyponitrous acid $(HON)_2$

The acid $(HON)_2$ can be isolated itself or as its salts, the hyponitrites. Reduction of aqueous $NaNO_2$ by sodium amalgam yields $Na_2N_2O_2$:

$$2\,NaNO_2 + 4\,Na + 2\,H_2O \rightarrow Na_2N_2O_2 + 4\,NaOH$$

from which yellow $Ag_2N_2O_2$ is precipitated by $AgNO_3$.
Treatment with ether-HCl crystallizes $(HON)_2$ as colorless, explosive leaflets of the weak acid which slowly decomposes even in the cold:

$$(HON)_2 \rightarrow N_2O + H_2O$$

The acid and its salts have *trans*-configurations:

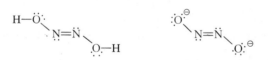

Isomerization by rotation around the N—N bond does not take place at room temperature.

12. Phosphorus and Arsenic

12.1. General

A transition from non-metal to metalloid to metal takes place in the Vth main group where the non-metallic nitrogen is followed by the metalloidal phosphorus and arsenic and the metallic antimony and bismuth. The emphasis in this section will be on phosphorus since arsenic chemistry strongly resembles that of phosphorus.

12.2. Bond Relationships in Phosphorus and Arsenic Compounds

The valence electron configuration of phosphorus in the ground state is:

The phosphorus atom can form three covalent bonds with the three half-occupied orbitals, or a P^{3-} ion. Intermediate bond types are also possible:

> covalent bonds: $\qquad\qquad$ PH_3, PCl_3, $P(C_6H_5)_3$, P_4
> covalent and ionic bonds: PH_2^- in $NaPH_2$
> ionic bond: $\qquad\qquad\quad$ P^{3-} in Na_3P

Analogous arsenic compounds are known. All have an electron octet.
The tendency to form π-bonds using the p-electrons decreases $N > P > As$. Most compounds of the lower elements polymerize, e.g., the gaseous compounds P_2, PN, and HCP, which correspond to N_2 and HCN:

$$:P\equiv P: \qquad :P\equiv N: \qquad H\!-\!C\equiv P:$$

The small internuclear distances suggest triple bonding. Strong heating of phosphorus vapor forms P_2 which polymerizes again to P_4 on cooling. Its dissociation energy of 468 kJ/mol is much larger than the 214 kJ/mol for the average PP single bond.

More stable than these high temperature species are the pyridine analogues C_5H_5P and C_5H_5As which are colorless liquids:

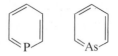

In AB_3 compounds ($A = P$, or As) the lone 3s- or 4s-electron pair which is located in an s-p-hybrid orbital with a large s-content can be utilized in the bonding of additional substituents. The valence angles lie between $90°$ and $105°$ (cf. p. 122). The hybridization changes to sp^3 when the electron pair is used in coordinate bonding, e.g., ($R = $ organic substituent):

$$PH_4^+ \quad PCl_4^+ \quad PR_4^+ \qquad AsH_4^+ \quad AsCl_4^+ \quad AsR_4^+$$

The tendency of AB_3 compounds to form onium ions decreases with increasing size of A. Numerous transition metal complexes having P- or As-containing ligands such as PF_3, PR_3 or P_4 are known.

Nitrogen chemistry differs from that of its higher homologs in the d-orbitals available for covalent and coordinate σ- and π-bonding.

Promotion of a 3s-electron of phosphorus to a $3d_{z^2}$ orbital creates the trigonal bipyramidal excited state $3sp^3d$ from which PF_5 and $P(C_6H_5)_5$ are derived. The same holds for arsenic (e.g., in AsF_5).

Compounds like $POCl_3$, $AsOF_3$, and PO_4^{3-} have π-bonds to oxygen from the sp^3-hybridized central atom originating from a $d_{x^2-y^2}$-orbital:

$$\begin{array}{ccc}
& \diagup Cl \\
O{=}P{-}Cl \\
& \diagdown Cl
\end{array}
\qquad
\begin{array}{ccc}
& \diagup F \\
O{=}As{-}F \\
& \diagdown F
\end{array}
\qquad
\begin{array}{c}
O^{\ominus} \\
| \\
O{=}P{-}O^{\ominus} \\
| \\
O^{\ominus}
\end{array}$$

The double bond in the phosphate ion is delocalized as in the isoelectronic sulfate ion (p. 201). Many tetrahedral phosphorus and arsenic compounds (e.g., phosphates and arsenates) are known.

The octahedral anions PF_6^- and AsF_6^- are formed when PF_5 and AsF_5 coordinate with fluoride ions. All six bonds are undistinguishable. The Lewis acidity of the unoccupied d-orbitals also appears in tetrahedrally coordinated phosphorus compounds where coordinate π-bonds from the substituents to the central atom are proposed, e.g., in PF_3 and $P_2O_7^{4-}$:

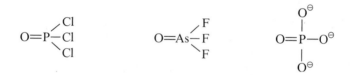

Evidence for the π-bonds is adduced from internuclear distances and valence angles (cf. p. 184). Some diphosphate anions, e.g., $Mg_2P_2O_7$, ZrP_2O_7, and SnP_2O_7 have linear POP bridges with complete delocalization of the electron pairs into d-orbitals of phosphorus (two 3-center π-bonds).

The coordination number at phosphorus may vary from one to six, and the oxidation state from -3 to $+5$:

Oxidation State:	-3	-2	-1	0	$+1$	$+2$	$+3$	$+4$	$+5$
N:	NH_3	N_2H_4	N_2H_2	N_2	N_2F_2	N_2F_4	N_2O_3	NO_2	N_2O_5
P:	PH_3	P_2H_4	$(PH)_x$	P_4	H_3PO_2	P_2F_4	P_2O_3	$H_4P_2O_6$	P_2O_5

However, the structures of analogous nitrogen, phosphorus and arsenic compounds need not be analogous (e.g., N_2O_3—P_2O_3—As_2O_3).

12.3. The Elements Phosphorus and Arsenic

12.3.1. Preparation

Phosphorus is a reactive element not found free in nature but as phosphates, e.g., phosphorite $Ca_5(PO_4)_3(OH, F, Cl)$, and apatite $Ca_5(PO_4)_3F$. Phosphorus is liberated from phosphates by reduction with coke, with sand as a slag former:

$$Ca_3(PO_4)_2 + 5C + 3SiO_2 \rightarrow 3CaSiO_3 + P_2\uparrow + 5CO$$

The reduction requires 1300 to 1450°. The vapor is predominantly P_2 which on condensation under water dimerizes to white phosphorus, P_4 (m.p. 44°C). Other phosphorus modifications are prepared from it.

Arsenic occurs naturally as sulfides and arsenides, e.g., arsenopyrite, FeAsS, and leucopyrite, $FeAs_2$, from which arsenic is sublimed by heating in a non-oxidizing atmosphere:

$$FeAsS \rightarrow FeS + As \qquad\qquad FeAs_2 \rightarrow FeAs + As$$

12.3.2. Modifications

White phosphorus, consisting of P_4 molecules, is a white, waxy, CS_2-soluble non-conductor, which spontaneously ignites in air, but can be stored under water. The bonding has been discussed (p. 96).

Polymeric, amorphous red phosphorus is formed exothermically when white phosphorus is heated to 180–350°C with traces of iodine as a catalyst. Heating to 450–550°C crystallyzes a polymeric insoluble phosphorus which is more dense and less reactive than white phosporus. Its melting at 620°C is accompanied by depolymerization and vaporization to give P_4-molecules.

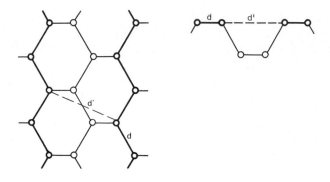

Fig. 96.   Structure of double layers in the lattice of black phosphorus. Equally marked atoms lie in a plane. The double layer consists of P_6-rings in the chair conformation.

The lattice of violet phosphorus is complex, consisting of mutually interconnected pentagonal channels.

Orthorhombic black phosphorus is formed when white phosphorus is heated under pressure (12 kbar at 200 °C). Black phosphorus is presumed to be the thermodynamically stable modification, It is insoluble, inert, and a semiconductor. Violet phosphorus is the stable modification from 550° to the melting point (620 °C).

Orthorhombic black phosphorus consists of staggered parallel double layers as shown in Fig. 96 in which the phosphorus atoms form zig-zag chains with valence angles of 96.6° and d = 2.22 Å. Each atom is additionally attached to an atom in the other half of the double layer. The smallest distance between two atoms in a layer which are separated by two other atoms in the other half of the double layer (d' = 3.31 Å) is smaller than the van der Waals distance (3.8 Å), indicating weak bonds. The ratio of distances is 1.49, corresponding to that in gray selenium, which is also a semiconductor (p. 209). The smallest internuclear distance between the double layers of black phosphorus is 3.59 Å, markedly smaller than the van der Waals distance, yet the crystals cleave parallel to the layers as in graphite.

Rhombic, black phosphorus changes reversibly at 83 kbar to rhombohedral, and at 111 kbar to a primitive, cubic lattice with distances of 2.38 Å, valence angles of 90° and coordination number of 6 for all atoms.

Three modifications of arsenic are known. The thermodynamically stable rhombohedral, gray arsenic forms glossy conducting metallic crystals consisting of puckered six-membered rings of arsenic atoms within layers. (Fig. 97). Each atom has three nearest neighbors at 2.51 Å in its own layer and three next-nearest neighbors at 3.15 Å in the neighboring layer to give a distorted octahedron. The van der Waals distance is 4.0 Å.

Gray arsenic sublimes at 610 °C with depolymerization without melting. Quenching the vapor which consists of As_4 molecules in CS_2 gives a solution of As_4 from

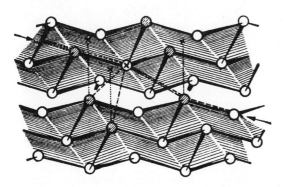

Fig. 97.   Structure and packing of double layers in the lattice of gray arsenic. The six neighbor atoms are crosshatched. (After H. Krebs, Grundzüge der anorganischen Kristallchemie, F. Enke Verlag, Stuttgart, 1968.)

which non-conducting, yellow arsenic crystallizes. This modification corresponds to white phosphorus. Yellow arsenic changes to gray arsenic, especially when exposed to light.

Exchange of P_4 and As_4 occurs at 1000 °C to give the mixed molecules AsP_3, As_2P_2 and As_3P.

12.4. Hydrides

The most important hydrides of phosphorus and arsenic are:

PH_3	H_2P—PH_2	AsH_3
Phosphine	Diphosphine	Arsine

Two additional arsenic hydrides are known; diarsine, stable at −100 °C, formed by decomposition of AsH_3 in a glow discharge, and triarsine, As_3H_5, which occurs in traces in the hydrolysis of an MgPAs alloy. Other phosphorus hydrides are known. The mean bond energy of the PP single bond (see p. 131) is larger than for the NN- or the AsAs-bond, and phosphorus like sulfur and silicon forms more chain and ring compounds.

The higher hydrides of phosphorus are based on chains and rings:

cyclo-Polyphosphines: $P_nH_n (n = 3, 4, \ldots)$

catena-Polyphosphines: $P_nH_{n+2} (n = 1, 2, \ldots)$

which are unstable and have only been identified spectroscopically. Many organophosphines are known, e.g., the cyclic series of puckered phenyl derivatives Ph_3P_3, Ph_4P_4, Ph_5P_5, and Ph_6P_6:

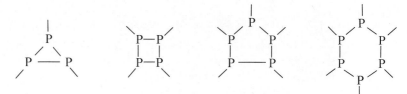

The catenaphosphines like Ph_5P_3 have zig-zag chains. Compounds containing As-As bonds are also known, e.g., As_2I_4, As_4S_3, $(CF_3As)_4$, and $(C_6H_5As)_6$.

Phosphine and Arsine:

Phosphine is a colorless, toxic gas with a characteristic odor prepared by hydrolysis of calcium phosphide:

$$Ca_3P_2 + 6H_2O \rightarrow 3Ca(OH)_2 + 2PH_3\uparrow$$

Diphosphine is generated as a by-product which is separated by fractional condensation or distillation.

Arsine likewise is a toxic gas generated in the hydrogenation of arsenic compounds:

$$AsCl_3 + 3LiAlH_4 \xrightarrow{-90\,°C} AsH_3 + 3LiCl + 3AlH_3$$
$$HAsO_4^{2-} + BH_4^- + 2H^+ \longrightarrow AsH_3\uparrow + B(OH)_4^-$$

The generation of AsH_3 from water-soluble arsenic compounds by nascent hydrogen (from Zn and dilute H_2SO_4) is used in testing for As.

Both PH_3 and AsH_3 are endothermic and burn in air to H_3PO_4 and As_2O_3. Upon heating, PH_3 decomposes partially and AsH_3 totally into the elements.

The PH_3 and AsH_3 molecules are trigonal-pyramidal like NH_3 but the valence angle decreases markedly in going from NH_3 to AsH_3 (p. 93). The free electron pair in PH_3 and AsH_3, unlike NH_3 is in an s-orbital, and the Lewis basicity diminishes from NH_3 to AsH_3. Protonation of the hydrides gives tetrahedral ions, NH_4^+, PH_4^+ and AsH_4^+ which for PH_3 and AsH_3 is accompanied by a large change in hybridization. Phosphonium and arsonium salts thus dissociate with greater ease to the hydrides. Phosphine and HI generate PH_4I which sublimes with dissociation at $80\,°C$ (the chloride sublimes at $-28\,°C$). Since PH_4I is hydrolytically cleaved in aqueous solution into PH_3, H^+ and I^-, PH_3 can be prepared from PH_4I and base:

$$PH_3 + HI \rightleftharpoons PH_4I \xrightarrow{H_2O} PH_3\uparrow + H_3O^+ + I^-$$

Condensing AsH_3 at 110K with a hydrogen halide forms the arsonium halides AsH_4Br and AsH_4I. Unsubstituted arsonium salts are unstable at $25\,°C$.

Silyl and germyl phosphine and arsine form from the hydrides in electric discharge:

$$SiH_4 + PH_3 \rightarrow H_3Si—PH_2 + H_2$$

Higher hydrides are formed likewise.

Organophosphines and -arsines are stronger Lewis bases than the parent hydrides. Triphenylphosphine, prepared from PCl_3 and phenylmagnesium bromide or phenyl lithium reacts with iodobenzene to give the tetraphenylphosphonium iodide:

$$PCl_3 + 3\,LiPh \xrightarrow{-3\,LiCl} Ph_3P \xrightarrow{+\,PhI} [Ph_4P]I$$

Triphenylarsine, made from $AsCl_3$, behaves similarly.
Organophosphines and -arsines are used as ligands in transition metal complexes. Colorless $NaPH_2$ is formed by bubbling PH_3 through a solution of sodium in liquid ammonia:

$$PH_3 + Na \xrightarrow{NH_3\ (l.)} NaPH_2 + \tfrac{1}{2}H_2$$

Arsine reacts analogously. The derivatives R_2PLi and R_2PK derived from R_2PCl are of value in organic synthesis for preparing PP-bonded compounds:

$$>\!P\!-\!Cl + M^{\oplus}\,P^{\ominus}\!< \;\rightarrow\; >\!P\!-\!P\!< + MCl$$

Phosphides, e.g., Na_3P, Ca_3P_2, etc., are prepared by melting phosphorus with the corresponding elements. Ionic, covalent (BP, SiP) and metallic phosphides (Fe_3P) are known.

Diphosphine:
The hydrazine analogue occurs as a by-product in the hydrolysis of Ca_3P_2, since this phosphide prepared from Ca and red phosphorus still contains some PP-bonds. Owing to its lower volatility, P_2H_4 can be separated as a colorless liquid from PH_3. Diphosphine ignites spontaneously, disproportionates in light and on heating to PH_3 and yellow hydrides of lesser hydrogen content. A polymeric phosphine, $(PH)_x$ which forms from triphosphine and higher hydrides decomposes on heating to PH_3 and elemental phosphorus.

12.5. Halides

Phosphorus reacts with halogens to form:

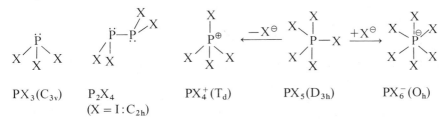

PX_3(C_{3v})   P_2X_4   PX_4^+(T_d)   PX_5(D_{3h})   PX_6^-(O_h)
                  (X = I : C_{2h})

The stability of the halides PX_3 and PX_5 decreases with increasing atomic weight of the halogen, and PI_5 is unknown. Mixed halides exist, as well as oxo- and thio-halides, e.g., $POCl_3$ and $PSCl_3$.

Arsenic also forms the halides AsX_3, As_2X_4, and AsX_5, as well as the ions AsX_2^+, AsX_4^-, AsX_4^+, and AsX_6^-.

Trihalides:

The colorless trichloride and bromide of phosphorus are synthesized from the elements like the analogous arsenic compounds; PI_3 forms red crystals. Colorless, gaseous PF_3 is obtained from PCl_3, and AsF_3 is prepared by fluorination of As_2O_3 with hydrogen fluoride:

$$PCl_3 + 3HF \rightarrow PF_3 + 3HCl$$
$$As_2O_3 + 6HF \rightarrow 2AsF_3 + 3H_2O$$

The trihalides are hydrolytically sensitive and form molecular gases, liquids or solids which are distillable and sublimable, except AsI_3 which crystallizes in a layer lattice. The valence angles are discussed on p. 122. The trihalides react with hydrides to release HX:

$$PCl_3 + 3H_2O \rightarrow P(OH)_3 + 3HCl$$
$$PCl_3 + 6NH_3 \rightarrow P(NH_2)_3 + 3NH_4Cl$$

The trihalides PX_3 or AsX_3 can act as either Lewis acids or bases, and PF_3 and PCl_3 are complex ligands, PCl_3 reacting with $Ni(CO)_4$ to give $Ni(PCl_3)_4$, and PF_3 giving $Fe(CO)_n(PF)_{5-n}$, with n = 0 to 5. Towards $(CH_3)_3N$, PCl_3 and $AsCl_3$ act as Lewis acids:

$$(CH_3)_3N + PCl_3(AsCl_3) \rightarrow (CH_3)_3\overset{\oplus}{N} - \overset{\ominus}{P}Cl_3(-\overset{\ominus}{A}sCl_3)$$

Liquid AsF_3 is conducting:

$$2AsF_3 \rightleftharpoons AsF_2^+ + AsF_4^-$$

and the ions can be isolated as salts:

$$AsF_3 + KF \rightarrow K[AsF_4]$$
$$AsF_3 + SbF_5 \rightarrow [AsF_2][SbF_6]$$
$$AsCl_3 + [(CH_3)_4N]Cl \rightarrow (CH_3)_4N[AsCl_4]$$

Oxygen and oxidizing agents convert PCl_3 to $POCl_3$, and heating with sulfur yields $PSCl_3$. The molecules are distorted tetrahedra. Unlike in NOF_3 (p. 147), the bonds PO and PS are double bonds as shown by the force constants:

	$P(SCH_3)_3$	$SPCl_3$	SPF_3	
f(PS)	2.56	4.89	5.21	[mdyn/Å]

The strengths of the PO and PS bonds depend on the electronegativity of the substituents as in the thionyl compounds (p. 246), with the greatest force constants and bond orders for the fluorides. For arsenic only $AsOF_3$ is known which is prepared from $AsCl_3$, As_2O_3 and F_2.

Tetrahalides:
The iodides P_2I_4 and As_2I_4 are synthesized directly from the elements using a deficiency of iodine which preserves some of the bonds in P_4 or gray arsenic:

$$P_4 + 4I_2 \xrightarrow{\text{CS}_2} 2P_2I_4 \text{ (red crystals)}$$

In the solid state the P_2I_4 molecules have the *trans*-structure (symmetry C_{2h}) shown on p. 313. The tetrachloride, fluoride and unsymmetrical difluoride, $H_2P{-}PF_2$, are also known.

Pentahalides:
Phosphorus reacts with excess Cl_2 or Br_2 to form PCl_5 or PBr_5, and PF_5 is obtained from PCl_5 and AsF_3. The only pentahalides of arsenic are AsF_5, and $AsCl_5$, accessible directly from the elements.
The pentafluorides are trigonal bipyramidal as is gaseous PCl_5. Solid PCl_5, recrystallized from nitrobenzene, however, is a salt, $[PCl_4^+]\,[PCl_6^-]$. Hence PCl_5 exhibits bond isomerism.
In the vapor phase, PBr_5 is completely dissociated into PBr_3 and Br_2, but it is a salt in the solid: $[PBr_4^+]Br^-$. The cations PX_4^+ are isoelectronic with the corresponding silicon halides and have a tetrahedral structure. The anions PX_6^- are regular octahedra with sp^3d^2-hybridization. In trigonal bipyramidal molecules like PF_5 the three equatorial and two axial fluorine atoms are inequivalent with different internuclear distances, $d(PF)$, and force constants, $f(PF)$. The ^{19}F NMR spectrum should thus display two signals in a $3:2$ ratio, each resonance appearing as a doublet owing to spin-spin coupling of $^{19}F(I = \frac{1}{2})$ and $^{31}P(I = \frac{1}{2})$. However,

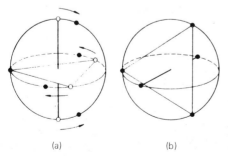

(a)   (b)

Fig. 98.   Pseudorotation in a trigonal-bipyramidal molecule. (a) ○ = original position of substituents; ● = position after exchange of places; (b) change in position of trigonal bipyramid after pseudorotation. Differences in equatorial and axial internuclear distances have been neglected.

only one doublet is observed. This contradiction has lead to the concept of pseudo-rotation, whereby equatorial and axial fluorine atoms rapidly exchange their places, and appear equivalent on a time average (Fig. 98). In this intramolecular rearrangement internuclear distances and valence angles change only slightly with a low energy barrier. Pseudorotation is observed for PF_5 even at $-120\,°C$. Pseudorotation is possible for all trigonal bipyramidal molecules, including SF_4 and ClF_3, but in mixed compounds like PCl_2F_3 with possible isomers I, II, and III, only structure I is observed with the most electronegative atoms occupying axial positions. Pseudorotation is less favorable in such molecules in which there are different space requirements for the bonding electron pairs (p. 122).

The pentafluorides are colorless, hydrolytically sensitive gases which are strong Lewis acids and fluoride ion acceptors (cf. p. 246):

$$PF_5 + (CH_3)_3N \rightarrow (CH_3)_3\overset{\oplus}{N} - \overset{\ominus}{P}F_5$$
$$PF_5 + (CH_3)_3P \rightarrow (CH_3)_3\overset{\oplus}{P} - \overset{\ominus}{P}F_5$$
$$2AsF_5 + 2HF \rightleftharpoons H_2F^+ + As_2F_{11}^- \text{ (in liquid HF)}$$

Phosphorus pentachloride accepts chloride ions to give PCl_6^-, and is decomposed by water via $POCl_3$ to H_3PO_4:

$$PCl_5 \xrightarrow[-2HCl]{+H_2O} POCl_3 \xrightarrow[-3HCl]{+3H_2O} PO(OH)_3$$

Upon heating, PCl_5 dissociates into PCl_3 and Cl_2, and thus behaves as a chlorinating agent:

$$P_2O_5 + 3PCl_5 \rightarrow 5POCl_3$$
$$SO_2 + PCl_5 \rightarrow SOCl_2 + POCl_3$$
$$ROH + PCl_5 \rightarrow RCl + POCl_3 + HCl$$

Monomeric and dimeric PCl_5 species are found in nonpolar solvents. In the vapor phase P_2Cl_{10} consisting of two PCl_6 octahedra with a common side (two shared Cl atoms) has been identified. However, PCl_5 dissolves in polar solvents to give $[PCl_4]^+ [PCl_6]^-$. Like $SbCl_5$, it reacts with chloride acceptors to form salts with the cation PCl_4^+. Arsenic pentachloride has recently been prepared, and salts with $AsCl_4^+$ and $AsCl_6^-$ ions are known:

$$AsCl_3 + Cl_2 + PCl_5 \rightarrow [AsCl_4]^+ [PCl_6]^-$$
$$[(C_2H_5)_4N]Cl + AsCl_3 + Cl_2 \rightarrow [(C_2H_5)_4N]^+ [AsCl_6]^-$$

No pentaiodides of phosphorus or arsenic are known.

12.6. Phosphoranes

Phosphoranes are substituted compounds derived from the hypothetical penta-hydride, PH_5, e.g., PF_5 or gaseous PCl_5. The pentahydride is not known because the hydrogen, unlike F, O, N, C, and Cl atoms (p. 199) cannot effect a sufficient contraction of the d_{z^2}-orbital to form an sp^3d-hybridized phosphorus atom. The following types of phosphoranes (R = organic radical) exist:

PF_5, PF_4R	$P(OR)_5$	PR_5	PCl_5
PF_3R_2, PF_2R_3			$RPCl_4$, R_2PCl_3
Fluoro-phosphoranes	Oxo-phosphoranes	Organic phosphoranes	Chloro-phosphoranes

Amino groups can also be bonded and mixed derivatives exist.

The d_{z^2}-orbital of phosphorus in these compounds participates primarily in for-ming axial bonds to atoms which contract this orbital, leaving the three equatorial positions for atoms lacking the contracting effect, as seen in the gaseous phosphor-anes PHF_4, PH_2F_3 and PH_3F_2 whose preparation starts with anhydrous hydrogen fluoride and a P—H compound:

$$HPO(OH)_2 + 4HF \rightarrow HPF_4 + 3H_2O$$
phosphorous acid

$$H_2PO(OH) \quad + 3HF \rightarrow H_2PF_3 + 2H_2O$$
hypophosphorous acid

$$H_2P—PH_2 \quad + 4HF \rightarrow H_3PF_2 + [PH_4]HF_2$$
diphosphine

Alkaline hydrolysis of PHF_4 and PH_2F_3 yields the starting materials or their anions. The tetrafluoride decomposes to PF_3 and HF but PH_2F_3 is stable to $100\,°C$. All the mixed derivatives PCl_nF_{5-n} are known.

Phosphorus halides serve as starting materials for organic phosphoranes:

$$PCl_3 \xrightarrow{PhMgBr} PPh_3 \xrightarrow{+PhI} [PPh_4^+]I^- \xrightarrow[-LiI]{+PhLi} PPh_5 \qquad Ph\!: C_6H_5$$

$$PCl_3 + RCl + AlCl_3 \rightarrow [RPCl_3]^+[AlCl_4^-] \xrightarrow[AsF_3]{HF\ or} RPF_4 \quad R\!: Alkyl$$

$$2R_3P + SF_4 \rightarrow 2R_3PF_2 + S_s. \qquad\qquad\qquad R\!: Organic\ Radical$$

$$2PF_5 + Ph_4Sn \rightarrow PhPF_4 + [Ph_3Sn]^+ PF_6^- \qquad\qquad Ph\!: C_6H_5$$

The phosphoranes are strong Lewis acids like PF_5 and PCl_5 and react with tertiary amines and fluoride ions to form octahedral $1:1$ adducts. Hexaorganophosphoranes with the anion PR_6^- have been synthesized.

The known arsenic compounds correspond to those of phosphorus, e.g., $As(C_6H_5)_5$ and $As(OCH_3)_5$, but their number is much smaller.

Phosphoranes are trigonal bipyramidal with the most electronegative substituents axial. Pseudorotation, discussed in the case of PF_5 (p. 315) is also encountered in other phosphoranes, e.g., (diethylamino)tetrafluorophosphorane, $(C_2H_5)_2NPF_4$, where the fluorine atoms are equivalent at room temperature in the NMR spectra, but on cooling to $-85\,°C$ reveal the presence of two fluorines in each position. Pseudorotation which leads to placing the less electronegative substituents into axial positions, e.g., in $(CH_3)_3PF_2$, is not observed.

12.7. Oxides

Phosphorus burns under a reduced pressure of oxygen to P_4O_6, and in an excess to P_4O_{10}. These oxides correspond to SO_2 and SO_3, but other phosphorus oxides are also known, as well as the analogous oxides As_2O_3 and As_2O_5.

In contrast to the endothermic nitrogen oxides, phosphorus and arsenic oxides are exothermic.

Phosphorus(III) Oxide:
Burning P_4 in an O_2-stream at diminished pressure and low temperature yields P_4O_6 which sublimes from the burning zone together with some P_4O_{10} and P_4 which can be seperated by repeated sublimation after converting P_4 into non-volatile red phosphorus by irradiation or with Hg vapor. The colorless, waxy phosphorus(III) oxide (m.p. $24\,°C$) consists in all phases as well as in solution of P_4O_6 molecules which can be derived from the P_4 tetrahedron by replacing the six PP-bonds with POP bridges:

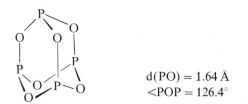

$$d(PO) = 1.64\,Å$$
$$<POP = 126.4°$$

The four P-atoms preserve their tetrahedral arrangement, but the large $\angle POP$ indicate that the lone pairs of the oxygen atoms are partly delocalized into the d-orbitals of the phosphorus.

Phosphorus(III) oxide is stable at $25\,°C$ in air but reacts with water:

$$P_4O_6 + 6H_2O \rightarrow 4H_3PO_3$$

as the anhydride of phosphorous acid. In hot water a complex reaction forms PH_3, H_3PO_4 and elemental phosphorus.

Phosphorus(V) Oxide:

Phosphorus burns in excess oxygen to P_4O_{10} which is produced commercially this way:

$$P_4 + 5O_2 \rightarrow P_4O_{10} \qquad \Delta H^\circ = -1493 \text{ kJ/mol } P_4O_{10}$$

The colorless, hexagonal crystals of P_4O_{10} are sublimed at red heat in an O_2 stream to oxidize the lower phosphorus oxides. Monomeric P_4O_{10} is derived from P_4O_6 in that each phosphorus carries a double-bonded oxygen atom:

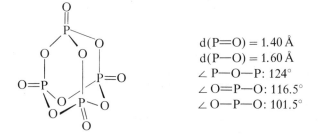

$d(P=O) = 1.40 \text{ Å}$
$d(P-O) = 1.60 \text{ Å}$
$\angle \, P-O-P: 124°$
$\angle \, O=P-O: 116.5°$
$\angle \, O-P-O: 101.5°$

The vigorous hydrolysis of P_4O_{10} proceeds stepwise via $H_4P_4O_{12}$ and $H_4P_2O_7$ to H_3PO_4:

$$P_4O_{10} \xrightarrow{+2H_2O} H_4P_4O_{12} \xrightarrow{+2H_2O} 2H_4P_2O_7 \xrightarrow{+2H_2O} 4H_3PO_4$$

In the first step two of the six POP-bridges are hydrolyzed, giving rise to the cyclic $(HPO_3)_4$:

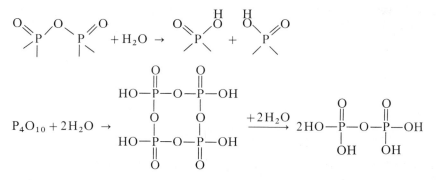

In a deficiency of water, P_4O_{10} is an efficient drying agent, forming a syrupy mixture of polyphosphoric acids. The anhydride N_2O_5 is obtained from HNO_3, Cl_2O_7 from $HClO_4$, SO_3 from H_2SO_4, and carbon suboxide, C_3O_2 from malonic acid with P_4O_{10}.

Heating P_4O_{10} in a sealed tube to $450\,^\circ$C forms a polymeric orthorhombic modification with a layered network structure of P_2O_5 units, like the isoelectronic $Si_2O_5^{2-}$ units in layered silicates (p. 354).

Phosphorus(III, V) Oxides:
The oxides, P_4O_6 and P_4O_{10} are the end-members of a series formed by stepwise addition of oxygen to the P_4O_6 molecule:

$$P_4O_6 \quad P_4O_7 \quad P_4O_8 \quad P_4O_9 \quad P_4O_{10}$$

The phosphorus(III, V) oxides are generated by thermal disproportionation of P_4O_6, or by controlled oxidation, or by reduction of P_4O_{10} with red phosphorus:

$$P_4O_6 + O_2 \xrightarrow{\;CCl_4\;} P_4O_7,\ P_4O_8$$

$$P_4O_6 \xrightarrow{\;150-450\,^\circ C\;} P_4O_7,\ P_4O_8,\ P_4O_9,\ P_{red}$$

$$P_4O_{10} + P_{red} \xrightarrow{\;450-525\,^\circ C\;} P_4O_8,\ P_4O_9$$

The latter two reactions are carried out under N_2 in a sealed tube to give mixtures of oxides separable by sublimation.
The P_4O_8 and P_4O_9 molecules are similar to P_4O_{10}, but lacking one or two double-bonded oxygen atoms.

Arsenic Oxides:
Arsenic burns to As_2O_3 which is also obtained in the hydrolysis of $AsCl_3$:

$$2\,As + \tfrac{3}{2}O_2 \quad \rightarrow \quad As_2O_3$$

$$2\,AsCl_3 + 3\,H_2O \quad \rightarrow \quad As_2O_3 + 6\,HCl$$

The product, which is the anhydride of H_3AsO_3, forms cubic crystals of monomeric As_4O_6 analogous to P_4O_6, also encountered in the vapor phase and in nitrobenzene solution.

Heating As_4O_6 to $>200\,^\circ$C converts it into polymeric, monoclinic As_2O_3.

Arsenic acid, H_3AsO_4, results when arsenic or As_2O_3 is dissolved in nitric acid, from which arsenic(V) oxide is formed on heating:

$$2\,H_3AsO_4 \xrightarrow{\;300\,^\circ C\;} As_2O_5 + 3\,H_2O$$

On heating As_2O_5 releases oxygen. It hydrolyzes forming H_3AsO_4 slowly. The structure of the oxide, which is amorphous, corresponds to the polymeric P_2O_5. Both oxides form solid solutions.

12.8. Sulfides

Red phosphorus reacts with molten sulfur to give various molecular sulfides depending on the molar ratio, e.g., P_4S_3, P_4S_5, P_4S_7, P_4S_9 and P_4S_{10} some of which are soluble in CS_2. The structure of P_4S_{10} corresponds to P_4O_{10}, i.e., has bridge and terminal sulfur atoms. In P_4S_9, one terminal sulfur atom is lacking (cf. P_4O_9). The other sulfides have structures shown in Figure 99.

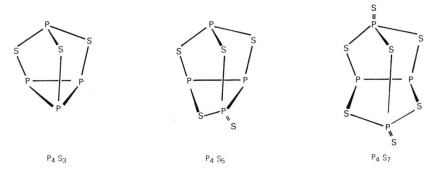

$P_4 S_3$ $P_4 S_5$ $P_4 S_7$

Fig. 99   Molecular structures of the phosphorus sulfides P_4S_3, P_4S_5 and P_4S_7.

The phosphorus sulfides are thermally stable heterocycles, even P_4S_3 which contains a three-membered ring and, nervertheless, distills without decomposition (m.p. 173 °C, b.p. 407 °C). The P_7^{3-} anion present in Sr_3P_{14} is isoelectronic and isostructural with P_4S_3.

However, P_4S_5 decomposes on melting to P_4S_3 and P_4S_7 (m.p. 308 °C, b.p. 529 °C), the latter like P_4S_{10} (m.p. 287 °C, b.p. 514 °C) being associated in the melt through the terminal sulfur atoms as reflected in the viscosity, surface tension, and entropy of evaporation. In the vapor phase P_4S_7 is monomeric, but P_4S_{10} dissociates to P_2S_5. No sulfide corresponding to P_4O_6 exists.

The sulfides ignite spontaneously and P_4S_{10} hydrolyzes to H_3PO_4 and H_2S, reacts with alcohols to give H_2S and diesters of dithiophosphoric acid, $HSPS(OR)_2$, and is desulfurized to P_4S_9 with PCl_3 with formation of $PSCl_3$.

Arsenic reacts with sulfur to form As_4S_3, As_4S_4, As_2S_3, and As_2S_5. The sulfides As_4S_4 and As_2S_3 occur in nature. Hydrogen sulfide precipitates As_2S_3 and As_2S_5 from aqueous hydrochloric acid solutions of As(III) and As(V), respectively. Monomeric As_4S_3 is analogous to P_4S_3, and As_4S_4 also has a cage structure similar to S_4N_4 (p. 235), with nitrogen atoms replaced by sulfur and sulfur replaced by arsenic. Polymeric As_2S_3 is monoclinic like As_2O_3.

12.9. Oxo Acids and Derivatives

For the oxo acids of phosphorus, their salts, esters and other derivatives the structural elements from which individual oxo-, thio-, halo-, etc., acids can be built will be considered first followed by the properties and reactions of individual compounds.

12.9.1. Oxo Acids with one Phosphorus Atom

The oxo acids contain the P—OH group which is acidic:

$$P—OH + H_2O \rightleftharpoons P—O^- + H_3O^+$$

In addition, the oxo acids contain P=O, and occasionally P—H groups:

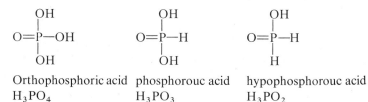

Orthophosphoric acid phosphorouc acid   hypophosphorouc acid
H_3PO_4 H_3PO_3 H_3PO_2

Unlike P—O—H, the P—H bond does not dissociate in water, and so is not neutralized or titratable, and hence H_3PO_3 is a diprotic and H_3PO_2 a monoprotic acid as emphasized by writing $H_2[HPO_3]$ for phosphorous and $H[H_2PO_2]$ for hypophosphorous acid.

The groups P—OH, P=O, and P—H can be prepared as follows:

$$P—O^- + H^+ \rightarrow P—OH$$

e.g., $Ca_3(PO_4)_2 + 3H_2SO_4 \rightarrow 2H_3PO_4 + 3CaSO_4\downarrow$

$$P—Cl + H_2O \rightarrow P—OH + HCl\uparrow$$

e.g., $POCl_3 + 3H_2O \rightarrow H_3PO_4 + 3HCl\uparrow$

$$PCl_3 + H_2O \rightarrow H_3PO_3 + 3HCl\uparrow$$

$$P—O—P + H_2O \rightarrow 2P—OH$$

e.g., $P_4O_{10} + 6H_2O \rightarrow 4H_3PO_4$

$$P_4O_6 + 6H_2O \rightarrow 4H_3PO_3$$

The free electron pair present at phosphorus in PCl_3 or P_4O_6 leads to isomerization:

$$>\ddot{P}—OH \rightarrow >\overset{\overset{\displaystyle H}{|}}{P}=O$$

Therefore, phosphorous acid, $HPO(OH)_2$, and not $P(OH)_3$ is obtained from PCl_3. The P—H group also forms in the disproportionation of P_4 in aqueous alkali solution:

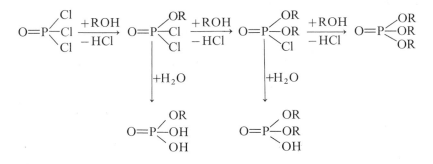

$$\text{>P—P< } + H_2O \xrightarrow{OH^-} \text{ >P—H} + O=\overset{\overset{\textstyle H}{\textstyle |}}{P}<$$

e.g., $P_4 + 6H_2O \xrightarrow{OH^-} 3H[H_2PO_2] + PH_3\uparrow$

This reaction corresponds to the disproportionation of Cl_2 to HCl and HOCl, and of S_8 to H_2S and thiosulfate in alkaline solution, in which nucleophilic degradation by hydroxyl ions takes place.

Orthophosphoric Acid:
The mineral $Ca_3(PO_4)_2$ is either reacted with dilute sulfuric acid, or reduced with coke to phosphorus which is burned to P_4O_{10} and then hydrolyzed to H_3PO_3 to give 85–90% solutions. Vacuum evaporation at 80° forms hydrogen-bonded, colorless layer crystals (m.p. 42 °C) of monomeric, anhydrous H_3PO_4. Phosphoric acid is also strongly hydrogen bonded in aqueous solution, accounting for its syrupy nature.
Orthophosphoric acid is a triprotic acid of medium strength (cf. Table 32, p. 161), forming dihydrogen phosphates, MH_2PO_4, hydrogen phosphates, M_2HPO_4, and orthophosphates, M_3PO_4. The orthophosphates contain the tetrahedral ion PO_4^{3-}, isoelectronic with SiO_4^{4-}, SO_4^{2-} and ClO_4^- (p. 202).
Orthophosphoric acid esters which play an important role in biological processes (e.g., adenosin triphosphate, ATP), are obtained in the condensation of alcohols with phosphoryl chloride, followed by hydrolysis:

$$O=P\overset{Cl}{\underset{Cl}{<}}Cl \xrightarrow[-HCl]{+ROH} O=P\overset{OR}{\underset{Cl}{<}}Cl \xrightarrow[-HCl]{+ROH} O=P\overset{OR}{\underset{Cl}{<}}OR \xrightarrow[-HCl]{+ROH} O=P\overset{OR}{\underset{OR}{<}}OR$$

$$\downarrow +H_2O \qquad\qquad \downarrow +H_2O$$

$$O=P\overset{OR}{\underset{OH}{<}}OH \qquad\qquad O=P\overset{OR}{\underset{OH}{<}}OR$$

Phosphorous Acid H_3PO_3:
Crystalline diprotic H_3PO_3 (m.p. 70 °C) is prepared by the hydrolysis of PCl_3 with hydrochloric acid and evaporation of the solution and forms hydrogen phosphites, $MH[HPO_3]$, and phosphites $M_2[HPO_3]$. H_3PO_3, as well as all other P—H and

P(III) compounds are strong reducing agents. Upon heating the acid disproportionates to phosphine:

$$4H_3PO_3 \text{ (l.)} \xrightarrow{200\,°C} PH_3\uparrow + 3H_3PO_4 \text{ (l.)}$$

Hypophosphorous Acid H_3PO_2:
Barium hypophosphite, formed when P_4 disproportionates in $Ba(OH)_2$ solution:

$$2P_4 + 3Ba(OH)_2 + 6H_2O \rightarrow 3Ba(H_2PO_2)_2 + 2PH_3\uparrow$$

liberates the acid with H_2SO_4 or by cation exchange. Crystalline H_3PO_2 is isolated
by evaporation.
Hypophosphorous acid in water is a moderately strong monoprotic acid and a
strong reducing agent, which as a solid disproportionates at 140 °C into PH_3 and
H_3PO_3, which in turn decomposes to PH_3 and H_3PO_4.

12.9.2. Condensed Phosphoric Acids

Condensed phosphates and their corresponding acids contain P—O—P units
formed by condensation:

$$P—OH + HO—P \rightarrow P—O—P + H_2O$$

$$P—OH + Cl—P \;\;\; \rightarrow P—O—P + HCl$$

The condensed phosphates correspond to the polysulfates (S—O—S) and poly-
silicates (Si—O—Si), which can form chain or ring structures.
The simplest condensed phosphorus(V) acid is $H_4P_2O_7$, formed by heating H_3PO_4
to $>200\,°C$, or by condensation of H_3PO_4 with $POCl_3$:

$$2H_3PO_4 \xrightarrow{-H_2O} \begin{array}{c} HO \\ | \\ HO-P-O-P-OH \\ \| \quad\quad \| \\ O \quad\quad\; O \end{array} \begin{array}{c} OH \\ | \\ \\ \end{array} \xleftarrow[180\,°C]{-HCl} \tfrac{5}{3}H_3PO_4 + \tfrac{1}{3}POCl_3$$

Diphosphates are obtained by heating hydrogen phosphates:

$$2M_2HPO_4 \rightarrow M_4P_2O_7 + H_2O$$

Higher condensed phosphates are prepared by dehydration of hydrogen phosphates
and dihydrogen phosphates:

$$2Na_2HPO_4 + NaH_2PO_4 \rightarrow Na_3P_3O_{10} + 2H_2O$$

$$n\,NaH_2PO_4 \rightarrow Na_nP_nO_{3n} + n\,H_2O$$

Chain terminating end groups are formed from HPO_4^{2-}; chain members from $H_2PO_4^-$; and cross linking groups from H_3PO_4:

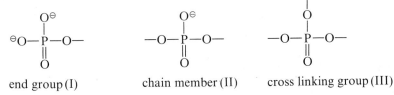

end group (I)   chain member (II)   cross linking group (III)

Condensed phosphates containing I and II consist of chains called polyphosphates (e.g., $Na_3P_3O_{10}$, sodium triphosphate) whose general formula is $M_nP_nO_{3n+1}$. Anions containing only II form rings called metaphosphates (e.g., $Na_3P_3O_9$) whose general formula is $M_nP_nO_{3n}$. Condensed phosphates containing some III are called ultraphosphates. Units I, II and II can be distinguished by their ^{31}P NMR spectra.

Hydrolysis of the P—O—P bridges yields POH groups. Condensed phosphate(V) is degraded in aqueous solution stepwise to orthophosphoric acid, and under carefully controlled conditions intermediates can be isolated, e.g. P_4O_{10} reacts with water at $0\,^\circ C$ in 70% yield to the cyclic tetrametaphosphoric acid, $H_4P_4O_{12}$, which can be isolated as the sodium salt.

Metaphosphates are salts of the polymeric acis $(HPO_3)_n$, where $n \geqslant 3$, e.g., sodium trimetaphosphate, $Na_3P_3O_9$, contains the cyclic, puckered anion $P_3O_9^{3-}$, isoelectronic with trimeric sulfur trioxide, S_3O_9:

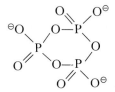

The $\measuredangle$P—O—P in which oxygen is common to two PO_4 tetrahedra, are 127° in $P_3O_9^{3-}$ and range $120\text{--}180^\circ$, since the oxygen lone pairs can be delocalized into the d-orbitals of phosphorus (p. 184).

Low molecular polyphosphates soften water by forming soluble complexes with Ca^{2+} and Mg^{2+} ions.

12.9.3. Peroxophosphoric Acids

Replacing OH groups in orthophosphoric acid by OOH gives monoperoxophosphoric, H_3PO_5, and diperoxophosphoric, H_3PO_6, acids. Both can be prepared from P_4O_{10} by reaction with aqueous hydrogen peroxide:

$$P\text{—}O\text{—}P + H_2O_2 \xrightarrow{-20\,^\circ C} P\text{—}OOH + HO\text{—}P$$

Peroxodiphosphoric acid, $H_4P_2O_8$, which contains a P—O—O—P bridge is formed like peroxodisulfuric acid, $H_2S_2O_8$, in the anodic oxidation of phosphate anions:

$$2 PO_4^{3-} \rightarrow P_2O_8^{4-} + 2e^-$$

The anion is isolated as $K_4P_2O_8$.

12.9.4. Thiophosphoric Acids

The thio acids, formed by replacing oxygen in the oxo acids, are unstable, but their anions can be isolated as salts:

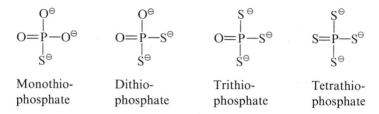

| Monothio-phosphate | Dithio-phosphate | Trithio-phosphate | Tetrathio-phosphate |

The four ions can be obtained by alkaline hydrolysis or thiolysis of P_4S_{10}:

$$P—S—P + OH^- \rightarrow P—S^- + {}^-O—P + H^+$$

$$P—S—P + SH^- \rightarrow P—S^- + {}^-S—P + H^+$$

e.g.: $P_4S_{10} + 6 Na_2S \xrightarrow{H_2O} 4 Na_3PS_4$

12.9.5. Halo- and Amidophosphoric Acids

Other monovalent radicals, e.g., —F, —Cl, —Br, —NH_2, or —N_3 can be used to replace OH-groups. Phosphoric acids with PP-bonds are also known.

12.9.6. Oxo- and Thio-Acids of Arsenic

The compounds in this group are:

Arsenites $MAsO_2$, or M_3AsO_3, salts of arsenous acid
Arsenates MH_2AsO_4, M_2HAsO_4, and M_3AsO_4, salts of arsenic acid
Thioarsenites M_3AsS_3
Thioarsenates M_3AsS_4

Only arsenic acid, $H_3AsO_4 \cdot \frac{1}{2} H_2O$, probably an oxonium salt, prepared by oxidation of As_2O_3 with nitric acid and evaporation, is stable. Arsenous acid is known only as aqueous solution of As_2O_3, from which the polymeric metaarsenites $MAsO_2$ (e.g., $KAsO_2$) and orthoarsenites (e.g., Ag_3AsO_3) are derived.

Thioarsenites and -arsenates are obtained from aqueous sulfide solutions of As_2S_3 and As_2S_5, respectively. The free acids liberated by addition of HCl decompose even at low temperatures into H_2S and the corresponding arsenic sulfide.

12.10. Phosphazenes

Phosphazenes, of which only the chlorophosphazenes will be treated here, contain the structural element —P=N—. The oligomeric phosphazenes are classified:

cyclic phosphazenes	chain phosphazenes
$(-PCl_2=N-)_n$ n = 3, 4, 5 ...	$R-(PCl_2=N)_n-R'$ n = 1, 2, 3...
	R: e.g., Cl; R': e.g., PCl_4

The formation of cyclic phosphazenes is based on the complex reaction of PCl_5 with NH_4Cl at 120–135 °C:

$$n\,PCl_5 + n\,NH_4Cl \rightarrow (-PCl_2=N-)_n + 4\,n\,HCl$$

Some reaction intermediates have been isolated in the following mechanism (compounds in parentheses are hypothetical):

$$NH_4Cl + PCl_5 \rightarrow [NH_4]^+[PCl_6]^- \xrightarrow{-3\,HCl} (HN=PCl_3)$$

$$\downarrow +2\,PCl_5$$

$$[Cl_3P-N=PCl_3]^+ [PCl_6]^-$$
m.p., 310–315 °C (decomp.)

$$+ NH_4Cl \;\Big|\; \begin{array}{l} -HCl \\ -PCl_5 \end{array}$$

$$[Cl_3P-N=PCl_2-N=PCl_3]^+ [PCl_6]^- \xleftarrow[-HCl]{+2\,PCl_5} (HN=PCl_2-N=PCl_3)$$
m.p. 228 °C

$$+ NH_4Cl \;\Big|\; \begin{array}{l} -3\,HCl \\ -PCl_5 \end{array}$$

$$(HN=PCl_2-N=PCl_2-N=PCl_3) \xrightarrow{-HCl}$$

m.p. 114 °C

Hexachlorocyclotri-phosphazatriene, $(PCl_2N)_3$, a colorless solid which can be vacuum- or steam-distilled, yields derivatives maintaining the six-membered ring, e.g., containing F, Br, alkyl, aryl, alkoxy, NH_2, N_3, etc., by Grignard or Friedel-Crafts alkylation, alcoholysis, amination, halogen exchange or with thiols.

Higher oligomeric $(PCl_2N)_n$ with n = 4 to 8, can be separated by vacuum distillation. Fluorophosphazenes, $(PF_2N)_n$, are known with $n \leq 17$.

The six-membered ring in $(PCl_2N)_3$ is nearly planar with $120°$ angles, and all internuclear distances, $d(PN)$, equal, but smaller than the sum of the single bond radii. The valence force constants, $f(PN)$, also indicate multiple bonds. As in benzene, several mesomeric structures must be considered in describing the ring.

However, in contrast to benzene, the 6 π-electrons are not completely delocalized but participate in three-center bonds which extend over the PNP-units. The nitrogen atoms are sp^2-hybridized and the distorted, tetrahedrally coordinated phosphorus atoms asymetrically sp-hybridized. Considering the three-fold axis (C_3) of the ring as the z-axis of the coordinate system, the π-bonds arise from overlap of the p_z-orbitals of nitrogen with each two d-orbitals of the neighboring phosphorus atoms (Fig. 100). This gives a π-bond order of 0.5 for each PN-bond. Two orthogonal phosphorus d-orbitals (d_{xz} and d_{yz}) at $90°$ are employed, so that maximum overlap takes place within the ring, but not exactly above and below the internuclear axis. This model explains the identical internuclear distances, stability and similarity of $(PCl_2N)_3$ and $(PCl_2N)_4$, since the number of three-center bonds may increase stepwise without marked change in properties as in an aromatic system. The larger rings are puckered, which is entirely compatible with the three-center model.

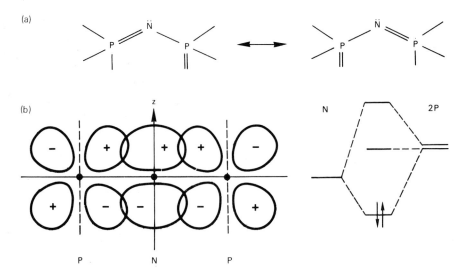

Fig. 100.   VB- and MO-model of the 3-center 2-electron π-bond in the cyclophosphazenes

Asymmetrical substitution changes valence angles and internuclear distances and puckers the ring, indicating incomplete delocalization of π-electrons. Isomerism (positional or geometrical) arises in mixed derivatives, as well as boat and chair conformations.

The cyclic phosphazenes are inorganic heterocycles. Such rings are formed by all non-metallic elements except the heavier halogens and the rare gases. Alternating sequences are most frequent, and condensed spiro-ring systems also exist.

13. Carbon

13.1. General

Carbon forms more compounds than any other non-metal except hydrogen in creating the basis for the field of organic chemistry. Moreover the special bonding character of the carbon atom imparts properties to these compounds which distinguish them from other non-metallic compounds.

Inorganic chemistry encompasses the modifications of carbon and its oxides, oxo acids, and chalcogenides. In addition, organometallic compounds such as the silicones (p. 358) and pentaphenylphosphorus (p. 317) are known.

13.2. Bond Relationships

The non-metallic elements of the IVth group are carbon, silicon, and germanium. The latter two are semi-conductors while the higher homologs tin and lead are metals. In the ^3P ground state, their atoms have the valence electron configuration s^2p^2, but the sp^3 excited state arising from promotion of one s-electron accounts for the formation of most compounds:

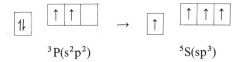

$$^3P(s^2p^2) \qquad ^5S(sp^3)$$

Only unstable carbenes are derived from the ^3P state, e.g., CF_2, p. 248 and CS, p. 341. The reason for their high reactivity is the free p-orbital (electron sextet) at the carbon atom.

The coordination number at carbon atoms with an electron octet may be 1 to 4:

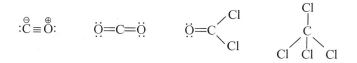

Coordination numbers >4 are only realized by many-center bonds, as in CH_5^+

found by mass spectroscopy which according to quantum mechanical calculations has C_s symmetry, or in $Al_2(CH_3)_6$ with bridging methyl groups:

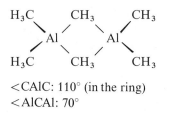

<CAlC: 110° (in the ring)
<AlCAl: 70°

Aluminum exhibits distorted tetrahedral coordination. Each bridging CH_3 group contributes a singly occupied sp^3-orbital which overlaps with two sp^3-orbitals of the aluminum atoms to form a 3-center 2-electron bond. Additional examples will be discussed in the section on carboranes (p. 375).

The electronegativity of carbon (2.5) prohibits the formation of ionic C^{4+} compounds with non-metals, and CF_4 and CO_2 are gases. Only singly-charged carbonium ions, R_3C^+, e.g., $[C(N_3)_3]^+$ $[SnCl_6]^-$ exist.

Carbon, however does react with strongly electropositive metals to form ionic carbides which contain C^{4-} (isoelectronic with N^{3-}, O^{2-}, F^- and Ne) and C_2^{2-} (isoelectronic with N_2 and CO) anions:

$$C^{4-} : Al_4C_3, Be_2C \qquad C_2^{2-} : CaC_2$$

Calcium carbide, CaC_2, has a distorted NaCl structure with parallel oriented C_2^{2-} groups in the anion positions.

The carbanions, R_3C^-, also exist, but carbon has little tendency to participate in ionic bonding. The weakly polar C—C and C—H bonds are the basis for organic chemistry.

The C—C bond is the strongest of any homonuclear non-metallic single bond, excepting that in H_2, as illustrated in Figure 62 (p. 131) and carbon atoms tend to form stable chains and rings. The C—H bond is also strong and with sp^3 hybridized carbon is only weakly polar. Alkanes, C_nH_{2n+2}, and alkyl groups, $—C_nH_{2n+1}$, are inert.

The sp^3 hybridized carbon atom differs from most non-metals in that its valence shell has no free electron pairs and no unoccupied orbitals. Attack by nucleophilic or electrophilic reagents is thus equally difficult. This fact underlies the uniqueness of carbon compounds, explaining the inertness of CF_4 and polytetrafluoroethylene (p. 245), whereas SiF_4, e.g., is easily hydrolyzed. In the sp^3 hybridized silicon atom, the 3d-orbitals are unoccupied, and a coordinate bond between H_2O and SiF_4 may form, providing a pathway for the generation of SiO_2:

$$SiF_4 + 2H_2O \rightarrow SiO_2 + 4HF$$

The analogous reaction of CF_4 to CO_2 and HF is thermodynamically favorable

($\Delta G° < 0$), but impeded kinetically, because H_2O finds no point of attack at the CF_4 molecule. Hence CF_4 is resistant towards hydrolysis like SF_6 (p. 230) and $FClO_3$ (p. 265) for the same reasons.

Carbon enters into multiple bonding with C, N, O, and S primarily, but also with Se, Te, P, and As. With larger partners the preference for single-bonded polymers increases, as can be seen from the thermal stability of the dichalcogenides of carbon (p. 202). The multiply bonded carbon atom is either sp or sp^2 hybridized:

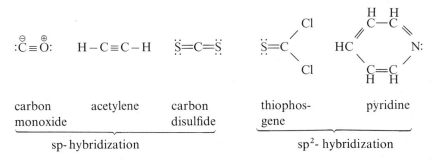

| carbon | acetylene | carbon | thiophos- | pyridine |
| monoxide | | disulfide | gene | |

sp- hybridization sp^2- hybridization

Delocalization of π-electrons in aromatic systems like benzene or the cyclopentadienyl anion is frequently encountered.

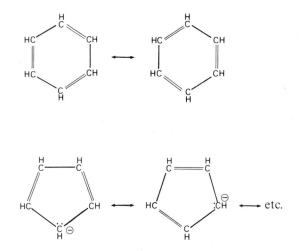

Fig. 101.   Valence structures for the molecules C_6H_6 (D_{6h}) and $C_5H_5^-$ (D_{5h}). In both molecules all C—C distances are equal as expressed by writing canonical structures.

Carbon p-orbitals perpendicular to the ring plane overlap leading to π-molecular orbitals extending over the ring. The ring plane is a node for these MO's. If $2n + 2$ (n = 1, 2, ect.) π-electrons are present, high stability is found.

The π-electron delocalization is expressed in the VB notation by writing several mesomeric structures.

No similarly stable π-electron systems for other non-metals are known, since delocalization is never complete (see p. 328 and p. 384).

13.3. Modifications of Carbon

Carbon crystallizes as graphite and diamond, with graphite the thermodynamically stable modification under normal conditions. Graphite consists of planar layers of fused six-membered rings containing σ- and π-bonds with van der Waals forces acting between the layers (Figure 102). Each sp^2 hybridized carbon has three equidistant neighbors bound by a σ-bond. The fourth electron is in a p-orbital perpendicular to the layer plane which overlaps with the p_{π}-orbitals of the three neighbors to create MO's which extend over the layer plane, giving equal internuclear distances. The delocalization leads to a large resonance energy and small nuclear distances (1.415 Å). Instead of the bond order of 1.3 expected from figure 102b, the internuclear distance corresponds to a value of 1.5 (see Fig. 63, p. 137).

Hexagonal graphite has staggered ABAB layers (as shown in Figure 103a). Its black color and electrical conductivity in the layer direction are related to the π-electrons within the layers, which form a quasi two-dimensional gas. The electrical conductivity in the perpendicular direction is smaller by a factor of 10^5.

The distance between the layers (3.35 Å) corresponds to van der Waals bonds, which explains the easy cleavage of graphite parallel to the planes.

Under high pressures graphite is converted into the denser diamond, but diamond synthesis employs temperatures of 1500–3000 K, catalysts like Fe, Cr, or Pt, as well as extreme pressures to achieve a sufficient reaction rate. Natural diamond may have been formed in similar geological processes. The sp^3 carbon atoms in the diamond network (Figure 103) are connected via σ-bonds as in CH_4.

Diamond is one of the hardest materials known. It is a non-conductor since all electrons are localized in two-center bonds.

Diamond burns to CO and CO_2 in air at 600–800 °C. At 1500 °C in the absence of air it is transformed into graphite.

The stable modifications of silicon and germanium as well as grey tin (α-Sn) are also in the cubic diamond form.

In contrast to diamond which is an insulator, pure silicon and germanium are semiconductors with weak, yet strongly temperature-dependent electrical conductivity related to the width of the energy gap between the valence and conduction bands which decreases from carbon to tin. The overlap of two tetrahedral orbitals creates a bonding and an antibonding MO. The bonding MO's of diamond form the valence band, the antibonding MO's the energetically higher conduction band. Since the valence band is completely occupied, electrical conductance is only possible when

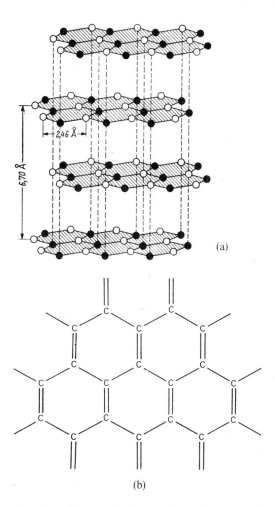

Fig. 102.   Structure of the graphite lattice (a), and representation of the bond relationships in the layers of the lattice with one mesomeric structure (b)

electrons are promoted to the empty conduction band. The energies required per mol are:

diamond: 670   Si: 105   Ge: 72   α-Sn: 8 [kJ/mol]

The high energy for diamond requires x-ray irradiation but ordinary heating suffices for silicon and germanium to bring about a marked decrease of electrical resistance.

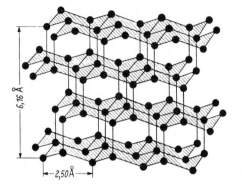

Fig. 103. The diamond lattice. The crosshatching is applied only for better visualization, and not to indicate a layer lattice. The structure is a three-dimensional network with identical bonds in all directions.

Surface Compounds:

The atoms on the surface of diamond and graphite lack neighboring atoms to utilize all their valence electrons, and hence exhibit enhaced reactivity.

Atoms or molecules can thus be chemisorbed onto such surfaces. Surface compounds form when the binding corresponds to chemical bonding in type and bond strength. Powdered diamond is hydrophobic, but becomes hydrophilic on treatment with oxidizing agents. Surface oxides are formed:

Heating the samples to 800 °C *in vacuo* releases the adsorbed species generating CO, CO_2, H_2O, etc. The resulting clean surface reacts with air at room temperature to form oxides but with Cl_2, F_2, and H_2 only at higher temperatures.

In graphite surface compounds are formed preferentially at the layer edges, since atoms within the layers lack unpaired electrons. Oxidation at 400 °C yields acidic surface oxides:

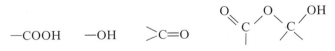

The OH groups can be titrated with base, e.g., 0.01 N NaOH or the hydrogen exchanged with other ions.

Surface compounds are also known with SiO_2 and other solids. They are of prac-

tical interest since they can change the mechanical, electrical and chemical properties of the solids.

13.4. Graphite Compounds

Graphite compounds retain the polymeric layer lattice. The bonding may be
covalent or ionic.

13.4.1. Covalent Graphite Compounds

Graphite reacts at 420–550 °C with fluorine without ignition to add F_2 across the
multiple bonds and forms compositions $CF_{0.68}$ to $CF_{0.99}$ containing sp^3 carbon
atoms:

$$>\!\!C\!\!=\!\!C\!\!< + \; F_2 \; \rightarrow \; \begin{array}{c} F \\ | \\ -\!C\!-\!C\!- \\ | \quad | \\ \quad F \end{array}$$

The σ-bonds remain intact, and in the limit the F : C ratio can become unity. The
electrical conductivity and black color disappear during fluorination. Preparations
with the highest fluorine content are colorless, transparent, non-conducting,
hydrophobic and inert. Upon heating they disproportionate into volatile carbon
fluorides and carbon black.

In the structure of $(CF)_n$, the fluorine atoms are presumably bonded on both sides of
puckered layers of carbon atoms. The internuclear distance, $d(CC) = 1.54$ Å,
corresponds to a single bond. The fluorinated layers are stacked irregularly at a
distance of 6.5 Å.

13.4.2. Ionic Graphite Compounds

The graphite lattice can react with reducing agents, e.g., the alkalis and alkaline
earths, and aluminum (a), or with oxidizing agents, e.g., the halogens or oxidizing
acids, (b):

$$\text{(a) } C_n + A \; \rightarrow \; C_n^- A^+ \qquad \text{(b) } C_n + B \; \rightarrow \; C_n^+ B^-$$

Potassium reacts at > 200 °C in the absence of air to form, depending on the
stoichiometry C_8K, $C_{24}K$, $C_{36}K$, $C_{48}K$, or $C_{60}K$. The compounds are bronze-
colored (C_8K) or bluish-gray, paramagnetic, and take the structures shown in
Fig. 104 in which potassium atoms are interspersed in graphite layers. In C_8K
each interlayer space is occupied; in the other phases each second, third, fourth,
or fifth layer. The potassium valence electrons are lost to the valence- and conductance bands of graphite, so that a partially ionic, partially metallic bonding

results. The conductivity of C_8K has a positive temperature effect, i.e., is metallic. It has a value 10 fold that of graphite in the direction of the layers, and 100 fold perpendicular to the layers.

Potassium graphite decomposes into its components on heating *in vacuo*.

Graphite reacts according to scheme (b) to give, e.g., C_8Cl, C_8Br, $C_{24}HSO_4 \cdot 2H_2O$ (blue) and $C_{24}HF_2 \cdot 2HF$.

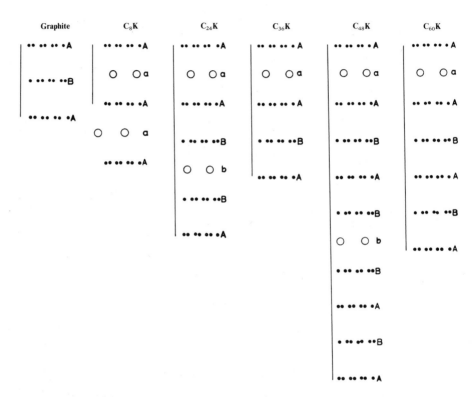

Fig. 104. Sequence of layers in potassium graphite compounds of varying composition (after H. Krebs, Grundzüge der anorganischen Kristallchemie, F. Enke Verlag, Stuttgart, 1968).

These compounds have a conductivity approaching that of aluminum by creation of positive holes in the valence band.

Metal halides, particularly chlorides, can be interspersed between the layers of graphite in the presence of elemental chlorine, e.g., $FeCl_3$ and Cl_2 react at $<300\,°C$ to give $C_n^+ \cdot FeCl_4^- \cdot mFeCl_3$ ($n \approx 27$, $m \approx 3$), which presumably has the structure shown in Fig. 105.

The interlayer distance is widened by the interspersed $FeCl_3$ to 9.40 Å.

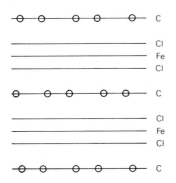

Fig. 105. Probable structure of graphite with interspersed $FeCl_3$.

The ionic graphite compounds are characterized by layer structures, an increase in interlayer spacing to ≤ 9.4 Å, the stepwise intercalation of the reaction species between the layers, and the hydrolytic sensitivity of the compounds.

13.5. Chalogenides of Carbon

13.5.1. Oxides

The following binary compounds are known:

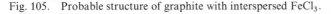

$$\overset{\ominus}{:}C\equiv\overset{\oplus}{O}: \qquad \ddot{O}=C=\ddot{O} \qquad \ddot{O}=C=C=C=\ddot{O}$$

$$CO \qquad\qquad CO_2 \qquad\qquad\qquad C_3O_2$$

The oxides C_6O_6 and $C_{12}O_9$ have been prepared, but will not be discussed.

Carbon monoxide is a colorless, poisonous gas, formed by burning carbon with insufficient oxygen:

$$C(s.) + O_2(g.) \rightarrow CO_2(g.) \qquad CO_2(g.) + C(s.) \rightleftharpoons 2CO(g.)$$

The second equilibrium, which plays a role in technical processes, is endothermic, so the position of the equilibrium moves to the side of CO with increasing temperature. Carbon monoxide is thermodynamically unstable, but the disproportionation into CO_2 and carbon is kinetically hindered.

Dehydration of formic acid with sulfuric acid produces the anhydride CO, which reacts with NaOH under pressure to give sodium formiate:

$$HCOOH \rightarrow CO + H_2O \qquad CO + NaOH \xrightarrow{160\,°C} HCOONa$$

Carbon monoxide is useful as a reducing agent and for bonding with transition metals in carbonyl complexes, e.g., with Ni to form $Ni(CO)_4$, with Fe to $Fe(CO)_5$, etc. The bonding of CO in carbonyls is analogous to that of N_2 in dinitrogen complexes since CO and N_2 are isosteres (p. 108) and both molecules have similar physical properties. Despite the formal charges, the dipole moment of CO is very small (0.12 D; $C \rightarrow O$) since the charges are compensated by the electronegativity difference. The bond order of 2.8 calculated from the valence force constant, and the dissociation energy of 1072 kJ/mol are, as for N_2, extraordinarily large.

The toxicity of CO arises from its rapid reaction with the iron in hemoglobin to form a stable carbonyl complex which is no longer suitable for O_2-transport.

Carbon dioxide is formed in the complete combustion of carbon, in the roasting of limestone, and in the decomposition of carbonates by acids. Like N_3^-, the CO_2-molecule contains two 3-center π-bonds.
Carbon dioxide is the thermodynamically stable oxide of carbon. The unstable peroxide, CO_3 is formed in the irradiation of a solution of O_3 in solid or liquid CO_2:

$$CO_2 + O_3 \xrightarrow{h \cdot \nu} CO_3 + O_2$$

The gas decomposes to CO_2 and O_2. The colorless gas tricarbon dioxide, C_3O_2 (b.p. $7°C$), is formed in the dehydration of malonic acid with P_4O_{10} *in vacuo*:

$$H_2C \underset{\diagdown COOH}{\overset{\diagup COOH}{}} \rightarrow O=C=C=C=O + 2H_2O$$

or by irradiation of CO:

$$4CO \rightarrow C_3O_2 + CO_2$$

Under normal pressure C_3O_2 polymerizes to a red product. Malonic acid is regenerated on hydrolysis.
The linear C_3O_2-molecule is derived from overlap of carbon sp hybrid orbitals with oxygen p-orbitals. Two 5-center π-bonds are superimposed over the four 2-center σ-bonds. The 5-center π-bonds are formed by overlap of five coplanar p-orbitals with π-symmetry leading to two bonding, one non-bonding, and two anti-bonding MO's, occupied by a total of six electrons. Of the 12 π-electrons, therefore, eight are in bonding states and four are non-bonding.
Analogous MO treatment demonstrates that for the hypothetical C_2O_2 and C_4O_2 molecules analogous structures would lead to an unstable triplet state.

13.5.2. Sulfides, Selenides, Tellurides

The CO_2 analogues CS_2 (colorless liquid) and CSe_2 (yellow liquid) are known, but CTe_2 has not been synthesized. Sulfur vapor and carbon react to form CS_2; CSe_2 is obtained from CH_2Cl_2 and liquid selenium:

$$C(s.) + S_2 \rightarrow CS_2$$

$$CH_2Cl_2 + 2Se(l.) \xrightarrow{600\,°C} CSe_2 + 2HCl$$

All the mixed dichalcogenides are known except COTe.
Decomposition of CS_2 vapor at 0.1 torr in a glow discharge produces CS which can be trapped in a matrix of CS_2 at $-190\,°C$ and detected by infrared spectroscopy. In reactivity CS (bond order 2.2) resembles CF_2, SiF_2, and BF (p. 248). It decomposes to:

$$3CS \rightarrow C_3S_2 + S$$

The red-brown liquid C_3S_2 corresponds to C_3O_2, but the Se- or Te-analogues are unknown.

13.5.3. Carbonic Acids and Carbonates

The acids H_2CO_3, H_2CS_3, and H_2CSe_3 will be discussed.
The anhydride of carbonic acid, CO_2, hydrolyzes slowly at pH 7:

$$CO_2 + H_2O \rightleftharpoons H_2CO_3$$

In aqueous solution CO_2 is physically dissolved and only loosely solvated. Only 0.2% is present as H_2CO_3, HCO_3^-, or CO_3^{2-}:

$$H_2CO_3 \rightleftharpoons H^+ + HCO_3^- \rightleftharpoons 2H^+ + CO_3^{2-}$$

Carbonic acid is a weak acid. The dissociation constants for the two equilibria above are $K_1 = 1.6 \cdot 10^{-4}$ and $K_2 = 4.7 \cdot 10^{-11}$. In alkaline solution CO_2 dissolves directly as hydrogen carbonate:

$$CO_2 + OH^- \rightleftharpoons HCO_3^-$$

Owing to its low concentration carbonic acid is completely dissociated, and H_2CO_3 cannot be isolated since CO_2 is released in the dehydration. Colorless crystals of an etherate $H_2CO_3 \cdot (CH_3)_2O$ (m.p. $-47\,°C$) are obtained from Na_2CO_3 at $-35\,°C$ with hydrogen chloride in dimethyl ether, which decompose above $-26\,°C$ into CO_2, H_2O, and the ether.
Bubbling CO_2 into alkali produces first carbonate and then hydrogen carbonate:

$$CO_2 + 2NaOH \rightarrow Na_2CO_3 + H_2O \xrightarrow{+CO_2} 2NaHCO_3$$

Difluorodioxocarbonate forms with CsF:

$$2\,CsF + CO_2 \xrightarrow{CH_3CN} Cs_2[CO_2F_2] \xrightarrow{100\,^\circ C} 2\,CsF + CO_2$$

The $CO_2F_2^{2-}$ anion is a derivative of the hypothetical orthocarbonic acid, H_4CO_4 [or $C(OH)_4$], which is only known in the form of esters. Trithio- and triseleno-carbonates are obtained from CS_2 or CSe_2:

$$CS_2 + SH^- + OH^- \rightarrow CS_3^{2-} + H_2O$$
$$CSe_2 + SeH^- + OH^- \rightarrow CSe_3^{2-} + H_2O$$

Mixed chalcogen carbonates may be prepared in this manner. The anions CX_3^{2-} (X=O, S, Se) are all trigonal-planar. The π-bond is delocalized (4-center π-bond):

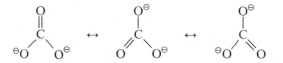

Unlike H_2CO_3, trithiocarbonic acid can be isolated as a red oil (m.p. $-27\,^\circ C$) by treating a suspension of $BaCS_3$ with hydrochloric acid. H_2CS_3 in water is a moderately strong acid ($K_1 = 1.2$ at 0°). The acid H_2CSe_3 is likewise prepared from $BaCSe_3$ in diethylether as a viscous red oil (decomp. $> -10\,^\circ C$).

Aqueous trithiocarbonates dissolve sulfur to form the perthiocarbonate ion, CS_4^{2-}. The free yellow solid perthiocarbonic acid, H_2CS_4, (m.p. $= -36.5\,^\circ C$) which decomposes into H_2S, CS_2 and sulfur is prepared from $(NH_4)_2CS_4$ and HCl gas in $(CH_3)_2O$ at $-78\,^\circ C$. The structures of H_2CS_4 and CS_4^{2-} are:

$$\text{H—S—S—}\underset{\underset{\displaystyle S}{\|}}{C}\text{—S—H} \qquad \text{and} \qquad {}^{\ominus}\text{S—S—}\underset{\underset{\displaystyle S}{\|}}{C}\text{—S}^{\ominus}$$

14. Silicon and Germanium

14.1. General

The transition from non-metallic to metallic behavior in the IVth main group of the periodic system takes place between germanium and tin. Carbon is a pure non-metal and silicon and germanium are semiconductors, but these elements are typical non-metals in their chemical behavior.

After oxygen, silicon is the next most abundant element contributing 27.5% to the earth's crust, but unlike carbon occurring only as oxides, SiO_2 and silicates. The Si—O bond is strong, and silicon has great affinity for oxygen.

Germanium and silicon are similar, like the pairs P/As, S/Se, and Cl/Br. Germanium is a relatively rare element ($<10^{-3}$ wt% of the earth's crust) and occurs mainly as thiogermanates.

14.2. Bond Relationships

The silicon atom has four valence electrons:

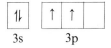

Few compounds are derived from this ^3P ground state:

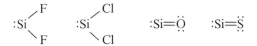

These unstable compounds are obtained only at high temperatures in equilibrium mixtures or isolated in a matrix (p. 248).

Four-valency originates from the excited state:

This configuration resembles that of excited carbon, but the orbitals in the third quantum shell are larger and more diffuse and d-orbitals are available for coordinate bond formation. These differences give rise to the divergent chemistries of carbon and silicon.

By analogy with the alkanes silicon and germanium form silanes, Si_nH_{2n+2}, and germanes, Ge_nH_{2n+2}. SiH_4 and GeH_4 are tetrahedral molecules like CH_4. Similar analogies hold for the halides Si_nX_{2n+2} and Ge_nX_{2n+2} (X=F, Cl, Br, I), but silicon and germanium are unable to form bonds corresponding to those in olefins (alkenes), alkynes, or aromatics since both atoms only rarely form sp^2 hybrids or p-p-π-bonds (e.g., in SiO, SiS). Multiple bonds between Si or Ge atoms have not been realized. The atoms apparently never form sp hybrids.

The instability of $H_2Si{=}SiH_2$ (silene) has the same rationale as for S_2 (p. 131) and P_2 (p. 307): the conversion of an Si=Si double bond to two Si—Si single bonds is exothermic so that Si_2H_4 formed as a reaction intermediate would polymerize to $(SiH_2)_x$. The polymerization of ethylene to the thermodynamically more stable polyethylene is also favorable, but requires a catalyst. Since silanes are more reactive than hydrocarbons, no catalyst is required. The higher reaction rate is related to the larger size of the silicon atom which is less well screened by its substituents.

No silicon analogues of the ketones, nitriles or carboxylic acids are known, however, many single-bonded analogues of ethers, thioethers, amines, and phosphines exist. Silicon takes coordination numbers from 1 to 6, with 4 being common (e.g., quartz, silicates, halo- and alkyl silanes). Values of 5 (rare) and 6 are made possible by 3d-orbital participation. In the octahedral SiF_6^{2-}, silicon is hybridized $sp^3d_{x^2-y^2}d_{z^2}$. Unlike CCl_4, $SiCl_4$ is a Lewis acid and reacts with tertiary amines:

$$SiF_4 + 2F^- \rightarrow SiF_6^{2-}$$
$$SiCl_4 + dipy \rightarrow SiCl_4(dipy)$$
$$SiCl_4 + 2py \rightarrow SiCl_4(py)_2$$

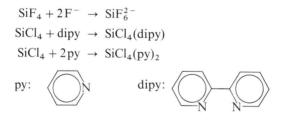

The d-orbitals also play a role in silicon reactivity. Unlike the inert alkanes, the silanes spontaneously ignite in air and are hydrolyzed, especially in the presence of OH^-. These reactions have no analogy in carbon chemistry and are related to the strongly accepting d-orbitals which make possible attack by nucleophilic reagents:

The intermediate is analogous to the $SiCl_4^-$-amine adducts.

These examples demonstrate that the analogy between carbon chemistry and that of Si and Ge is only a formal one.

The d-orbitals not only behave as σ-acceptors but can also engage in coordinate π-bonds, e.g., in SiO_2 which unlike CO_2 is a high polymer of sp^3 hybridized silicon atoms in a siloxane network:

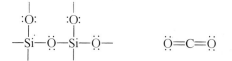

Several stable, but interconvertible modifications (quartz, tridymite, cristobalite, etc.) of SiO_2 are known. The m.p. of quartz (1986 K) is a result of the high Si—O bond energy. The $\angle$ Si—O—Si is 142° for α-quartz, 150° for α-cristobalite, and 152° for β-cristobalite, indicating delocalization of oxygen electron pairs into silicon d-orbitals:

$$Si-\overset{..}{\underset{..}{O}}-Si \leftrightarrow Si\overset{\ominus}{=}\overset{\oplus}{O}-Si \leftrightarrow Si\overset{\ominus}{=}\overset{2\oplus}{O}=Si^{\ominus}$$

The Si—O bond is thus multiple, explaining the high thermal stability of the silicates and silicones.

Table 51   Average Bond Energies (in kJ/mol at 25 °C)

	—H	—F	—Cl	—O—
C	415	486	327	343
Si	320	540	360	368
Ge	289	465	339	—

Silicon forms π-bonds with fluorine, oxygen and nitrogen, e.g., in disiloxane, $(SiH_3)_2O$ (p. 184), trisilylamine, $(SiH_3)_3N$, (p. 283), silyl isocyanate, H_3SiNCS (p. 284), and SiF_4, as shown by comparing the bond energies of C, Si, and Ge with H, F, Cl, and O (Table 51). Bond energies decrease down a group (cf. CH > SiH > GeH), but in the fluorides, chlorides and oxygen compounds a maximum is observed at silicon. An explanation is found in the mesomeric structures:

$$\overset{\diagdown}{\underset{\diagup}{Si}}-\overset{..}{\underset{..}{F}} \leftrightarrow \overset{\diagdown}{\underset{\diagup}{Si}}\overset{\ominus}{=}\overset{\oplus}{\underset{..}{F}}$$

These π-bonds influence the valence angles and the bond energies, and lead to a shortening of internuclear distances.

Support for this concept comes from the weak basicity of $(SiH_3)_3N$ in which the electron pair seems no longer available, since $(SiH_3)_3N$ does not form an adduct with $(CH_3)_3B$ as $(CH_3)_3N$ does. The π-donor ability towards Si is:

$$N > O > F; \qquad N > P; \qquad O > S; \qquad F > Cl > Br > I$$

Germanium tends less to form coordinate π-bonds, as seen in the similar basicity of $[Ge(CH_3)_3]_3N$ with the tertiary alkylamines.

The higher reactivity of silanes and germanes derives from the lesser Si—H and Ge—H bond energies (Table 51). Likewise, the Si—Si (197 kJ/mol) and Ge—Ge (163 kJ/mol) bonds are considerably weaker than the C—C bond (331 kJ/mol), and the tendency for catenation decreases in the series C > Si > Ge.

Unlike the C—H bond, the Si—H bond is considerably polarized owing to the lower electronegativity (Allred-Rochow) of silicon:

$$H: 2.2 \qquad C: 2.5 \qquad Si: 1.7 \qquad Ge: 2.0$$

Silanes thus react with positively polarized hydrogen atoms with elimination of H_2:

$$\geq Si-H + H-O-H \ \rightarrow \ \geq Si-OH + H_2$$
$$\geq Si-H + H-Cl \ \rightarrow \ \geq Si-Cl + H_2$$

The electronegativity of Si does not permit the formation of cations, Si^{4+}; SiF_4 is a gas, and SiO_2 forms a network lattice, but with large ionic contributions in both cases.

14.3. The Elements

Silicon and germanium crystallize in a diamond lattice (p. 336), but under high pressure other, higher density modifications are realized. No graphite analogues are known for either element.

Elemental silicon is won from the endothermic reduction of quartz with carbon, or calcium carbide in an electric furnace:

$$SiO_2 + 2C \ \rightarrow \ Si + 2CO\uparrow \qquad SiO_2 + CaC_2 \ \rightarrow \ Si + 2CO\uparrow + Ca\uparrow$$

The exothermic reduction with active metals are suitable laboratory preparations:

$$3SiO_2 + 4Al \ \rightarrow \ 3Si + 2Al_2O_3$$

Reduction of the tetrahalides (X═F or Cl) with aluminum or zinc yields the element, as does the pyrolysis of silane:

$$SiH_4 \ \rightarrow \ Si + 2H_2$$

High purity silicon is prepared in this manner.

Germanium is prepared by reduction of the oxide with hydrogen:

$$GeO_2 + 2H_2 \ \rightarrow \ Ge + 2H_2O$$

The oxide is obtained by hydrolysis of $GeCl_4$.

Further purification of silicon and germanium is afforded by zone melting. Silicon crystallizes in hard, brittle, dark grey octahedra (m.p. 1413 °C). The brown powder resulting in the reduction of SiO_2 with magnesium metal can be recrystallized from liquid aluminum. Silicon is a poor electrical conductor whose conductivity increases with temperature (p. 335) or by the presence of impurities; e.g., doping with 1 ppm boron increases conductivity 10^5 fold. Silicon dissolves in alkali with hydrogen evolution, but is stable towards acids, except a mixture of HNO_3 and HF. On exposure to air, especially during grinding, surface oxides form which reduce its reactivity (passivation). Finely powdered silicon is pyrophoric and hydrolyzes to form SiO_2 and H_2.

Germanium crystallizes in grayish-white, brittle octahedra (m.p. 959 °C) which are resistant towards non-oxidizing acids. With oxidizing acids, GeO_2 forms.

14.4. Hydrides

Silicon forms silanes Si_nH_{2n+2}, characterized to $Si_{15}H_{32}$, and polymeric hydrides $SiH_{0.7-0.9}$. The silanes are colorles gases (n = 1, 2), liquids (n > 3), or solids, sensitive to oxidation and hydrolysis. They spontaneously ignite to SiO_2 and H_2O[14]. The germanes are known to Ge_9H_{20} as air-sensitive gases (GeH_4) liquids or solids.

Preparation:

The mono-, di-, and trisilanes can be prepared from the corresponding chlorides with $LiAlH_4$ in ether:

$$SiCl_4 + LiAlH_4 \rightarrow SiH_4\uparrow + LiCl + AlCl_3$$

A silane mixture is obtained in the hydrolysis of Mg_2Si in warm 20% phosphoric acid:

$$Mg_2Si + 4H^+ \rightarrow SiH_4 + 2Mg^{2+}$$

Magnesium silicide is prepared at 1100 °C from the elements under argon. The hydrolysis product contains silanes to n = 15, with concentrations decreasing with increasing molecular weight. The higher silanes (n > 3) include branched chains. Separation is possible by vacuum distillation or by gas chromatography. On a technical scale SiH_4 is prepared from $SiCl_4$ and LiH in a LiCl—KCl melt. Higher silanes are synthesized by silent electrical discharge:

$$SiH_4 \rightarrow SiH_3 + H \qquad\qquad 2SiH_3 \rightarrow Si_2H_6$$
$$SiH_3 \rightarrow SiH_2 + H \qquad\qquad SiH_2 + Si_2H_6 \rightarrow Si_3H_8$$

Mixed Si—Ge hydrides can be prepared in this manner by starting with a SiH_4—GeH_4 mixture.

---

14 As a cyclic analogue of the cycloalkanes, Si_5H_{10} has been synthesized.

Germanes are prepared by reduction of GeO_2 with $NaBH_4$ in acidic solution or by hydrolysis of Mg_2Ge in aqueous acid or with NH_4Br in liquid ammonia. Higher germanes are obtained by decomposing GeH_4 in a silent electric discharge.

Reactions of Silanes:
Silanes decompose to the elements on heating, providing pure silicon and higher silanes. Unlike the alkanes, silanes have negative hydrogen, $[\overset{\delta+}{Si} - \overset{\delta-}{H}]$, which favors the alkali catalyzed hydrolysis:

$$SiH_4 + 2H_2O \rightarrow SiO_2 + 4H_2$$

Silanes undergo hydrogen-halogen exchange:

$$SiH_4 + HCl \rightarrow SiH_3Cl + H_2$$
$$3Si_2H_6 + BCl_3 \rightarrow 3Si_2H_5Cl + \tfrac{1}{2}B_2H_6$$

Partially halogenated silanes undergo condensation:

$$2SiH_3Cl + H_2O \rightarrow (SiH_3)_2O + 2HCl$$
$$3SiH_3I + 4NH_3 \rightarrow (SiH_3)_3N + 3NH_4I$$
$$4SiH_3I + 3N_2H_4 \rightarrow (SiH_3)_2N - N(SiH_3)_2 + 2N_2H_6I_2$$
$$3SiH_3Br + 3KPH_2 \xrightarrow{\ -100\,°C\ } (SiH_3)_3P + 2PH_3 + 3KBr$$

Silyl salts are prepared in monoglyme ($H_3C-O-CH_2-CH_2--O-CH_3$):

$$SiH_4 + K \rightarrow KSiH_3 + \tfrac{1}{2}H_2$$
$$Si_2H_6 + KH \rightarrow KSiH_3 + SiH_4$$

The potassium salt crystallizes in a NaCl lattice, and reacts with chlorides, releasing KCl:

$$KSiH_3 + (CH_3)_2AsCl \rightarrow H_3Si-As(CH_3)_2 + KCl$$

The germanes react similarly. Germane, GeH_4, is attacked by O_2 on heating to GeO_2 and H_2O, but is stable toward 30% NaOH. Like SiH_4 it forms germyl halides, GeH_3X, and metal salts, e.g., $KGeH_3$, obtained from potassium metal and GeH_4 in liquid NH_3.

14.5. Halides

The silicon halides are derived from the silanes by replacement of hydrogen with halogen. SiF_4 and $SiCl_4$ are the best investigated compounds. The higher halides $(SiX_2)_n$ and $(SiX)_n$ are less thoroughly characterized.
The germanium halides GeX_4, Ge_2X_6, and GeX_2 are known.

Fluorides:
Gaseous tetrafluorosilane can be prepared by fluorination of silicon or SiO_2, or more easily:

$$2CaF_2 + 2H_2SO_4 + SiO_2 \rightarrow SiF_4\uparrow + 2CaSO_4 + 2H_2O$$

Inasmuch as SiF_4 hydrolyzes very easily, using concentrated sulfuric acid removes the water formed thus preventing the hydrolysis of SiF_4. In the laboratory it is prepared from $BaSiF_6$:

$$BaSiF_6 \xrightarrow[\text{vacuum}]{300-350\,°C} BaF_2 + SiF_4\uparrow$$

The hydrolysis of SiF_4 is an equiliburium:

$$SiF_4 + 2H_2O \rightleftharpoons SiO_2 + 4HF$$

The main product in the vapor phase with excess SiF_4 is hexafluorodisiloxane, SiF_3OSiF_3, a colorless gas. With small amounts of liquid water, SiF_4 gives the oxonium salt of hexafluorosilicic acid:

$$3SiF_4 + 6H_2O \rightarrow 2(H_3O)_2SiF_6 + SiO_2$$

This salt in water is a strong acid and converts hydroxides and carbonates to the corresponding hexafluorosilicates.
Tetrafluorogermane is prepared by heating $BaGeF_6$ obtained from BaF_2, GeO_2 and HF. Gaseous GeF_4 fumes like SiF_4 when exposed to air, hydrolyzing with liquid water to GeO_2 and $(H_3O)_2GeF_6$.
The tetrafluorides are Lewis acids which form 1 : 1 or 1 : 2 adducts with donors:

$$SiF_4(NH_3)_2, \quad SiF_4[OS(CH_3)_2]_2, \quad GeF_4(SH_2)_2, \quad GeF_4(NCCH_3).$$

The monomeric silicon difluoride has been discussed (p. 248). The higher perfluorosilanes are derived from it, or from ZnF_2 and the corresponding perchloropolysilanes.
Germanium difluoride is more stable than SiF_2 since the stability of the oxidation state $+2$ increases from C to Pb. The colorless crystals of GeF_2 contain fluorine-bridged units to give four-coordinated, pseudotrigonal bipyramids with a lone pair at germanium. The difluoride is made by the reduction of GeF_4 with germanium at 150–300 °C, or from germanium and HF at 225 °C:

$$Ge(s.) + GeF_4 \rightleftharpoons 2GeF_2 \qquad Ge + 2HF \rightarrow GeF_2 + H_2$$

Chlorides:
Silicon tetrachloride is formed by chlorination of silicon or its compounds, e.g., by chlorination of a ferrosilicon alloy:

$$Fe/Si + 3.5 Cl_2 \xrightarrow{>400\,°C} SiCl_4 + FeCl_3$$

Ferrosilicon is prepared from the elements and reacts faster than silicon. The tetrachloride is a colorless liquid whose hydrolysis in the vapor phase proceeds only in presence of HCl acceptors, but rapidly with liquid water:

$$SiCl_4 \xrightarrow{+H_2O} SiCl_3OH \xrightarrow{+SiCl_4} SiCl_3\text{—}O\text{—}SiCl_3 \xrightarrow{+3H_2O} 2 SiO_2 + 6 HCl$$

tetrachloro-   trichloro-   hexachloro-
silane   silanol   disiloxane

The intermediates can be isolated.

The tetrachloride is a weaker Lewis acid than SiF_4, and while adducts with tertiary amines form, no $SiCl_6^{2-}$ ion exists.

Ammonolysis also proceeds with HCl elimination, e.g., with NH_3 to form hexachlorodisilazane:

$$2 SiCl_4 + 3 NH_3 \rightarrow SiCl_3\text{—}NH\text{—}SiCl_3 + 2 NH_4Cl$$

Cyclohexachlorotrisilazane, $(SiCl_2NH)_3$, is also isolated.

The higher chlorosilanes, $Si_nX_{2n+2}(n \leqq 6)$ are generated in the chlorination of $CaSi_2$ and in an electric discharge through $SiCl_4$ using silicon electrodes.

Higher chlorosilanes are formed in the disproportionation of the lower ones catalyzed by $(CH_3)_3N$:

$$4 Si_2Cl_6 \rightarrow Si_5Cl_{12} + 3 SiCl_4$$

$$3 Si_3Cl_8 \rightarrow Si_5Cl_{12} + 2 Si_2Cl_6$$

$$5 Si_2Cl_6 \rightarrow Si_6Cl_{14} + 4 SiCl_4$$

Passing $SiCl_4$ vapor over silicon at 1000 °C forms $SiCl_2$ in an equilibrium:

$$Si(s.) + SiCl_4(g.) \rightleftharpoons 2 SiCl_2(g.) \qquad K_p(1615\ K) = 1$$

On quenching, $SiCl_2$ polymerizes to $(SiCl_2)_n$, and higher perchlorosilanes to Si_6Cl_{14} are known. The carbenoid $SiCl_2$ inserts into the E—Cl bond of BCl_3, CCl_4, or PCl_3 to form $Cl_2B\text{—}SiCl_3$, $Cl_3C\text{—}SiCl_3$, and $Cl_2P\text{—}SiCl_3$, respectively.

Germanium tetrachloride is prepared from the elements or GeO_2 and hydrochloric acid. The colorless, fuming liquid hydrolyzes to GeO_2.

The dichloride occurs in the reduction of $GeCl_4$ with Ge at $>680\,°C$ and in the dissociation of trichlorogermane:

$$GeHCl_3 \rightleftharpoons GeCl_2 + HCl$$

The colorless crystals of $GeCl_2$ are hydrolyzed by liquid water to hydrous germanium(II) hydroxide, and react with air to GeO_2 and $GeCl_4$.

Other Silicon Halides:
The bromide and iodide, synthesized from the elements, are colorless, moisture-sensitive and less thermally stable than SiF_4 or $SiCl_4$. Mixed halides are also known.
The chlorosilanes SiH_3Cl, SiH_2Cl_2 and $SiHCl_3$ are obtained:

$$Si + 3\,HCl \xrightarrow[300\,°C]{CuCl_2} SiHCl_3 + H_2$$

$$SiH_4 + HCl \longrightarrow SiH_3Cl + H_2$$

$$SiCl_4 + H_2 \longrightarrow SiHCl_3 + HCl$$

$$2\,SiHCl_3 \xrightarrow[300\,°C]{AlCl_3} SiCl_4 + SiH_2Cl_2$$

14.6. Oxides

Mono- and dioxides are known:

$$SiO \quad SiO_2 \qquad GeO \quad GeO_2$$

The polymeric dioxides occur in several modifications while the monoxides are monomeric in the vapor phase but polymerize on solidification.
Silicon dioxide occurs naturally in many crystalline forms, e.g., quartz (rock crystal, amethyst, etc.), tridymite, cristobalite, and the high pressure forms coesite and stishovite; and amorphous or microcrystalline forms frequently containing water such as opal (chalcedony, agate, flint, etc.) and silica gel.
The following equilibria exist between the modifications which are thermodynamically stable at normal pressure:

$$\underset{\text{(trigonal)}}{\alpha\text{-Quartz}} \underset{575\,°C}{\rightleftharpoons} \underset{\text{(hexagonal)}}{\beta\text{-Quartz}} \underset{867\,°C}{\rightleftharpoons} \underset{\text{(hexagonal)}}{\beta\text{-Tridymite}} \underset{1470\,°C}{\rightleftharpoons} \underset{\text{(cubic)}}{\beta\text{-Cristobalite}} \underset{1713\,°C}{\rightleftharpoons} \text{melt}$$

α-Tridymite and α-cristobalite are metastable.

All SiO_2 modifications except stishovite have tetrahedral silicon atoms linked by oxygen into networks (cf. p. 345 and the discussion of valence angles, p. 184). The Si—O bonds are strongly polar and very stable owing to the multiple bond contribution (p. 355). In stishovite which is formed at 1300°C and 120 kbar, crystallizing in a rutile (TiO_2) lattice, the Si atoms are octahedrally coordinated by two additional oxygen atoms.

The low rate of phase transition stems from the complex networks in SiO_2 modifications since not only a change in crystal symmetry is involved. Thus cristobalite is obtained from an SiO_2 melt only by slow cooling. On rapid cooling, the melt solidifies to a glass which is metastable and crystallizes on annealing at $1000\,^\circ C$. Silicon dioxide is resistant to acids (excepting HF) and dilute alkali, but metasilicates form on melting with alkali metal hydroxides:

$$SiO_2 + 2\,NaOH \rightarrow Na_2SiO_3 + H_2O$$

Finely-divided or freshly prepared SiO_2 is more reactive, as it contains silanol groups and releases water on aging or heating to form siloxane groups, Si—O—Si. Gaseous SiO is generated by heating SiO_2 with silicon in a $3:1$ ratio *in vacuo* at $1000–1300\,^\circ C$:

$$Si(s.) + SiO_2(s.) \rightleftharpoons 2\,SiO(g.)$$

On slow cooling SiO disproportionates to Si and SiO_2, but on quenching a black-brown polymeric $(SiO)_n$ is obtained which is glassy or fibrous. The surface of the glassy SiO in air is covered with SiO_2 which renders it passive. Fibrous SiO is pyrophoric.

Monomeric SiO has been isolated along with $(SiO)_2$ and $(SiO)_3$ in a matrix at low temperatures. The dissociation energy of the monomer is 715 kJ/mol, corresponding to double bonding. Polymerization creates additional SiO and SiSi bonds.

The hexagonal, cristobalite-like germanium dioxide formed in the hydrolysis of $GeCl_4$ with aqueous NH_3 gives an acidic reaction in water. Germanates form on GeO_2 dissolution in alkali. On heating to $380\,^\circ C$ hexagonal GeO_2 converts to the insoluble tetragonal form which crystallizes in the rutile lattice (octahedrally coordinated germanium).

Reduction by heating with Ge gives monomeric GeO. Polymeric GeO arises from the reduction of GeO_2 with H_3PO_2 in aqueous HCl. The intermediate hydrous $Ge(OH)_2$ is dehydrated at $650\,^\circ C$ to GeO. Germanium monoxide reacts with HCl to give $GeHCl_3$ and H_2O at $175\,^\circ C$.

14.7. Oxo Acids, Silicates and Germanates

Silicon:
The simplest oxo acid of silicon, by analogy with the acids of the neighboring non-metals should have the formula H_4SiO_4:

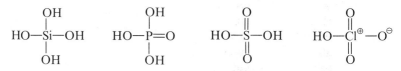

Orthosilicic acid, $Si(OH)_4$, is known, but unlike the other acids cannot be isolated. Hydrolysis of $SiCl_4$ in excess water at pH = 3.2 yields a solution of $Si(OH)_4$, a weak acid that spontaneously polymerizes with loss of water to oligo- and poly-silicic acids:

$$(HO)_3SiOH + HOSi(OH)_3 \xrightarrow{-H_2O} (HO)_3Si—O—Si(OH)_3 \rightarrow \text{etc.}$$

Condensation of $Si(OH)_4$ can occur in one dimension with the formation of chains and rings, or in two dimensions in layers, or in three dimensions to give an insoluble network of SiO_2:

$$n\,Si(OH)_4 \rightleftharpoons (SiO_2)_n\downarrow + 2n\,H_2O$$

The intermediates which arise in the condensation are unstable, but salts of these acids are known among natural silicates.

The condensation is reversible; e.g., SiO_2 dissolves in water to form $Si(OH)_4$, but the saturation concentration is only $7 \cdot 10^{-5}$ mol/l with solid quartz in excess. The solubility of amorphous SiO_2, however, is $2 \cdot 10^{-3}$ mol/l.

Silicates are formed by melting SiO_2 with oxides, hydroxides or carbonates. Depending on the ratio, siloxane bridges are cleaved and anions generated:

$$\geqslant Si—O—Si\leqslant + Na_2O \rightarrow \geqslant Si—O^- Na^+ + Na^+\ {}^-O—Si\leqslant$$

Sodium and potassium silicates are available in aqueous solutions as "waterglass". They are obtained by melting powdered quartz and carbonate at $1300\,°C$ and contain hydrogen silicates like MH_3SiO_4 and $M_2H_2SiO_4$. The solutions are alkaline owing to hydrolysis:

$$SiO_2 + 2\,Na_2CO_3 \rightarrow Na_4SiO_4 + 2\,CO_2\uparrow$$

$$Na_4SiO_4 + 3\,H_2O \rightleftharpoons 4\,Na^+ + 3\,OH^- + H_3SiO_4^-$$

Acidification of aqueous silicate solutions generates free silicic acids which condense to precipitate a gel, $SiO_2 \cdot aq$, which contains silanol groups and can be dehydrated on heating. The simple anions present in silicate minerals are shown in Figure 106.

Orthosilicates have the tetrahedral anion SiO_4^{4-}, isoelectronic with sulfate, as in olivine, Mg_2SiO_4, and garnet, $Ca_3Al_2[SiO_4]_3$. The disilicates include thortveitite, $Sc_2[Si_2O_7]$, and barysilite, $Pb_3[Si_2O_7]$. The anions of thortveitite have linear siloxane bridges.

In the three cyclosilicates, the SiO_4-tetrahedra are cross-linked via common oxygen atoms so that 6, 8 and 12-membered rings with alternating Si—O units are generated. The metasilicates, $[SiO_3]_n^{2n-}$, include α-wollastonite, $Ca_3[Si_3O_9]$, and beryl, $Be_3Al_2[Si_6O_{18}]$. In addition, bicyclic anions also exist.

The disilicate ion is the first of the chain anions, $(SiO_3)_n^{2n-}$ (cf. Fig. 106), which include β-wollastonite, $Ca[SiO_3]$ and enstatite, $Mg[SiO_3]$. These minerals consist of parallel arrangements of negatively charged chain molecules, between which cations are interspersed for

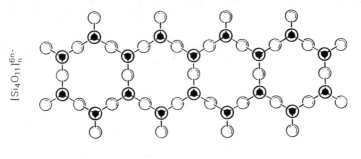
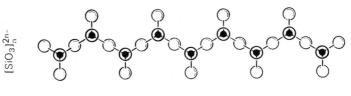
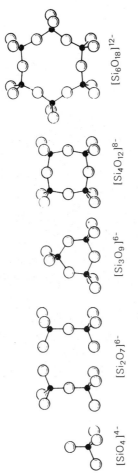
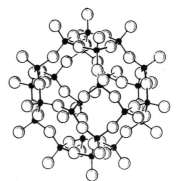
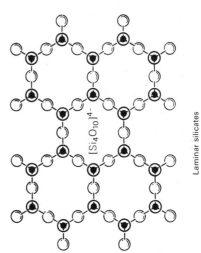

Fig. 106. Structures of Silicate Anions.

electroneutrality. The size and charge of the cations are of secondary importance, thus metasilicates with mixed cations, as in diopside, $CaMg[Si_2O_6]$, and spodumen, $LiAl[Si_2O_6]$, exist.

Linking $(SiO_3)_n$-chains via common oxygen atoms forms band structures which lead to two-dimensional phyllosilicates, as pictured in Figure 106.

A dependence of internuclear distance, $d(SiO)$, on oxygen atom position is observed in the alkali and alkaline earth metal salts of these oligo- and polysilicate ions, e.g., in $(SiO_2)_n$ $\bar{d}(SiO)$ is 1.60 Å, while in the siloxane bridges of silicates it is 1.64 Å, and in the terminal, negatively charged oxygen atoms it is 1.58 Å, indicating the participation in $(p \rightarrow d)$ π-bonding with distribution of charge along the whole framework.

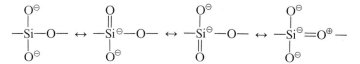

The model is also supported by the valence force constants, $f(Si—O)$.

Three-dimensional linking of SiO_4-tetrahedra gives rise to quartz, tridymite, and cristobalite in which all of the oxygen atoms are linked in siloxane bonds and no

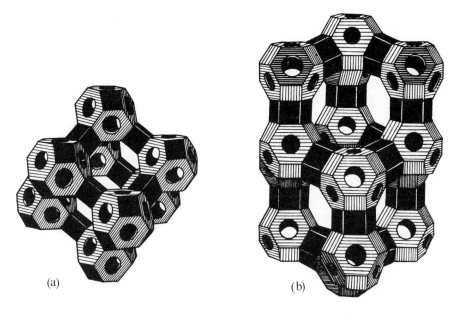

(a)   (b)

Fig. 107. Schematic representations of the lattice of two zeolites (a: type A, b: type X) from O. Grubner, P. Jiru and M. Ralek, Molekularsiebe, Deutscher Verlag der Wissenschaften, Berlin, 1968).

cations are required. Replacement of up to 50% of the silicon atoms by isoelectronic Al^- ions gives the network aluminosilicates which require cations for electro-neutrality, as in orthoclase, $K[AlSi_3O_8]$, and anorthite, $Ca[Al_2Si_2O_8]$. The cations are found in lattice holes formed by the SiO_4 and AlO_4 tetrahedra.

Aluminosilicates are also derived from the oligo- and polysilicate ions, discussed above.

The aluminosilicate zeolites have large-pore structures, as in the mineral faujasite, $NaCa_{0.5}[Al_2Si_5O_{14}] \cdot 10H_2O$, whose structure contains the basket-like units shown for the three-dimensional silicates in Fig. 106. These units are interlinked to form channels as pictured in Fig. 107, in which the cations as well as the water molecules are located. These cations are only electrostatically bound, and thus are freely mobile and exchangeable. Zeolites therefore serve as cation exchangers. The chambers in the faujasite structure, shown in Fig. 107, are accessible by orifices of certain dimensions, e.g., the inner diameters of the cages are 6.6 Å and 11.6 Å, but the respective orifices only 2.5 Å and 7.5 Å. Natural zeolites serve as water insoluble absorbents, as well as molecular sieves. Dehydrating the zeolite *in vacuo* at 400 °C gives hygroscopic material, suitable for drying of gases and solvents, which can be regenerated by heating *in vacuo*. Drying takes place because only the H_2O mole-cule fits into the cages, while larger solvent molecules are not absorbed. Similarly H_2O can be removed from O_2, N_2, Cl_2 and rare gases, in which polarity may play a role. The sieve effect permits the removal of O_2 from argon.

Zeolites are used in the gas chromatographic separation of *ortho-* and *para-*H_2, or H_2, HD, and D_2, etc.

Synthetic molecular sieves, manufactured by hydrothermal synthesis from silicates and aluminates, are custom-tailored for aperture and size of cage, e.g., the Linde "Sieve A", $Na_{12}[Al_{12}Si_{12}O_{48}] \cdot 27H_2O$. Practical use requires that the lattice is not damaged or destroyed during dehydration.

Germanium:

Like silicon, germanium does not form stable oxo acids, but germanates and polygermanates obtained by dissolving GeO_2 in alkali or by melting GeO_2 with metal oxides. Depending on mixing ratios and reaction conditions, orthogermanates, M_4GeO_4, metagermantes, M_2GeO_3, containing chain anions, and polygermanantes, $M_2Ge_2O_5$, are formed. Oligogermanates are also known.

14.8. Glasses

Silicon dioxide and silicates solidify when cooled rapidly in a glassy state. Such silicate glasses are of paramount importance.

A glass has the densest atomic packing with ordering in small domains, but no extended order as in crystals. The comparison of quartz crystals vs. glass is illustra-ted in Fig. 108.

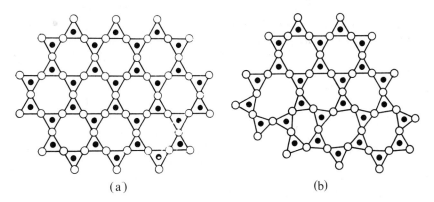

(a)   (b)

Fig. 108.   Schematic, two-dimensional representation of cross-linking of SiO_4 tetrahedra in a) crystalline, and b) glassy SiO_2.

Glass formation is always a consequence of inhibited crystallization. Glasses are formed by solidification of melts, or condensation of vapors. Orientation of the particles in the melt for crystal formation must be inhibited. Various complex polymers are in equilibrium with each other in the SiO_2 or silicon anion melts.

Three-dimensional network glasses are not restricted to SiO_2 and silicates, but are also encountered in the oxides B_2O_3, GeO_2, P_2O_5, As_2O_3 and As_2O_5. Such glasses do not melt sharply, but soften to liquids continiously. Annealing can lead to crystallization.

Glasses are also formed in melts of small molecules, e.g., molten KNO_3 containing 30–47 mol% $Ca(NO_3)_2$ solidifies as a glass, as the cationic charges and sizes interfere with crystallization.

The highly associated hydrogen-bonded liquids, such as concentrated sulfuric or phosphoric acids, glycerine and other alcohols undergo glassy solidification as the molecules lack the necessary mobility for crystal growth.

Glasses are richer in energy than the corresponding crystals as they retain some of the heat of melting, which is released upon crystallization.

14.9. Organosilicon Compounds

Silicon and germanium form compounds:

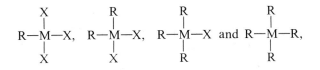

where M = Si or Ge, R = alkyl or aryl, and X = H, Cl, OR, NR$_2$, etc. Silicon tetra-chloride is alkylated or arylated by Li-, Zn-, Hg- or Al-organics, or with Grignard reagents:

$$SiCl_4 + 4RMgX \rightarrow R_4Si + 2MgX_2 + 2MgCl_2 \qquad X = Cl, Br$$

The Si—C bond is thermally and chemically stable, e.g., tetramethylsilane pyrolyzes only at >650°C, and resists hydrolysis by dilute sodium hydroxide, although the silicon atom is vulnerable to the attack by OH$^-$ ions via its d-orbitals, and the conversion to SiO$_2$ and CH$_4$ is thermodynamically feasible. The lack of reaction aries from the low polarity of the Si—C bond as shown below (Δx_E = difference in electronegativities):

	Si—F	Si—CF$_3$	Si—Cl	Si—CH$_3$	Si—H	Ge—H
Δx_E:	2.4	1.8	1.1	0.8	0.5	0.2

The tetrafluoride, SiF$_4$, hydrolyzes vigorously and the Si—CF$_3$ bond is hydrolyzed by cold water; SiCl$_4$ reacts slowly with liquid water, (CH$_3$)$_4$Si is resistant to dilute alkali; SiH$_4$ requires base catalysis, and GeH$_4$ is stable even to 30% alkali. The hydrolysis rate is thus a function of bond polarity. Easy access of the nucleophile to the center atom is not sufficient to initiate the reaction. The inertness of the Si—CH$_3$ bond has practical significance for the silicones which are polymeric siloxanes:

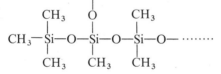

The starting material for silicone polymers are the methylchlorosilanes, which are prepared from elemental silicon in presence of copper metal as catalyst (Rochow synthesis):

$$6CH_3Cl + 3Si \xrightarrow{\text{290–340°C}} CH_3SiCl_3 + (CH_3)_2SiCl_2 + (CH_3)_3SiCl$$

Hydrolysis of these chlorosilanes leads to mono-, di-, or tri-functional silanols which condense like orthosilicic acid to form chemically and thermally stable siloxane bridges:

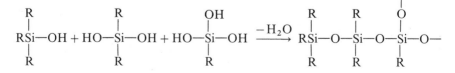

The silanols are chain terminating groups, silanediols are chain members, and silanetriols are chain branching groups. Mixing the three components permits adjustment of the degree of polymerization to give fluids of various viscosities, oils, or elastomers, which owing to their stability find application as lubricating and insulating materials, gaskets, laquers, etc. The $\angle$ SiOSi in the organopolysiloxanes are 130–140° and the internuclear distances, d(SiO), are 1.64 Å, indicating weaker bonds than those in SiO_2.

If methylchlorosilanes are hydrolyzed in aqueous HF, the corresponding methylfluorosilanes are formed via silanol intermediates:

$$CH_3SiCl_3 + 3\,HF \xrightarrow{H_2O} CH_3SiF_3 + 3\,HCl$$

Gaseous CH_3SiF_3 reacts with aqueous KF to give a hydrolysis-resistant methylpentafluorosilicate:

$$CH_3SiF_3 + 2\,KF \xrightarrow{H_2O} K_2[CH_3SiF_5]$$

14.10. Other Silicon Compounds

Silicon Carbide:
Reduction of SiO_2 with coke in an electric furnace gives dark colored silicon carbide, SiC, rather than elemental silicon. Pure, colorless SiC crystallyzes in the diamond lattice, and like Si is a semi-conductor. It is thermally stable and acid resistant, and has a hardness like diamond. Silicon carbide is used as an abrasive under the name "carborundum" and in resistance units for electrical furnaces.

Silicon Sulfide:
The sulfides SiS and SiS_2 as well as selenides and tellurides corresponding to the oxides SiO and SiO_2 are known. The disulfide is formed by fusing the elements, or by double decomposition of SiO_2 and Al_2S_3 at 1100 °C. In contrast to the three-dimensional structure of SiO_2, colorless SiS_2 consists of chains of tetrahedral silicon atoms in spirocyclic linkages:

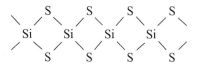

Owing to the weaker donor strength of the sulfur, the SiS bonds are single without (p $\rightarrow$ d) π contribution and SiS_2 is more reactive than SiO_2, hydrolyzing to $SiO_2 \cdot$ aq and H_2S. The chain of SiS_2 units gives rise to fibrous colorless crystals.

Heating SiS_2 *in vacuo* to 850°C or passing CS_2 vapor at 2 torr and 1000°C over elemental silicon forms monomeric SiS which condenses on cold surfaces as a red glass, $(SiS)_n$.

Silicides:

Silicon forms silicides with strong electropositive metals in which the metal cations are bound to monomeric or polymeric anions as in Mg_2Si, Ca_2Si, CaSi, and $CaSi_2$. The calcium silicide, Ca_2Si, contains Si^{4-} ions, CaSi has Ca^{2+} and Si_n^{2n-} chains (isoelectronic with sulfur chains of comparable length), and $CaSi_2$ has a layer structure, consisting of parallel puckered Si_n^{n-} layers interspersed with Ca^{2+} ions. The silicon layers consist of fused Si_6 rings which are also present in diamond lattice (Figure 103).

Chlorination of $CaSi_2$ yields a highly reactive silicon:

$$CaSi_2 + Cl_2 \xrightarrow{20-40°C} 2\,Si(s.) + CaCl_2$$

The product ignites under water and burns to SiO_2 and H_2, forms chlorosilanes Si_nCl_{2n+2} (chains), Si_nCl_{2n} (rings), and Si_nCl_{2n-2} (bicyclics, e.g., $Si_{10}Cl_{18}$) via $(SiCl)_n$ with excess Cl_2, and undergoes alcoholysis in methanol to form $Si(OCH_3)_4$ and H_2.

The layered, graphite-like $(SiCl)_n$ which is also accessible from $CaSi_2$ and ICl, can be fluorinated with SbF_3 to $(SiF)_n$, and hydrogenated with $LiAlH_4$ to $(SiH)_n$. Amorphous brown hydrides $SiH_{0.7-0.9}$ are also generated from $CaSi_2$ and CaSi by action of acids. In contrast, Ca_2Si forms SiH_4.

15. Boron

15.1. General

Boron is the only non-metal in the IIIrd group, and its chemistry differs completely from that of its higher homologs or other non-metals. Some similarities with silicon are noted.

Boron is relatively rare; however, it is concentrated in the earth's crust in extensive deposits, exclusively as mixed oxides such as borax, $Na_2B_4O_5(OH)_4 \cdot 8H_2O$, and kernite, $Na_2B_4O_5(OH)_4 \cdot 2H_2O$.

Boron consists of ^{10}B (19.6%) and ^{11}B (80.4%)[15] with nuclear spins of $I = \frac{3}{2}$ for ^{11}B and $I = 3$ for ^{10}B. ^1H- and ^{11}B-NMR-spectroscopy is important for the structure elucidation of boranes.

Boron compounds are used in glass compositions, soaps, cleansing agents and in soldering and welding.

15.2. Bond Relationships

The boron atom has an electron configuration of $2s^2p$. Only unstable compounds are derived from this ^2P state, e.g., boron monofluoride, BF, discussed on p. 248. Most boron compounds are based on the excited ^4P state of the boron atom which has three unpaired electrons and one unoccupied orbital:

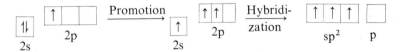

Boron is thus a trivalent element, forming BCl_3, $B(CH_3)_3$, $B(OH)_3$, etc., in which the central atom is sp^2-hybridized with trigonal planar coordination and an unoccupied p_z orbital of similar orbital energy which can accept two electrons. Hence, BX_3 compounds are Lewis acids, resembling SiX_4, in which the d-orbitals are responsible for the Lewis acidity.

15 The isotope ratio $^{11}B/^{10}B$ varies depending on the origin of boron ores between 3.93 and 4.14, or 79.7 to 80.6% ^{11}B.

The relative strength of the BX_3 Lewis acids is indicated by the enthalpy change in adduct formation, as with trimethylamine:

$$(CH_3)_3N + BX_3 \rightleftharpoons (CH_3)_3\overset{\oplus}{N}-\overset{\ominus}{B}X_3$$

The sequence of enthalpies is:

$$BH_3 > BBr_3 > BCl_3 > BF_3 > BH_2(CH_3) > BH(CH_3)_2 > B(CH_3)_3$$

The greatest acidity is expected for BF_3 whose extremely polar bonds ($\Delta x_E = 2.1$) should result in a highly positive charge at the boron atom, and increase its acceptor strength. The fluorine atoms, however, have free electron pairs in orbitals with the same symmetry as the p-acceptor orbital of boron with which they can overlap to give a coordinate (p → p) π-bond and diminish the partial charges on all atoms (p. 111). SCF calculations give an occupation with 0.3 electron for the p_π-orbital of boron and partial charges of 1.4e at the boron and $-0.47e$ at each fluorine atom. Since Cl- and Br-atoms are poorer donors, the above Lewis acidity sequence results. In BH_3, which arises in the dissociation of diborane, π-bonding is not possible, and BH_3 is the strongest BX_3 Lewis acid. The lesser acidity of the organic substituted boranes arises from steric hindrance. Larger alkyl groups reduce acceptor strength drastically.

The occupation of the vacant p-orbital in BX_3-compounds by reaction with σ-donors is accompanied by a change in boron hybridization from sp^2 to sp^3 to give T_d symmetry as in BF_4^-. The complexes may be neutral or anionic or cationic:

$$BH_4^- \quad [(CH_3)_3N\,BH_3] \quad [(C_5H_5N)_2BH_2]^+ \quad [(C_5H_5N)_4B]^{3+}$$

Salts with isolated B^{3+} ions do not exist, since the energy required cannot be compensated by forming an ionic lattice.

The BX_3 compounds hydrolyze to form orthoboric acid:

$$BX_3 + 3H_2O \rightarrow B(OH)_3 + 3HX$$
$$X: F, Cl, OCH_3, CH_3, N(CH_3)_2, I, H(B_2H_6)$$

In contrast, the tetrahedral ions BX_4^- are resistant to hydrolysis like the isoelectronic CX_4 compounds, e.g., $NaBF_4$ dissolves in water without decomposition, KBH_4 is stable in air, and the complex $BCl_3N(CH_3)_3$ is stable in boiling water. The hydrolysis of BX_3 compounds goes through an intermediate H_2O complex which is not possible in BX_4^- compounds whose valence orbitals are occupied:

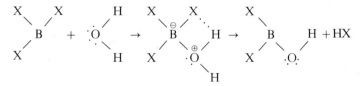

The intermediate BF_3OH_2 can be isolated from the BF_3 reaction.
The amminolysis of the boron trihalides to $B(NR_2)_3$ proceeds similarly, and BX_3NHR_2, BX_2NR_2, and $BX(NR_2)_2$ intermediates can be isolated.

Comparsion of the Chemistry of Boron with that of the other Non-Metals:
Boron has more valence orbitals than valence electrons. It can only form three 2-center 2-electron bonds plus one additional coordinate bond, nevertheless, boron assumes coordination numbers from 1 to 8, unique among the non-metals:

Coordination number:	1	2	3	4	5 6 7 8
Compound:	BF	HOBO	BF_3	BF_4^-	crystalline boron, boranes, carboranes

Coordination numbers 5, 6, 7 and 8 are realized by the formation of many-center bonds which will be treated in the sections on elemental boron and boranes.
Boron has little tendency to form double bonds. Boron-boron double bonds do not exist, and $B{=}O$ bonds are only present in high temperature molecules such as $HO{-}B{=}O$. Boron also forms few catenated chains and rings with the $>B{-}B<$ unit.
The electropositive boron atom ($X_E = 2.0$ on the Allred-Rochow scale) forms strong, polar B—F, B—O, B—Cl, and B—N bonds. The hydrogen is negatively polarized in the B—H bond, and the boranes resemble the silanes rather than the alkanes in their reactions, such as hydrolysis, and oxidation in air. Their structure, however, is different from other non-metal hydrides.
The only stable boron oxide, B_2O_3, is similar to SiO_2 in being a network solid, and reacts with metal oxides and hydroxides to form borates containing B—O—B bonds, which like the Si—O—Si bonds are stabilized by double bond contributions:

In contrast to the silicic acids, several boric acids can be isolated.

15.3. Elemental Boron

Elemental boron crystallizes in three-dimensional networks in which certain structural subunits of the unit cells can be recognized (cf. Table 52).

Preparation:
Large scale reduction of B_2O_3 or $B(OH)_3$ with Mg leads to products of low purity (ca. 90%).

Table 52   Modifications of Boron

Name	Atoms per Unit Cell	Structural Units
α-rhombohedral boron	12	One B_{12}-icosahedron
β-rhombohedral boran	105	One B_{84}-unit, two B_{10}-groups and one B-atom
α-tetragonal boron	50	four B_{12}-icosohedra and two B-atoms
β-tetragonal boron		

Reduction of BCl_3 or BBr_3 with H_2 or pyrolysis of B_2H_6 or BI_3 yields pure boron.

$$2 BBr_3 + 3 H_2 \rightarrow 2 B(s.) + 6 HBr$$

Pyrolysis of BI_3 on a tantalum wire at 800–1100 °C produces crystals of α-rhombohedral boron.

Depending on the temperature and other reaction conditions, BBr_3 reduction produces amorphous boron (800–1250 °C), α-rhombohedral boron (1000–1250 °C), α-tetragonal boron (1100–1300 °C) or the thermodynamically stable β-rhombohedral boron (>1330 °C) which is also obtained by crystallizing a boron melt (m.p. 2250 ± 50 °C).

Depending on the modification, boron forms dark red, brown, or glossy black, semiconducting crystals, which are extremely hard [β-rhombohedral boron = 9.3 (Mohs' scale)].

The various modifications are monotropic. Their relationships in the phase diagram are uncertain since all transitions are kinetically hindered owing to the complex lattice structures. Hot boron is corrosive, forming borides with metals, and reduces oxides, even CO or SiO_2, making preparation of high purity boron difficult. Boron at low temperatures is inert to attack by HF or HCl, but it is oxidized to boric acid by HNO_3 or aqua regia. Fusing with alkali hydroxide produces the corresponding borates and hydrogen.

Crystal Structures:

Of all crystalline modifications α-rhombohedral boron has the simplest structure with a unit cell containing 12 equivalent atoms in an icosahedron as shown in Fig. 109. Each atom has five nearest neighbors located in the base of a pentagonal pyramid of which it forms the apex. Icosahedra are quasi-double pyramids with their base planes turned to each other at an $\angle = 36°$ from coincidence. A C_5 symmetry axis passes through each boron atom, which lie at 1.77 Å from one another. In the α-rhombohedral boron, the B_{12}-icosahedra are linked to each other so that six of the 12 atoms, marked "r" in Figure 109(a), have next-nearest neighbors at a distance of 1.7 Å which belong to an adjacent icosahedron and are located on a C_5 axis of the first icosahedron. The six boron atoms marked "e" in the figure have two additional neighbors belonging to different icosahedra, which form an equilateral triangle having an edge of 2.0 Å (Figure 109(b)). Hence, half the atoms have seven and half six coordination.

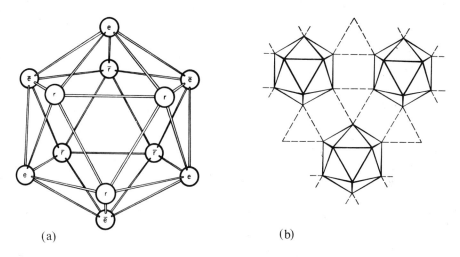

(a)   (b)

Fig. 109.   (a) B_{12}-icosahedron, (b) linking of icosahedra in α-rhombohedral boron. After E. L. Muetterties, The Chemistry of Boron and Its Compounds, J. Wiley & Sons, New York 1967.

The unit cell of α-tetragonal boron shown in Figure 110 contains four icosahedra which are linked directly or via tetrahedral boron atoms with other icosahedra. The internuclear distances are 1.81 Å within the icosahedra, 1.68 Å between icosahedra, and only 1.60 Å to the tetrahedral boron atoms.

The structure of the thermodynamically stable β-rhombohedral boron is complex, consisting of symmetrical units as in the α-form with 20% of the atoms having eight-coordination.

Figure 110 shows much free space between the icosahedra, permitting foreign atoms to be incorporated such as the metals Al or Ga and the non-metals C, Si, N or P to form boron-rich metal borides, as well as boron carbides, silicides, nitrides and phosphides without distortion of the icosahedra. The cross-liking is responsible for the hardness of boron and its derivatives.

Bond Mechanisms:

The atoms in crystalline boron have coordination numbers of 4 to 8. Since boron has only three valence electrons, all atoms must participate in many-center bonds

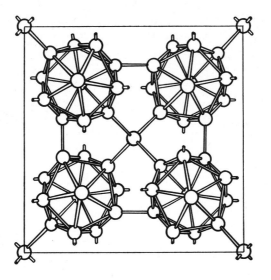

Fig. 110.   Structure of α-tetragonal boron. The bisphenoidal array of four icosahedra in one unit cell is projected down the tetragonal *c*-axis. The distortion of some of the inter-icosahedral bonds from the preferred pentagonal pyramidal coordination geometry is evident in the projection. After E. L. Muetterties, The Chemistry of Boron and Its Compounds, J. Wiley & Sons, New York 1967.

(cf. HF_2^- p. 172, I_3^- p. 253). From the four orbitals available in boron a maximum of eight atoms can be bonded with 3-center bonds as shown in Figure 111.

In (a) the atomic orbitals of three symmetrically arranged sp^3-hybridized boron atoms overlap. Two of the orbitals are half-occupied, and the third is empty, hence, a bonding electron pair results. The two anti-bonding orbitals remain empty.

In (b) the p-orbital of the central atom overlaps the hybrid orbitals of both neighbor atoms to give a bonding, a nonbonding and an antibonding orbital. The two electrons occupy the bonding MO only. An open 3-center 2-electron bond results in (b) and a closed three-center bond in (a). Both bond types are encountered in certain boranes, and in α-rhombohedral boron between the icosahedra. However, within the B_{12}-subunits many-center bonds of a higher order must be assumed in order to arrive at a satisfactory explanation for the stability of the icosahedra.

Each boron atom of the icosahedron has four atomic orbitals, one of which is used to make a terminal bond, and the remaining three used within the B_{12} framework. The MO treatment takes into account 36 atomic orbitals within the icosahedron. For optimum overlap the boron atoms are assumed to be sp-hybridized with the sp-orbital axes on the C_5 axis through each boron atom and with one radial lobe pointing to the center of the icosahedron. Two p-orbitals remain whose axes are tangential to the nearly spherical icosahedron. The twelve sp-hybrid orbitals overlap in the center of the icosahedron and the 24 p-orbitals on the surface. As

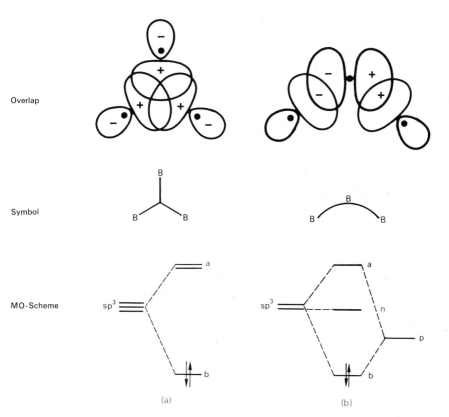

Fig. 111.   Two types of two-center bonds between three boron atoms, as they are present in boranes, carboranes and boron modifications.

shown in the diagram (Figure 112) 13 bonding and 23 antibonding MO's are formed, and the B_{12}-icosahedron achieves maximum stability with 26 valence electrons. Thus of 36 electrons of the 12 boron atoms, 10 remain for terminal bonding. In agreement, $B_{12}H_{12}$ does not exist since only 10 electrons are available for two-center bonds to hydrogen atoms, however, the stable ion $B_{12}H_{12}^{2-}$ (p. 374) containing the B_{12}-icosahedron is known, in which the two additional electrons allow each boron atom to bind one hydrogen.

Accordingly, only half the bonds between the icosahedra in α-rhombohedral boron can be two-center, requiring six electrons per icosahedron. These bonds originate from the atoms marked "r" in Fig. 109(a). The remaining four electrons make it possible for the other six atoms to enter into six closed, three-center bonds to each two adjacent icosahedra.

The single atoms in α-tetragonal boron, owing to their small internuclear distances,

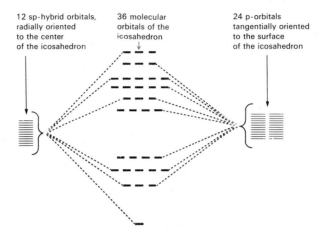

| 12 sp-hybrid orbitals, radially oriented to the center of the icosahedron | 36 molecular orbitals of the icosahedron | 24 p-orbitals tangentially oriented to the surface of the icosahedron |

Fig. 112.   Molecular orbital diagram for the many-center bonds in a B_{12}-icosahedron.

are apparently bonded to adjacent icosahedra via two-center bonds. The direct bonds between icosahedra must, therefore, be partly two- and partly three- center bonds.

15.4. Boranes and Hydridoborates

15.4.1. General

Boron forms at least 17 binary compounds with hydrogen, none analogous to the hydrides of C, Si, N, P, O, or S. Two series are known:

$$B_nH_{n+4}: B_2H_6, B_4H_8, B_5H_9, B_6H_{10}, B_8H_{12}, B_{10}H_{14}, B_{16}H_{20}, B_{18}H_{22}^*$$

$$B_nH_{n+6}: B_4H_{10}, B_5H_{11}, B_6H_{12}, B_9H_{15}^*, B_{10}H_{16}$$

The starred compounds occur in two isomeric forms, and other stable (e.g., B_8H_{18} and $B_{20}H_{16}$) and unstable boranes exist. The parent borane, BH_3, is unstable but plays a role as intermediate.

The nomenclature of the boranes is illustrated with a few examples:

BH_3: Borane (3) B_2H_6: Diborane (6) B_4H_{10}: Tetraborane (10)

The prefix indicates the number of boron atoms, while the numeral in parentheses is the number of hydrogens. When there is no ambiguity as in diborane, the number may be omitted.

Hydridoborates are salts which contain boron-hydrogen anions, the simplest being BH_4^- (boranate ion). Complex ions with from 2 to 20 boron atoms also exist.

15.4.2. Diborane

Gaseous diborane is easy to prepare by hydrogenating BX_3 compounds:

$$4\,BCl_3 + 3\,LiAlH_4 \xrightarrow{\text{ether}} 2\,B_2H_6\uparrow + 3\,LiCl + 3\,AlCl_3$$

$$4\,BF_3 + 3\,NaBH_4 \xrightarrow[\text{ether}]{25\,°C} 2\,B_2H_6\uparrow + 3\,NaBF_4$$

Lithium or sodium hydride can also be employed. Hydridoborates decompose with hydrogen chloride or phosphoric acid:

$$2\,LiBH_4 + 2\,HCl \rightarrow B_2H_6 + 2\,H_2 + 2\,LiCl$$

In these reactions the dimeric B_2H_6 is formed in equilibrium with BH_3:

$$2\,BH_3 \rightleftharpoons B_2H_6$$

The dissociation energy of B_2H_6 is about 150 kJ/mole. The equilibrium at 25 °C is completely on the side of the dimer. In the absence of a suitable donor the highly Lewis acidic BH_3-molecule (p. 362), reacts with itself to give each boron atom an electron octet by formation of many center bonds:

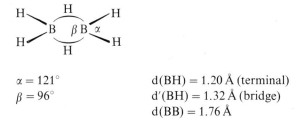

$\alpha = 121°$ $d(BH) = 1.20\ \text{Å (terminal)}$
$\beta = 96°$ $d'(BH) = 1.32\ \text{Å (bridge)}$
                                         $d(BB) = 1.76\ \text{Å}$

The boron atoms are tetrahedral with terminal angles indicating sp^3- to sp^2-hybridization. Models of bridge formation can be constructed using either hybridization. Starting with sp^3-hybridized boron and considering two-center bonds to the terminal hydrogen atoms based on their internuclear distances, one sp^3-orbital remains available on each boron atom for the formation of the bridge bonds by overlap with the hydrogen ls-orbitals as illustrated in Figure 113 to form 3-center 2-electron bonds. The sp^3-orbitals overlap not only with the s-orbital, but also with each other to form a partial B—B bond, as reflected in the small internuclear distance, $d(BB)$, corresponding to a B—B single bond. These relationships are mirrored in the MO diagram in Figure 114, showing the formation of a bonding, an antibonding and (for the B—H—B bridge) a non-bonding MO.

Two electrons per BHB-bridge suffice to fill the bonding MO. Additional electrons would weaken the bond.

The electron distribution in B_2H_6 according to this model corresponds to that in C_2H_4 with sp^3-hydridized carbon atoms and two arced bonds between them (p. 95).

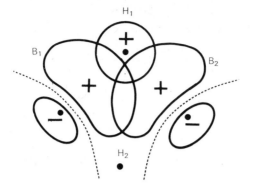

Fig. 113. Overlap of the hydrogen ls-orbital with two sp³ hybrid orbitals of the boron atoms in diborane.

The large terminal angles are rationalized by the theory of electron pair repulsion (p. 117) since the electrons in the bridge bonds require less space.

From the position of the second hydrogen in Fig. 113, it is apparent that the BHB bridges are so close that all six atomic orbitals overlap ($4 \times sp^3$ and $2 \times s$) and, therefore, the interaction can be considered as 4-center 4-electron bonding.

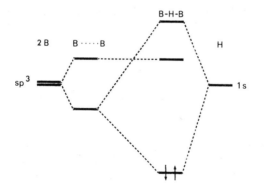

Fig. 114.   MO-Diagram for the bridge bonds in diborane B_2H_6.

Just as the σ-π-model is an alternative for the arc bond model for C_2H_4 (p. 95), an alternative model based on sp^2-hybridized boron atoms involves the overlap of six atomic orbitals: two boron sp^2-hybrid orbitals along the internuclear axis, a ls orbital from each hydrogen atom above and below it, and two boron p-orbitals in the same plane which have π-symmetry with the internuclear axis (Figure 115). MO calculation for this model likewise gives two bonding and four antibonding MO's so that four electrons fill the bonding orbitals.

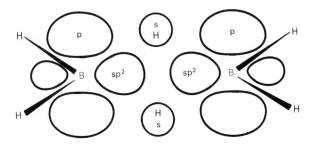

Fig. 115.   Orbital overlap in the B_2H_6 molecule with sp^2-hybridization of the boron atoms.

The bridge bonds can be cleaved by Lewis bases to give borane adducts:

$$B_2H_6 + 2D \rightarrow 2\overset{\oplus}{D}-\overset{\ominus}{B}H_3$$

$$D: CO, PH_3, PF_3, PR_3, NR_3, (CH_3)_2S$$

Borane carbonyl H_3BCO is formed at $-45°C$. The salt $Li_3[P(BH_3)_4]$ in which four BH_3-molecules are bound to a phosphide ion is prepared by reaction of H_3BPH_3 with $LiBH_4$ or B_2H_6/C_4H_9Li. The liquid adduct $H_3BS(CH_3)_2$ can be used to store BH_3. Stronger Lewis bases can displace weaker ones like dimethyl sulfide.

Diborane reacts with hydride ions in analogy to its reaction with Lewis bases:

$$B_2H_6 + 2MH \xrightarrow{\text{ether}} 2MBH_4$$

$$M: Li, Na, K$$

15.4.3. Higher Boranes

The colorless, hydrolytically sensitive higher boranes ignite in air and range in thermal stability from iso-B_9H_{15} unstable at $-30°C$ to $B_{20}H_{16}$ which melts at $197°C$. Tetraborane(10) is a gas, but the remaining boranes are liquids or solids.
Preparation:
The syntheses are based on mutual rearrangements following unknown reaction mechanisms:
a) Controlled pyrolysis of B_2H_6 or other boranes

$$2B_2H_6 \xrightarrow{120°C} B_4H_{10} + H_2$$

$$5B_2H_6 \xrightarrow{200-240°C} 2B_5H_9 + 6H_2$$

$$2i\text{-}B_9H_{15} \xrightarrow{>-30°C} B_8H_{12} + B_{10}H_{14} + 2H_2$$

B_4H_{10}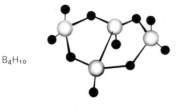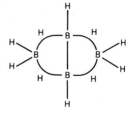

B_5H_9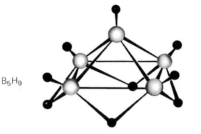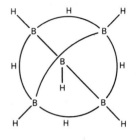

B_6H_{10}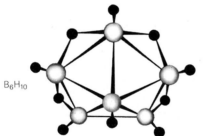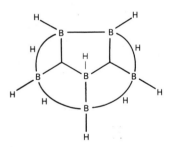

$B_{20}H_{16}$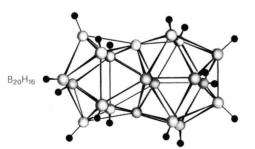

Fig. 116.   Structures and structural formulae of the boranes B_4H_{10}, B_5H_9, B_6H_{10} and $B_{20}H_{16}$.

b) Lewis base-catalyzed rearrangements

$$4\,B_4H_{10} \xrightarrow{(CH_3)_3N} 3\,B_2H_6 + 2\,B_5H_9 + 2\,H_2$$

$$B_5H_{11} \xrightarrow{(CH_2)_6N_4} B_9H_{15}$$

c) Acid decomposition of Mg_3B_2 to generate B_2H_6 with B_4H_{10}, B_5H_9, B_5H_{11}, B_6H_{10} and $B_{10}H_{14}$.

d) Loss of hydrogen in a glow discharge

$$2\,B_5H_9 \rightarrow B_{10}H_{16} + H_2$$

$$2\,B_{10}H_{14} \rightarrow B_{20}H_{16} + 6\,H_2$$

e) Protonation of hydridoborates

$$K\,B_9H_{15} + HCl\,(liquid) \xrightarrow{-80\,^\circ C} B_9H_{15} + KCl + \tfrac{1}{2}H_2$$

$$B_{20}H_{18}{}^{2-} \xrightarrow{+H^+, H_2O} n\text{-}B_{18}H_{22} \text{ and } i\text{-}B_{18}H_{22}$$

Procedures a) and b) find most universal use.

Structures:
The complicated borane molecules often display high symmetry. They can be constructed from the following elements:

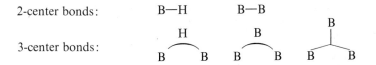

The structures of B_4H_{10}, B_5H_9, B_6H_{10}, and $B_{20}H_{16}$ are shown in Fig. 116. Topological formulae are also given for the first three molecules[16].
Substitution products are derived from the higher boranes, in which the terminal hydrogen atoms are replaced by halogen atoms, organic substituents, etc.

15.4.4. Hydridoborates

The parent boranate ion, BH_4^-, is isoelectronic with CH_4 and NH_4^+, and like them is tetrahedral. Boranates are generated by hydrogenation of BX_3 compounds, or double decomposition:

16 The bonding in boranes is discussed in K. Wade, Electron Deficient Compounds, Nelson, London (1971).

$$2\,LiH + B_2H_6 \rightarrow 2\,LiBH_4$$

$$4\,NaH + B(OCH_3)_3 \rightarrow NaBH_4 + 3\,NaOCH_3$$

$$4\,LiH + BF_3 \rightarrow LiBH_4 + 3\,LiF$$

$$AlCl_3 + 3\,NaBH_4 \rightarrow Al(BH_3)_3 + 3\,NaCl$$

Diborane is very sensitive toward oxygen and water, but the salts $NaBH_4$ and KBH_4 resist decomposition in air. $Al(BH_3)_3$ is a volatile liquid.

Sodium borohydride is a water soluble hydrogenation agent. It reacts with B_2H_6 to form various higher hydridoborates:

$$2\,NaBH_4 + B_2H_6 \xrightarrow{\;0\,°C\;} 2\,NaB_2H_7$$

$$NaBH_4 + B_2H_6 \xrightarrow[2\;bar]{25-35\,°C} NaB_3H_8 + H_2$$

$$2\,NaBH_4 + 5\,B_2H_6 \xrightarrow[(C_2H_5)_3N]{180\,°C} Na_2B_{12}H_{12} + 13\,H_2$$

The higher hydridoborates with charges between -1 and -4 contain the same structural elements as the boranes. The following nomenclature is used:

$B_2H_7^-$: Heptahydridodiborate (-1) $B_6H_6^{2-}$: Hexahydridohexaborate (-2)

Of the ca. 20 hydridoborates, the ions $(BH)_n^{2-}$ with $(n = 6$ to $12)$ contain closed, quasi-aromatic, polyhedral boron atom frameworks, especially the thermally and hydrolytically stable $B_{10}H_{10}{}^{2-}$ and $B_{12}H_{12}{}^{2-}$.
The $B_{12}H_{12}{}^{2-}$ salts are generated by heating diborane/boranate mixtures in the presence of Lewis bases, while $B_{10}H_{10}{}^{2-}$ is obtained from $B_{10}H_{14}$ in boiling xylene:

$$B_{10}H_{14} + 2(C_2H_5)_3N \xrightarrow{\;-H_2\;} B_{10}H_{12}(NR_3)_2 \rightarrow (R_3NH)_2B_{10}H_{10}$$

An alternative method is tetraethylammonium boranate thermolysis:

$$10(C_2H_5)_4NBH_4 \xrightarrow{\;190\,°C\;} [(C_2H_5)_4N]_2B_{10}H_{10} + 8(C_2H_5)_3N + 11\,H_2 + 8\,C_2H_6$$

Heating the products in sodium hydroxide gives $Na_2B_{10}H_{10}$. The structures of the $B_{10}H_{10}{}^{2-}$ and $B_{12}H_{12}{}^{2-}$ ions are shown in Figure 117. In $B_{10}H_{10}{}^{2-}$, the boron atoms form a polyhedron of two square pyramids, while $B_{12}H_{12}{}^{2-}$ contains a B_{12} icosahedron. Each boron atom carries a terminal hydrogen atom attached by a 2-center bond. In $B_{12}H_{12}{}^{2-}$ the borons and hydrogens are equivalent. The proton or ^{11}B-NMR spectrum of an isotopically pure ^{11}B sample consists of a doublet arising from spin-spin coupling between boron and adjacent hydrogen nuclei.

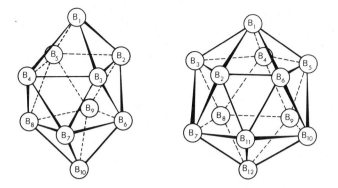

Fig. 117. Structures of the B_n frameworks in the ions $B_{10}H_{10}^{2-}$ and $B_{12}H_{12}^{2-}$. The atoms are numbered for identifying substituents.

In $B_{10}H_{10}^{2-}$ the two apical and eight equatorial boron atoms are equivalent, and the ^{11}B-NMR spectrum of an isotopically pure ^{11}B sample consists of two doublets with an intensity ratio of 1:4.

The $B_{10}H_{10}^{2-}$ and $B_{12}H_{12}^{2-}$ ions are chemically similar and distinguished from other hydridoborates by their stability. The water soluble alkali metal salts are kinetically stable to 600° and 800 °C, respectively, and resist alkali or 3 N HCl at 100 °C, and even mild oxidizing agents, despite the thermodynamically favored hydrolysis to boric acid and hydrogen.

The origin of the stability lies in the quasi-aromatic electron delocalization in the symmetrical B_n-framework. Boron removal with the resulting symmetry reduction requires high activation energy. Closed structures are given the prefix "*closo*", open structures the prefix "*nido*" (= nest), e.g., $B_{12}H_{12}^{2-}$ is dodecahydro*closo*-dodecaborate-(-2)-ion. The *nido*-boranes include B_4H_{10}, B_5H_9, and B_6H_{10}, but not $B_{20}H_{16}$ (cf. Fig. 116).

As in other aromatic systems, substitution reactions on $B_{10}H_{10}^{2-}$ and $B_{12}H_{12}^{2-}$ maintain the B_n-framework. Both ions react with Cl_2, Br_2 and I_2 to form halogenated derivatives (e.g., $B_{12}Cl_{12}^{2-}$, $B_{12}Br_{12}^{2-}$, $B_{12}I_{12}^{2-}$). The $B_{10}H_{10}^{2-}$ anion reacts stepwise with nitrous acid and NaBH$_4$ to yield the colorless, sublimable inner diazonium salt $B_{10}H_8(N_2)_2$, with two N_2 molecules bound to the apical boron atoms which undergoes displacement reactions at these positions with NH$_3$, pyridine, CO, H$_2$S, CH$_3$CN, etc.

15.5. Carboranes

Replacement of framework boron atoms in the boranes and hydridoborates by other non-metals gives rise to carbo-, phospho-, thioboranes, etc. Of greatest importance are the carboranes.

The best investigated carboranes, $C_2B_{n-2}H_n$, derived formally from the *closo*-hydridoborates, $(BH)_n^{2-}$, by the replacement of two BH^- groups by isoelectronic CH groups to give closed, neutral, isomeric molecules, e.g., dicarbo-*closo*-decacarborane, $C_2B_{10}H_{12}$, derived from the icosahedral $B_{12}H_{12}^{2-}$ ion. Fig. 118 shows the three possible isomers for $C_2H_{10}H_{12}$.

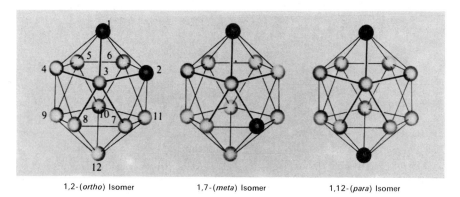

1,2-(*ortho*) Isomer   1,7-(*meta*) Isomer   1,12-(*para*) Isomer

Fig. 118.   The three isomeric forms of the carborane $C_2B_{10}H_{12}$.

The carbon atoms can be 1,2-(*ortho*-), 1,7-(*meta*-), or 1,12-(*para*-). The numbering begins at one apex, counting the next plane of five atoms clockwise, continues likewise for the next plane, finishing with the other apex (cf. Fig. 118).

The 1,2-$C_2B_{10}H_{12}$ isomer is formed in the reaction of $B_{10}H_{14}$ with acetylene in the presence of donors (D), like dialkyl sulfides, nitriles, phosphines, tertiary amines, etc.:

$$B_{10}H_{14} + RC \equiv CR \xrightarrow{D} R-\underset{\underset{\textstyle B_{10}H_{10}}{\diagdown O \diagup}}{C-C}-R + 2H_2$$

The 1,2-$C_2B_{10}H_{12}$ isomer is stable to ca. 470 °C, and transforms into the 1,7-isomer at >470 °C, which in turn changes at 615 °C to yield the thermodynamically stable 1,12-isomer.

The mechanism for 1,2 to 1,7-isomer rearrangement proceeds through a cubic-octahedral intermediate in which the framework atoms are displaced only slightly, as in Figure 119. The C_2B_{10} framework decomposes at >630 °C.

The carboranes $C_2B_{10}H_{12}$ like the *closo*-hydridoborates resist attack by boiling water, or by oxidizing or reducing agents, but nucleophilic reagents degrade the C_2B_{10} cage, and carboranes with 9 to 6 boron atoms are formed, e.g., in the formation of the *nido*-carborane $C_2B_9H_{13}$:

$$C_2B_{10}H_{12} \xrightarrow{\text{CH}_3\text{O}^{\ominus}/\text{CH}_3\text{OH}} C_2B_9H_{\overline{12}} \begin{cases} \xrightarrow{\text{H}^+} C_2B_9H_{13} \\ \xrightarrow{\text{NaH}} C_2B_9H_{11}{}^{2-} \end{cases}$$

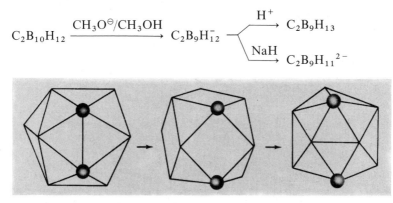

Fig. 119.   Mechanism for the conversion of $1,2\text{-}C_2B_{10}H_{12}$ into $1,7\text{-}C_2B_{10}H_{12}$. After K.Wade, Electron Deficient Compounds, Nelson, London, 1971.

The C_2B_9-framework is derived from the C_2B_{10} icosahedron by loss of one apical (Fig. 117) boron atom to leave unoccupied atomic orbitals at the vacant site which can coordinate with metal atoms like the cyclopentadienyl anion, $C_5H_5^-$. An Fe^{III} complex, $[Fe(C_5H_5)C_2B_9H_{11}]$ is obtained from $FeCl_2$, $Na_2C_2B_9H_{11}$, and NaC_5H_5. Complexes of Fe, Co, and Ni with two $C_2B_9H_{11}$-ligands analogous to ferrocene, are portrayed in Figure 120.

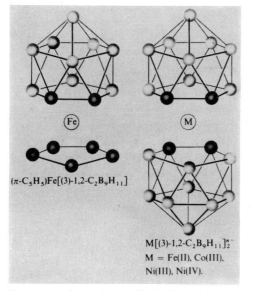

$(\pi\text{-}C_5H_5)Fe[(3)\text{-}1,2\text{-}C_2B_9H_{11}]$

$M[(3)\text{-}1,2\text{-}C_2B_9H_{11}]_2^{n-}$
$M = Fe(II), Co(III),$
$Ni(III), Ni(IV).$

Fig. 120.   Structures of complexes with the ligand $1,2\text{-}C_2B_9H_{11}^{2-}$. The missing apex of the ligand is numbered 3. (After K.Wade, Electron Deficient Compounds, Nelson, London, 1971).

15.6. Halides

The binary boron halides BX_3 and B_2X_4 ($X = F$, Cl, Br, I) are known together with halides of lower halogen content.

Trihalides:
Both BF_3 and BCl_3 are manufactured in quantity. Suitable syntheses are:

$$6\,NaBF_4 + B_2O_3 + 9\,H_2SO_4 \;\rightarrow\; 8\,BF_3\uparrow + 6\,NaHSO_4 + 3\,H_3OHSO_4$$

$$B_2O_3 + 3\,CaF_2 + 6\,H_2SO_4 \;\rightarrow\; 2\,BF_3\uparrow + 3\,CaSO_4\downarrow + 3\,H_3OHSO_4$$

$$B_2O_3 + 3\,C + 3\,Cl_2 \;\xrightarrow{530\,°C}\; 2\,BCl_3 + 3\,CO$$

$$BF_3 + AlCl_3 \;\rightarrow\; BCl_3 + AlF_3$$

$$BF_3 + AlBr_3 \;\rightarrow\; BBr_3 + AlF_3$$

$$B_2O_3 + 3\,C + 3\,Br_2 \;\rightarrow\; 2\,BBr_3 + 3\,CO$$

$$LiBH_4(NaBH_4) + 2\,I_2 \;\xrightarrow{130\,°C}\; BI_3 + LiI + 2\,H_2$$

The colorless trihalides are gaseous (BF_3), liquid (BCl_3, BBr_3) or solid (BI_3). The molecules are of trigonal planar symmetry D_{3h} with zero dipole moment. Unlike AlF_3, BF_3 does not form ionic crystals, although the electronegativity difference between B and F is 2.1. The ionization energy required to form a B^{3+} cation is not compensated by the electron affinities of the fluorine atoms and the lattice energy of an hypothetical ionic BF_3.

As the acid halide of boric acid, BX_3 reacts with hydrogen compounds:

$$>\!B\!-\!X + H\!-\!E\!-\; \rightarrow\; >\!B\!-\!E\!- + HX \qquad E: O, N, S$$

Hydrolysis gives boric acid, $B(OH)_3$, and HX, and BCl_3 and BBr_3 are esterified by alcohols and react with secondary amines to form amides, $B(NR_2)_3$, and with CH_3SH, to form $CH_3S\!-\!BCl_2$.

Ligand exchange results in equilibrium between pairs of trihalides, e.g., BCl_3 and BBr_3:

$$BCl_3 + BBr_3 \;\rightleftharpoons\; BBrCl_2 + BBr_2Cl$$

Trialkylboranes react with the BF_3-etherate to give alkyldifluoroboranes:

$$BR_3 + 2\,[BF_3 \cdot O(C_2H_5)_2] \;\rightarrow\; 3\,BRF_2 + 2\,(C_2H_5)_2O$$

Likewise, BF_3 and $B(OH)_3$ vapor react to give $BF_2(OH)$ and B_2O_3 reacts with BX_3 to form $B_3O_3X_3$, containing a six-membered alternating B_3O_3 ring ($X = F$, Cl). Boron trihalides act as Lewis acids with amines, ethers, nitriles, ketones and other donors. The adducts can form colorless, sublimable crystals [$BCl_3N(CH_3)_3$ (m.p. 243 °C)] or liquids [$BF_3O(CH_3)_2$ (b.p. 126 °C)]. The tetrahaloborate anions, BX_4^-, are generated from BX_3 and ionic halides:

$$BF_3 + KF \rightarrow KBF_4$$

$$4\,BF_3 + 3\,KCl \rightarrow 3\,KBF_4 + BCl_3$$

Small amounts of water converts BF_3 to a solution of oxonium tetrafluoroborate:

$$4\,BF_3 + 6\,H_2O \rightarrow 3\,H_3O^+ + 3\,BF_4^- + B(OH)_3$$

Additional anions like $BF_3(OH)^-$ are also present. The stability of the haloborates increases with cationic size, e.g., $NaBF_4$ decomposes at 384 °C, $CsBF_4$ melts without decomposition at 550 °C. Isolation of the hydrolysis-sensitive ions BCl_4^-, BBr_4^-, BI_4^- requires large cations.

The Lewis acidity of the BX_3 makes them catalysts in organic syntheses, e.g., in the nitration of aromatic compounds with N_2O_5 the formation of nitronium ions is catalyzed by BF_3:

$$O_2N{-}O{-}NO_2 + BF_3 \rightarrow NO_2^{\oplus} + F_3\overset{\ominus}{B}{-}ONO_2$$

Subhalides:
The diboron tetrahalides have 2-center BB-single bonds. B_2F_4 and B_2Cl_4 crystallize as planar B_2X_4 molecules of symmetry D_{2h} with 120° valence angles. Liquid or gaseous B_2Cl_4 has the lower D_{2d} symmetry, however, with the two halves of the molecule twisted by 90°. The barrier to rotation around the B—B axis (7.5 kJ/mol) is compensated for by the lattice energy of the solid. Gaseous B_2F_4 is generated in the fluorination of boron monoxide.

$$2(BO)_n + 2n\,SF_4 \rightarrow n\,B_2F_4 + 2n\,SOF_2,$$

Liquid B_2Cl_4 is obtained in the reduction of BCl_3 with gaseous copper metal:

$$2\,BCl_3 + 2\,Cu \rightarrow B_2Cl_4 + 2\,CuCl$$

Reduction of BCl_3 in an electrical discharge between copper electrodes and with the discharge tube filled with copper wire is also a suitable method.
The difunctional Lewis acid B_2Cl_4 acts as the acid chloride of hypodiboric acid, $B_2(OH)_4$, and undergoes disproportionation:

$$B_2Cl_4 \rightarrow BCl_3 + \frac{1}{n}(BCl)_n$$

Chlorides containing closed B_n frameworks similar to the *closo*-hydridoborates, e.g., light-yellow B_4Cl_4, purple B_8Cl_8, yellow-orange B_9Cl_9, and red, paramagnetic $B_{12}Cl_{11}$ are formed. The boron atoms in B_4Cl_4 form a tetrahedron and in B_8Cl_8 a distorted square antiprism (Fig. 121). The terminal chlorine atoms are bonded by two-center bonds. In B_4Cl_4 this requires four of the 12 valence electrons, leaving eight electrons to form four closed three-center bonds within the B_4-cage, or one per tetrahedral face of three sp^3-hybrid boron atoms.

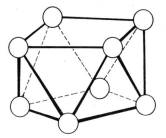

Fig. 121.   Structure of the boron framework in B_8Cl_8 (square antiprism).

15.7. Oxygen Compounds of Boron

The thermally and chemically stable oxygen compounds include boron trioxide, B_2O_3, orthoboric acid, $B(OH)_3$, metaboric acids, $(HBO_2)_n$, and the monomeric and polymeric borates.

Boron Acids:
Boric acid, H_3BO_3, occurs naturally, but is manufactured from borates by acid hydrolysis, e.g., from borax with hydrochloric or sulfuric acid:

$$Na_2B_4O_5(OH)_4 + 2HCl + 3H_2O \rightarrow 4H_3BO_3 + 2NaCl$$

Boric acid is a moderately soluble monoprotic acid, almost undissociated and monomeric in solution. The pK ($= 9.2$) has been derived for the equilibrium:

$$B(OH)_3 + 2H_2O \rightleftharpoons H_3O^+ + B(OH)_4^-$$

Boric acid forms a layer lattice in which planar $B(OH)_3$-molecules are hydrogen-bonded in two-dimensional layers with van der Waals forces acting between the layers (Fig. 122). The scale-like appearence of the colorless crystals reflects this structure.

Like ortho acids of other non-metals, boric acid releases water on heating and undergoes intermolecular condensation via oligo- and polyboric acids to boron trioxide:

$$3B(OH)_3 \xrightarrow[-3H_2O]{90-130\,°C} B_3O_3(OH)_3 \xrightarrow{140-160\,°C} (HBO_2)_n \xrightarrow[-H_2O]{400\,°C} B_2O_3$$

Orthorhombic cyclotriboric acid (α-metaboric acid) whose structure, shown in Figure 123, is based on the planar B_3O_3 rings frequently encountered in borate chemistry, is formed first. The molecules are bound into layers by hydrogen bonds.

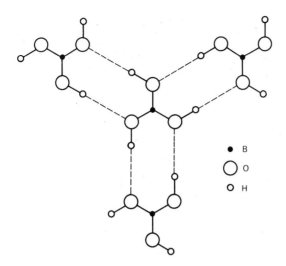

Fig. 122. Binding of planar B(OH)$_3$ molecules by linear OHO hydrogen bonds in the layers of crystalline orthoboric acid.

Two other metaboric acid modifications (β: monoclinic and γ: cubic) arise upon further heating in which the boroxole rings are cross-linked to each other via oxygen atoms. Boron is partially four-coordinated in β-HBO$_2$ and totally in γ-HBO$_2$. Dehydration of both produces the glassy anhydride B$_2$O$_3$ (m.p. 450 °C).

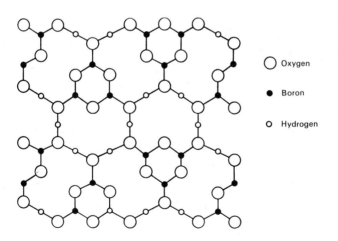

Fig. 123. Structure of crystalline cyclo-triboric acid, B$_3$O$_3$(OH)$_3$.

Crystalline B_2O_3 consists of planar chains of triangular BO_3 units in which each oxygen atom is held in common by two triangles. The chains are interlinked to bands, and these form an infinite, three-dimensional network. Gaseous, V-shaped $O{=}B{-}O{-}B{=}O$ molecules with sp-hybridized boron are released on heating.

Boron trioxide hydrolyzes to regenerate $B(OH)_3$ and undergoes alcoholysis to give esters $B(OR)_3$. In alkali B_2O_3 dissolves to give borates. With HF it forms BF_3. Aqueous borate solutions contain the anion $B(OH)_4^-$ at high pH, but in weak base where both $B(OH)_3$ and $B(OH)_4^-$ are present (see equation on p. 380), condensation takes place:

$$2\,B(OH)_3 + B(OH)_4^- \;\rightleftharpoons\; B_3O_3(OH)_4^- + 3\,H_2O$$

The trimeric anion is derived from the cyclo-triboric acid by addition of one OH^- ion and is the predominant species, but $B_3O_3(OH)_5^{2-}$, $B_4O_5(OH)_4^{2-}$, and $B_5O_6(OH)_4^-$ are also present. These polyanions hydrolyze in neutral and acidic solution to regenerate $B(OH)_3$. The structures of these anions, which are also found in crystalline borates, are:

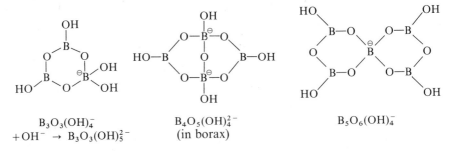

$B_3O_3(OH)_4^-$
$+OH^- \rightarrow B_3O_3(OH)_5^{2-}$

$B_4O_5(OH)_4^{2-}$
(in borax)

$B_5O_6(OH)_4^-$

In addition to $B(OH)_4^-$ $B(OH)_3$ also condenses in aqueous solution with other polyhydroxyl compounds like glycerine, mannitol, etc., to generate oxonium ions:

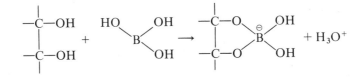

These, unlike boric acid, can be tritrated with aqueous sodium hydroxide.

Borates:
Monomeric, oligomeric and polyborates, containing the trigonal BO_3 and/or the tetrahedral BO_4 units, are known.

Anhydrous borates are formed in fusing $B(OH)_3$ with metal oxides, hydroxides, or carbonates, but almost all naturally occurring or borates recrystallized from

aqueous solutions, however, are hydrated, containing structural water (OH-groups) or water of crystallization (H_2O molecules).
Monoborates consist of the anions:

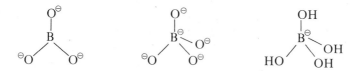

The anions $B_2O_5^{4-}$ and $B_2O(OH)_6^{2-}$ found in the diborates can be constructed from these units. In the bleaching agent sodium peroxoborate, $Na_2[B_2(O_2)_2(OH)_4]6\,H_2O$, prepared form $B(OH)_3$ and Na_2O_2, the boron atoms are connected by two peroxo groups.
The oligomeric borates contain the boroxole ring B_3O_3, either isolated as in $Na_3B_3O_6$, the sodium salt of the cyclotriboric acid, or condensed with a second ring as in borax, or spirocyclic via a shared boron atom bonded to a second ring (cf. formulae on p. 382). The higher metaborates, e.g., $Ca(BO_2)_2$, contain catenated anions, which may be interbonded in layers:

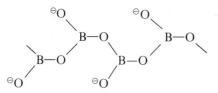

In the mineral turmaline, a complex aluminosilicate with a 10% boron content which is the most abundant boron containing mineral, a three-dimensional lattice is reached. Boron also forms rings with sulfur, e.g., B_2S_2, B_2S_3, B_3S_3, B_2S_4 and B_4S_2 rings are known, the latter containing B—B bonds, with each boron atom carrying a monovalent substituent.

15.8. Boron-Nitrogen Compounds

In the $>$B—N$<$ bond, a Lewis acid is directly adjacent to a Lewis base, giving the possibility for coordinate (p → p) π-bonding:

$$>\text{B}—\ddot{\text{N}}< \leftrightarrow >\overset{\ominus}{\text{B}}=\overset{\oplus}{\text{N}}<$$
    (a)   (b)

The unit (b) is an isostere of the C=C double bond in olefins, and hence similarities may be expected, as for borazine and boron nitride.

Borazine (HB—NH)$_3$:

The simplest boron-nitrogen compound, the aminoborane, H_2B—NH_2, is unstable to polymerization, but substituted derivatives, R_2B—NR'_2, are known (R = an organic substituent, R' = H, Cl, or an organic substituent).

Condensation of B—Cl with N—H compounds forms BN-bonds, e.g., 1,3,5-tri-chloroborazine is obtained from BCl_3 and NH_4Cl and reduced to borazine:

$$3\,BCl_3 + 3\,NH_4Cl \xrightarrow[C_6H_5Cl]{150\,°C} B_3N_3Cl_3H_3 \xrightarrow[(n-C_4H_9)_2O]{NaBH_4} B_3N_3H_6$$

Borazine is a colorless liquid with an aromatic odor. It resembles its isostere benzene in melting and boiling point, density, parachor, surface tension, zero dipole moment and molecular structure:

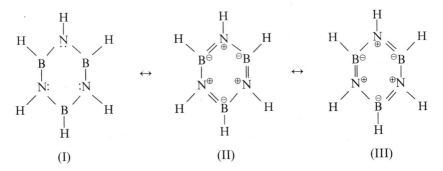

(I)   (II)   (III)

All internuclear distances [d(BN) = 1.436 Å], and valence angles (120°) are identical. The ring is planar (D$_{3h}$ symmetry). MO calculations indicate that the p$_\pi$-orbitals of the sp^2-boron atoms are occupied with only 0.5 electrons (in benzene: 1.0), i.e., structure I has considerable weight. The BN bond is thus polar, and differs from the C=C bond in its chemical behavior. Owing to the σ-electron polarization, the nitrogen atoms, despite their positive formal charge, are negative (-0.23e) and the boron atoms positive ($+0.32$e). Accordingly, borazine is thus more reactive than benzene, e.g., adding three moles of HCl to give $B_3N_3H_9Cl_3$ which can be reduced with $NaBH_4$ to the cyclohexane analogue, $B_3N_3H_{12}$ (m.p. 204 °C).

The thermolysis of borazine at 380 °C yields the biphenyl and naphthalene analogues:

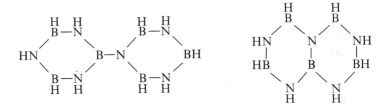

Boron Nitride $(BN)_n$:

The reduction of B_2O_3 with KCN at 1100 °C in a graphite crucible yields rhombohedral BN. Fusing $B(OH)_3$ with $(H_2N)_2CO$ in a NH_3 atmosphere at 500–950 °C, and subsequent annealing at 1800 °C gives hexagonal BN. Both BN modifications have graphite structures, but the planar $(BN)_n$ layers are differently stacked, as shown in Fig. 124.

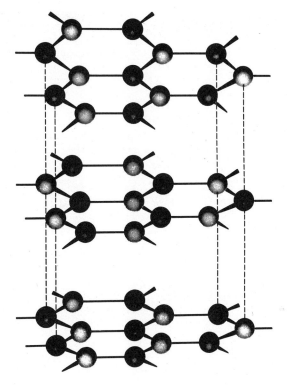

Fig. 124.   Layered structure of hexagonal boron nitride, $(BN)_n$ (After E. L. Muetterties, The Chemistry of Boron and its Compounds, J. Wiley, New York, 1967).

All atoms in the layers are sp^3-hybridized. The BN internuclear distance (1.446 Å) resembles that in borazine, and participation of coordinate $(p \rightarrow p)$ π-bonds must be assumed. Unlike graphite, hexagonal BN is colorless and an electrical insulator, with a gap width of 440 kJ/mol arising from the splitting of the π-bonding system into two sub bands owing to the polarity of the BN bond. In graphite the π-bonding system overlaps with the valence and conduction bands created by the overlap of σ-orbitals. Boron nitride becomes a conductor only at very high temperatures. Boron nitride is thermally and chemically stable, reacting with fluorine to give BF_3 and N_2, and with HF to give NH_4BF_4.

Like graphite which transforms into diamond at high pressure, two high pressure modifications are obtained from hexagonal BN:

$$\text{BN (hex.)} \begin{cases} \xrightarrow[\text{2000–4000 K}]{\text{70–120 kbar}} & \text{BN (cubic, Zincblende lattice)} \\ \\ \xrightarrow[\text{300–2000 K}]{\text{120–130 kbar}} & \text{BN (hexagonal, Wurtzite-lattice)} \end{cases}$$

These transitions are accompanied by a change in hybridization of all atoms from sp^2 to sp^3 and the formation of coordinate σ-bonds between the now puckered layers:

$$>\overset{\ominus}{B}=\overset{\oplus}{\ddot{N}}< \;\leftrightarrow\; >B-\ddot{N}< \;\rightarrow\; -\overset{|}{\underset{|}{\overset{\ominus}{B}}}-\overset{|}{\underset{|}{\overset{\oplus}{N}}}-$$

Cubic boron nitride is as hard as diamond and crystallizes in the same lattice with BN internuclear distances of 1.56 Å. At $>1390°$ this "inorganic diamond" returns to "inorganic graphite" (BN hexagonal).

Subject Index

Formula Index

Formulae are listed alphabetically as they appear in the text and the name for the element follows the alphabetized symbol. Compounds, ions, salts of each element are arranged alphanumerically.